Lecture Notes in Civil Engineering 929

Lecture Notes in Civil Engineering (LNCE) publishes the latest developments in Civil Engineering—quickly, informally and in top quality. Though original research reported in proceedings and post-proceedings represents the core of LNCE, edited volumes of exceptionally high quality and interest may also be considered for publication. Volumes published in LNCE embrace all aspects and subfields of, as well as new challenges in, Civil Engineering. Topics in the series include:

- Construction and Structural Mechanics
- Building Materials
- Concrete, Steel and Timber Structures
- Geotechnical Engineering
- Earthquake Engineering
- Coastal Engineering
- Ocean and Offshore Engineering; Ships and Floating Structures
- Hydraulics, Hydrology and Water Resources Engineering
- Environmental Engineering and Sustainability
- Structural Health and Monitoring
- Surveying and Geographical Information Systems
- Indoor Environments
- Transportation and Traffic
- Risk Analysis
- Safety and Security

To submit a proposal or request further information, please contact the appropriate Springer Editor:

- Pierpaolo Riva at pierpaolo.riva@springer.com (Europe and Americas);
- Swati Meherishi at swati.meherishi@springer.com (Asia—except China, Australia, and New Zealand);
- Wayne Hu at wayne.hu@springer.com (China).

All books in the series now indexed by Scopus and EI Compendex database!

Zbigniew Perkowski · Damian Beben ·
Zbigniew Zembaty · Maria Rossella Massimino ·
Jun He · Adam Rak
Editors

Environmental Challenges in Civil Engineering IV

Selected Contributions to the 7th International Conference on Environmental Challenges in Civil Engineering, April 20–22, 2026 Opole, Poland

Springer

Editors
Zbigniew Perkowski
Faculty of Civil Engineering and Architecture
Opole University of Technology
Opole, Poland

Zbigniew Zembaty
Faculty of Civil Engineering and Architecture
Opole University of Technology
Opole, Poland

Jun He
Changsha University of Science
and Technology
Changsha, China

Damian Beben
Faculty of Civil Engineering and Architecture
Opole University of Technology
Opole, Poland

Maria Rossella Massimino
Department of Civil-Industrial Engineering
and Architecture
University of Catania
Catania, Italy

Adam Rak
Faculty of Civil Engineering and Architecture
Opole University of Technology
Opole, Poland

ISSN 2366-2557 ISSN 2366-2565 (electronic)
Lecture Notes in Civil Engineering
ISBN 978-3-032-39514-6 ISBN 978-3-032-39515-3 (eBook)
https://doi.org/10.1007/978-3-032-39515-3

This Springer imprint is published by the registered company Springer Nature Switzerland AG
The registered company address is: Gewerbestrasse 11, 6330 Cham, Switzerland

Preface

Dear Readers,

We present to you a monographic collection of works titled *Environmental Challenges in Civil Engineering IV*, comprising original papers presented at the 7th International Scientific Conference on Environmental Challenges in Civil Engineering (ECCE 2026), held in Opole, Poland, on April 20–22, 2026.

A series of ECCE Conferences has been held cyclically every two years since 2014. In 2026, the event was organized as a KreativEU Network Conference by the Opole Branch of the Polish Association of Civil Engineers and Technicians (PACET), the Faculty of Civil Engineering and Architecture of the Opole University of Technology (a member of KreativEU Alliance), the Opole District Chamber of Civil Engineers, and the Commission of Civil Engineering of the Katowice Branch of the Polish Academy of Sciences. It received honorary patronage from Prof. Maria Kaszyńska, the Chairwoman of PACET, Prof. Marcin Lorenc, the Rector of the Opole University of Technology, Mr. Szymon Ogłaza, the Marshal of the Opole Voivodeship, and Mrs. Aleksandra Drescher, the President of the Provincial Fund for Environmental Protection and Water Management in Opole.

This year's conference was attended by scientific staff from Poland, Germany, Lithuania, the Netherlands, Portugal, Romania, Slovakia, and Ukraine. The event was also attended by numerous engineers of the Opole Branch of PACET and the Opole District Chamber of Civil Engineers.

The papers printed in this collection demonstrate that the conference, to the great satisfaction of its organizers, has become a platform for exchanging ideas between science and construction practice. The presented materials reflect the current dynamics of changes in the civil engineering sector, where traditional engineering approaches meet the imperatives of environmental protection, sustainable development, and digital transformation.

Each paper submitted to the ECCE 2026 conference underwent a two-stage review process. First, it was pre-screened by the scientific committee for compliance with the conference scope and then reviewed by independent reviewer in a double-blind peer-review process.

On the first day of the conference, three plenary lectures were presented. We would like to especially thank the authors of the plenary lectures for their valuable contribution, i.e.:

- Piotr Podkowa, Seweryn Stęplowski, Cementownia "Odra" S.A., Poland: *Implementation of a Modern Calciner in the Production Line of Cementownia "Odra" S.A.*
- Mariusz Pustelnik, Mostopol Sp. z o.o., Poland: *Flood in Southern Poland in 2024: Immediate and Final Reconstruction of Bridge Structures*
- Maurizio Scarciglia[1], Anna Szczegielniak[2], Luca Trabattoni[2]; [1]Breda University of Applied Sciences, The Netherlands; [2]Opole University of Technology, Poland:

Degrowth, Demographic Change, and Spatial Justice: A Joint Educational Model for Rural-Urban Planning

The remaining papers were presented on the second and third days of the meeting during six sessions.

This monograph comprises twenty chapters, which were developed from papers selected by the editors and, crucially, extensively debated by the conference participants. The thematic scope of the chapters is inherently interdisciplinary. This aligns precisely with the intentions of the conference's Scientific Committee, reflecting the fact that in modern sustainable construction and its allied disciplines—which, among others, focus on social development, environmental protection, and the preservation of heritage—integrating diverse perspectives constitutes standard practice. The selected papers are organized into five parts, systematically progressing from a 'micro' scale, beginning with materials engineering, to a 'macro' scale, culminating in spatial planning. However, across all levels, the core conference theme of sustainable construction and architecture remains a central thread. Consequently, the sections of the monograph are structured as follows:

- Part I: *Sustainable Building Materials, Environmental Engineering* —seven chapters.
- Part II: *Sustainable Additive Manufacturing in Concrete Technology* —three chapters.
- Part III: *Infrastructure, Structural Engineering* —three chapters.
- Part IV: *Diagnostics and Preservation of Built Heritage*—four chapters.
- Part V: *Social Spatial Justice, Spatial Planning*—three chapters.

At this point, we also want to thank all the authors of the submitted papers for their contributions and the conference's sponsors, who presented their latest achievements and projects in construction and environmental protection during their special sessions. We also want to thank the members of the Scientific and Organizing Committees (listed in the following sections of the monograph), our honorary patrons, and the volunteers. Without all of you, this book would not have been possible. Detailed information about the conference can be found on the website: https://ecce.po.edu.pl/.

Inviting you to read this monograph

The Organizing Committee of the ECCE 2026 Conference

Chairman

Wiesław Baran	Opole University of Technology/Opole Branch of the Polish Association of Civil Engineers and Technicians, Poland

Members

Damian Bęben	Opole University of Technology, Poland
Zbigniew Perkowski	Opole University of Technology/Commission of Civil Engineering, Katowice Branch of the Polish Academy of Sciences, Poland
Adam Rak	Opole University of Technology/Opole District Chamber of Civil Engineers, Poland

Secretaries

Jan Czmiel	Doctoral School of Opole University of Technology, Poland
Małgorzata Godyń	Opole District Chamber of Civil Engineers, Poland
Krzysztof Irek	Opole University of Technology, Poland
Joanna Kubielas-Walasek	Doctoral School of Opole University of Technology, Poland

The Scientific Committee of the ECCE 2026 Conference

Chairman

Zbigniew Perkowski	Opole University of Technology, Poland

Members

Glenda Abate	University of Catania, Italy
Dariusz Bajno	Wrocław University of Science and Technology, Poland
Damian Bęben	Opole University of Technology, Poland
Mirosław Bogdan	Opole University of Technology, Poland
Piotr Bońkowski	Opole University of Technology, Poland
Giovanni Bosco	University of L'Aquila, Italy
Petra Bujňáková	University of Žilina, Slovakia
Inês Domingues Serrano	Polytechnic University of Tomar, Portugal
Jacek Domski	Koszalin University of Technology, Poland
Łukasz Drobiec	Silesian University of Technology, Poland
Stefania Grzeszczyk	Opole University of Technology, Poland
Piotr Górski	Opole University of Technology, Poland
Anna Halicka	Lublin University of Technology, Poland
Anna Hoła	Wrocław University of Science and Technology, Poland
Jun He	Changsha University of Science and Technology, China
Ming Hu	University of Notre Dame, IN, USA
Elżbieta Janowska-Renkas	Opole University of Technology, Poland
Andrzej Kadłuczka	Cracow University of Technology, Poland
Justyna Kleszcz	Opole University of Technology, Poland
Seweryn Kokot	Opole University of Technology, Poland
Jarosław Konior	Wrocław University of Science and Technology, Poland
Marcin Koniorczyk	Łódź University of Technology, Poland
Peter Koteš	University of Žilina, Slovakia
Tomasz Krykowski	Silesian University of Technology, Poland
Renato de Leone	University of Camerino, Italy
Andrzej Marynowicz	Opole University of Technology, Poland
Maria Rossella Massimino	University of Catania, Italy
Silviu-Marian Miloiu	Valahia University of Târgovişte, Romania
Manuela Neri	University of Brescia, Italy
Lucia Nováková	Trnava University, Slovakia
Miguel José Oliveira	University of Algarve, Portugal
Adam Rak	Opole University of Technology, Poland
Andrea Segalini	University of Parma, Italy
Halil Sezen	Ohio State University, OH, USA
Olexandr Shymanovskyi	V.N. Shimanovsky Ukrainian Institute of Steel Construction, Ukraine

Elisa da Silva	University of Algarve, Portugal
Grzegorz Sławomir Świt	Kielce University of Technology, Poland
Luca Trabattoni	Opole University of Technology, Poland), University of Pavia, Italy
Jan Vaslestad	GeoAnlegg AS, Multiconsult, Norway
Amer Wadi	Higher Colleges of Technology, United Arab Emirates
Robert Wójcik	University of Warmia and Mazury, Poland
Jerzy Wyrwał	Opole University of Technology, Poland
Zbigniew Zembaty	Opole University of Technology, Poland

Scientific Secretaries

Dominka Bysiec	Opole University of Technology, Poland
Mariusz Czabak	Opole University of Technology, Poland
Krystian Jurowski	Opole University of Technology, Poland
Alina Kaleta-Jurowska	Opole University of Technology, Poland

Contents

About the Editors

Assoc. Prof. Zbigniew Perkowski, Ph.D., D.Sc. Opole University of Technology

Faculty of Civil Engineering and Architecture

Katowicka 48, 45-061 Opole, Poland

E-mail: `z.perkowski@po.edu.pl`

Zbigniew Perkowski is an employee of the Department of Building Materials Engineering at the Faculty of Civil Engineering and Architecture of the Opole University of Technology (Poland). He has been working at the Opole University of Technology since 1999, where he is currently the vice-chairman of the University Council. He teaches Theory of Elasticity and Plasticity, Thermomechanics, Strength of Materials, and Concrete Structures. He is an author or co-author of 3 books, a patent in the field of concrete structures, and over 70 publications in proceedings of domestic and international conferences, as well as articles in peer-reviewed scientific journals, including the *International Journal of Heat and Mass Transfer*, *Journal of Building Physics*, *Engineering Structures*, *Measurement*, and *Archives of Civil and Mechanical Engineering*. He served on the scientific boards of several international and domestic conferences. He belongs to the Section of Building Physics of the Committee of Civil Engineering and Hydroengineering of the Polish Academy of Sciences. He has been the chairman of the Commission of Civil Engineering at the Katowice Branch of the Polish Academy of Sciences since 2015. In 2013, he received the Professor Wacław Żenczykowski Award from the Polish Association of Civil Engineers and Technicians (PACET) for his achievements in research on damage mechanics in civil engineering. Composite structures, building physics, pavement friction, assessment of structural members through digital image correlation, and inverse problems in engineering are his most important current research interests.

Prof. Damian Bęben, Ph.D., D.Sc. Opole University of Technology

Faculty of Civil Engineering and Architecture

Katowicka 48, 45-061 Opole, Poland

E-mail: d.beben@po.edu.pl

Damian Bęben is the chairman of the Scientific Council of the civil engineering, geodesy, and transport discipline at the Opole University of Technology (Poland). He is an author and co-author of 3 books and over 200 publications at national and international conferences, as well as in the peer-reviewed scientific journals indexed in the Journal Citation Reports. He has h-indices of 17 and 20 in the Web of Science Core Collection and Scopus, respectively. According to Stanford University and Elsevier, he is among the top 2% of researchers worldwide. He has been a reviewer in many international scientific journals, the National Centre for Research and Development, the Polish Ministry of Science and Higher Education, and the Foundation of Polish Science. He is a member of the International Association for Bridge Maintenance and Safety (IABMAS); International Association for Life-Cycle Civil Engineering (IALCCE); International Association of Computer Science and Information Technology (IACSIT); Transportation Research Board (TRB) of the National Academies, Committee on Subsurface Soil-Structure Interaction (AFS40). He was a scholarship holder of the Foundation for Polish Science for Young Prominent Scientists, a scientific scholarship for outstanding young scientists awarded by the Ministry of Science and Higher Education; 2011 Outstanding Reviewer for the *Journal of Bridge Engineering* (ASCE); and a recipient of a PhD scholarship from the European Social Fund for a PhD. Hi is an associate editor of *Vibration* and *Frontiers in Built Environment* (Bridge Engineering section). His most important research interests are the analysis of bridges, including soil-steel composite structures; underground and buried structures; the durability of engineering structures; field load tests of structures; and environmental protection in transportation engineering.

Prof. Zbigniew Zembaty, Ph.D., D.Sc. Opole University of Technology

Faculty of Civil Engineering and Architecture

Katowicka 48, 45-061 Opole, Poland

E-mail: z.zembaty@po.edu.pl

Zbigniew Zembaty works at the Faculty of Civil Engineering and Architecture of the Opole University of Technology (Poland). He is an author or co-author of a book and about 100 other publications, including about 75 covered by Scopus, with about 1000 citations and respective h-index 21 (Web of Science Core Collection: 18). Listed among the top 2% of researchers in the World by Stanford/Elsevier Index (https://elsevier.digitalcommonsdata.com/datasets/btchxktzyw/6). The main areas of scientific research cover civil engineering, structural dynamics, random vibrations, inverse problems, structural health monitoring, seismic engineering, and geophysics. He is the co-editor of three Springer books: *Seismic Behaviour and Design of Irregular and Complex Civil Structures II, III, & IV* (2016, 2020, 2022). He cooperates with Milan Polytechnic, Trento and Pavia Universities (Italy), and Israel Institute of Technology, including participation in the doctoral schools at Trento University and Technion-Haifa, with lectures on random vibrations. He participates in the work of scientific committees at numerous international conferences, including the World Conference on Earthquake Engineering (16th WCEE, Santiago, Chile, 2017). Current active research focuses on the rotational components of seismic ground motion. He is a member of the special international group IWGoRS, which deals with 6-component ground-motion seismology and seismic engineering. More information: https://z.zembaty.po.opole.pl/.

Prof. Maria Rossella Massimino, Ph.D., D.Sc. University of Catania

Department of Civil Engineering and Architecture

Geotechnical Section, Building 13, Via Santa Sofia 64, 95125 Catania, Italy

E-mail: maria.massimino@unict.it

Maria Rossella Rita Massimino is a full professor of geotechnics at the Department of Civil Engineering and Architecture (DICAR) at Catania University (Italy). She teaches Soil Mechanics, Foundations, and FEM Geotechnical Design. She serves as the reference lecturer for several Erasmus Agreements. She is the author and co-author of over 170 publications presented at national and international conferences and in peer-reviewed scientific journals. She has an h-index of 22 and 1277 citations, according to the Scopus database. She has attended numerous national and international conferences as an invited and keynote speaker. She leads several national and international research projects. She is a member of the Editorial Committees of several international journals indexed on Scopus. She is a member of the Italian Geotechnical Society (AGI) and the Italian National Group of Geotechnical Engineering (GNIG). She participates in the Research Doctorate College on the Assessment and Mitigation of Urban and Territorial Risks at Catania University. She was a member of the European Technical Committee on Application of Eurocode 8 (ERTC-12). She received an ECOLEADER grant, the International Geosynthetics Society IGS Student Award, and the Italian *Troiano Award* for the best Ph.D. thesis conducted in Italy during 1999–2001. Her main research interests include foundations, soil-structure interactions, soil dynamics, local site response, and environmental protection by new geomaterials.

Prof. Jun He, Ph.D. Changsha University of Science and Technology
Faculty of Civil and Environmental Engineering
No. 960, Wanjiali South Road, 410114 Changsha, China
E-mail: hejun@csust.edu.cn

Jun He is a professor and doctoral supervisor whose distinguished academic credentials include a European Union Marie Skłodowska-Curie Fellowship and the Hunan Provincial Outstanding Youth Fund Award. His research background comprises postdoctoral work at Waseda University, Japan, and visiting scholarships at the University of Oklahoma, USA, and Heriot-Watt University, UK. He currently serves as the director of the Hunan Provincial Key Laboratory of Bridge and Building Green Construction and Maintenance and the chair of the Technical Committee on Mechanical Performance and Design Methods of Composite Bridges at the World Transport Congress (WTC).

He is also an active member of several leading professional organizations, including IABSE, IABMAS, ASCE, and JSCE.

He has secured and directed multiple research initiatives, including EU international collaborative projects, four National Natural Science Foundation of China (NSFC) grants, and two Hunan Provincial Natural Science Foundation awards, in addition to contributing to more than 10 national research programs such as the National 973 and 863 Programs. His scholarly contributions encompass over 200 high-quality publications, including more than 120 SCI-indexed papers in renowned journals such as the *Journal of Structural Engineering* (ASCE) and *Engineering Structures*. His research outcomes have been integrated into industry standards, design guidelines, and textbooks. According to data from Stanford University and Elsevier, he ranks among the top 2% of most-cited researchers worldwide.

His work has been recognized with awards, including the First Prize of Hunan Provincial Science and Technology Progress Award and the Second Prize of Hunan Provincial Teaching Achievement Award. He frequently serves on organizing committees and as a session chair at international conferences and is acknowledged as an outstanding reviewer for top-tier journals such as *Composite Structures* and the *Journal of Structural Engineering* (ASCE). His research program focuses on the mechanical behavior and design methodologies of steel-concrete composite bridges; the application of advanced materials in composite bridge systems; the development of sustainable structures featuring demountable and replaceable components for rapid assembly; and reliability and durability assessment of composite bridges.

Assoc. Prof. Adam Rak, Ph.D., D.Sc. Opole University of Technology

Faculty of Civil Engineering and Architecture

Katowicka 48, 45-061 Opole, Poland

E-mail: a.rak@po.edu.pl

Adam Rak is a specialist in environmental engineering, water management, and construction project management. Since 2004, he has been affiliated with the Opole University of Technology, Poland, where he serves as an associate professor. He is the director of the postgraduate program *Cost Estimation and Management of Construction Project* and a member of the Construction Project Engineering Section of the Committee of Civil Engineering and Hydroengineering of the Polish Academy of Sciences. His research focuses on applying mathematical methods to model technological processes in environmental engineering and construction, developing numerical models to optimize supervision and control procedures in investment processes, and conducting analyses related to environmental protection, sustainable building, and environmental conditions that affect the preparation and implementation of investment projects. He is the author or co-author of 185 publications and studies. They include two books (*Modelling Changes in Water Quality in a Mountain Retention Reservoir and Construction Investment Projects*: *Environmental Conditions for Preparation and Implementation*), international scientific journals (*Polish Journal of Environmental Studies*, *International Journal of Engineering Research and Application*, *International Journal of Water Sciences*, *Open Journal of Architectural Design*, *Water Supply – Water Practice and Technology*, and *Wulfenia Journal*) and chapters in scientific monographs, including *Cutting Edge Research in New Technologies and Environmental Engineering III* (Taylor & Francis Group). He is an expert and consultant for international consulting companies. He is the author or co-author of over 100 studies, concepts, and reports concerning water and wastewater management, agricultural reuse of wastewater, and environmental impact assessments of investment projects. He has served as an expert and coordinator on numerous infrastructure and environmental projects of both European and national significance, including water supply systems, wastewater treatment, and the preparation of large investment projects. He has supervised approximately 180 engineering and master's theses, as well as doctoral dissertations, and has served as a reviewer in doctoral and habilitation proceedings. His scientific and expert work integrates theoretical research with engineering practice in the preparation and execution of large infrastructure projects related to water management, sewerage systems, and environmental protection.

Sustainable Building Materials, Environmental Engineering

Characterization and Alkaline Activation of Linz-Donawitz (LD) Steel Slag for Use as a Sustainable Cementitious Binder

Rezan Javed(✉), Henrik Funke, and Sandra Gelbrich

Chemnitz University of Technology, 09126 Chemnitz, Germany
rezan.javed@mb.tu-chemnitz.de

Abstract. This investigation focuses on characterizing Linz-Donawitz (LD) slag before and after processing to assess its suitability as an alkali-activated binding material. The purpose is to explore its viability as a substitute for traditional cement in construction, thereby helping reduce CO_2 emissions associated with cement manufacturing. The mineralogical, chemical, and physical properties of the LD slag were thoroughly examined before and after processing. During processing, the material was leached. The LD slag was combined with alkali silicates and alkali hydroxides to initiate the activation process. The assessment focused on characterizing the binder's mechanical performance, setting behavior, and microstructural development to determine the influence of treatment (leaching) and the type of alkaline activator on the material's overall reactivity and suitability as a cementitious substitute. Modified mixtures were prepared by varying both the type and concentration of alkali activators. The mechanical properties, including four-point bending, tensile strength, and compressive strength, were evaluated after 28, 56, and 112 days. Among all analyzed grain sizes of untreated LD slag, the highest bending tensile strength was observed for $d_{80} < 63$ µm, reaching 2.1 MPa, while the maximum compressive strength was measured at 12.8 MPa. In contrast, the treated LD slag exhibited a maximum bending tensile strength of 1.4 MPa and compressive strength of 7.4 MPa, suggesting that the treatment process influenced the strength. In addition, the introduction of additional constituents altered the material's reactivity and structural development. Further confirmation of the morphological changes was obtained through Scanning Electron Microscopy (SEM) analysis. Based on experimental results, untreated LD slag was adopted as an aggregate in industrial-scale mixtures along with other constituents. A reference area was prepared to demonstrate the potential for various indoor and outdoor flooring applications.

Keywords: Linz-Donawitz (LD) Slag · Alkali Activated Materials (AAM) · Secondary Cementitious Materials (SCMs).

1 Introduction

The production of steel consumes substantial amounts of iron ore, limestone, fuel, water, air, and energy, while producing several types of slag as by-products. Steelmaking slags are classified by production process into Blast Furnace (BF), Ladle Furnace (LF), Electric

Z. Perkowski et al. (Eds.): ECCE 2026, LNCE 929, pp. 3–18, 2027.
https://doi.org/10.1007/978-3-032-39515-3_1

Arc Furnace (EAF), and Linz-Donawitz (LD) slags. These slags differ in their mineralogical and chemical compositions, which affect their potential applications. LD slag is produced as a by-product of the Linz-Donawitz (LD) process, which converts molten pig iron into crude steel. Its primary constituents include free lime, metallic and non-metallic iron, and calcium silicates, resulting in a highly basic material. In the steelmaking process, molten iron from the Blast Furnace (BF) is transferred to the Linz-Donawitz (LD) converter, where it undergoes oxidation to remove carbon and other impurities, such as phosphorus and silicon. Lime is added to the converter, and the resulting waste, primarily composed of calcium silicate and other oxide impurities, is skimmed off as slag. The slag is then subjected to open quenching, in which it is cooled with water and air [1].

According to the EUROSLAG Association, the European Union generated around 38 million tons of steel slag in 2024, with approximately 92% (35 million tons) recycled and used in the industrial and construction sectors [2]. The disposal of large quantities of waste slag presents significant challenges due to environmental concerns and limited space within steel plants. Employing cost-effective and environmentally friendly techniques to utilize integrated steel plant waste can substantially reduce production costs. Reducing global warming and preserving natural resources have become key environmental priorities in the modern era. Reusing these wastes as substitutes for raw materials helps conserve valuable natural resources, such as dolomite, which is used as a flux in steelmaking. This approach not only minimizes the environmental impacts associated with mining but also decreases the need for slag disposal space [3, 4].

The main goal is to partially or fully replace Portland cement clinker with composite materials, a strategy that has gained considerable attention in recent decades due to the high CO_2 emissions associated with conventional clinker production. The reactive precursors most widely used in alkali-activated binders include granulated blast-furnace slag (GBFS), fly ash, and silica fume. Alkali-activated binders have increasingly been recognized as effective for specialty concretes due to their superior properties, including enhanced chemical and thermal resistance. These binders are produced by activating inorganic, non-metallic, predominantly amorphous aluminosilicate materials with alkaline solutions, a process that simultaneously dissolves and polymerizes the precursors, forming a cohesive binder.

LD slag was selected for this study, similar to other commonly used by-products in alkali-activated materials. The chemical composition of LD slag can vary depending on the specific steelmaking process, operational conditions, and any additives used. However, it generally contains significant amounts of CaO, Fe, SiO_2, and Mn, which contribute to its potential as a precursor in alkali-activated binder systems. LD slag is predominantly crystalline due to its high basicity and the relatively slow cooling conditions during and after steelmaking, which favor the formation of stable calcium silicate and ferrite phases. The main crystalline phases identified in LD slag include calcium silicates, primarily belite (Ca_2SiO_4) and minor alite (Ca_3SiO_5), which contribute to its latent hydraulic reactivity. Additionally, calcium ferrites such as srebrodolskite ($Ca_2Fe_2O_5$) and iron-rich oxides, including wüstite (FeO) and magnetite (Fe_3O_4), are commonly present, providing structural stability to the slag matrix. Free lime (CaO) may also occur as discrete crystals, influencing both the reactivity and the volumetric behavior of the material. While the crystalline phases dominate, a minor amorphous fraction is often present, which can affect the overall chemical reactivity during applications such as cementitious or alkali-activated systems. These factors also present significant engineering challenges in the development of efficient geopolymer concrete technology, as outlined below:

- Low hydraulic activity of binder, for instance, due to a higher proportion of crystalline phase compared to the amorphous phase.
- Heterogeneous mineral composition depending on the melt type.
- Complications with volumetric stability (expansion and cracking of the product related to, e.g., CaO and MgO).
- Excessive iron oxides and a deficiency of reactive silica and aluminum.
- Relatively high activator demand.
- The high calcium content causes excessively rapid setting, which hinders the development of a strong geopolymer structure.
- Crystalline phases exhibit a highly ordered and thermodynamically stable lattice structure. The strong, well-organized atomic bonds within this structure reduce their susceptibility to dissolution in alkaline environments. Consequently, the limited dissolution of crystalline phases leads to reduced chemical reactivity.

A detailed investigation of LD slag was undertaken in this study, beginning with preliminary characterization. The research subsequently aimed to optimize the alkali activation process to enhance the material's mechanical properties and explore its potential as a raw material for construction applications.

2 Material and Methods

2.1 Material Characterization and Development

Three distinct LD slag types, sourced from Nickelhütte Aue GmbH, were analyzed, each differing exclusively in their elemental composition. In terms of physical appearance, LD slag is characterized as a dark gray to black, angular, and rough-textured granular material consisting of irregularly shaped grains with a wide distribution of grain sizes. Initially, particle size distribution analysis was carried out by using Bettersizer S3 Plus. The particle-size distribution outcomes of all LD slag variants were very similar. Therefore, only one representative result is shown in Fig. 1. The x-axis represents particle size in μm, while the y-axis represents the cumulative percentage, indicating the fraction of particles smaller than a given size. In the graph, the grey bars (histograms) depict the differential particle-size distribution, showing the number of particles in each size range. The blue curve represents the cumulative distribution. The most important points are D10, D50, and D90. D10 = 2.5 μm indicates that 10% volume consists of particles smaller than 2.5 μm. Furthermore, D50 (median) = 20.1 μm signifies that 50% of the volume has a particle size below 20.1 μm. In addition, a D90 of 732.4 μm indicates that 90% of the volume is below 732.4 μm.

To characterize the LD slag, grains of different sizes were obtained by sieving, with sizes ranging from 500 μm to $d_{90} < 10$ μm. Grains with a size of up to 63 μm were separated using a RETSCH AS 400 sieve shaker, as shown in Fig. 2. The process of obtaining a further finer grain size involved collaboration with a third-party facility.

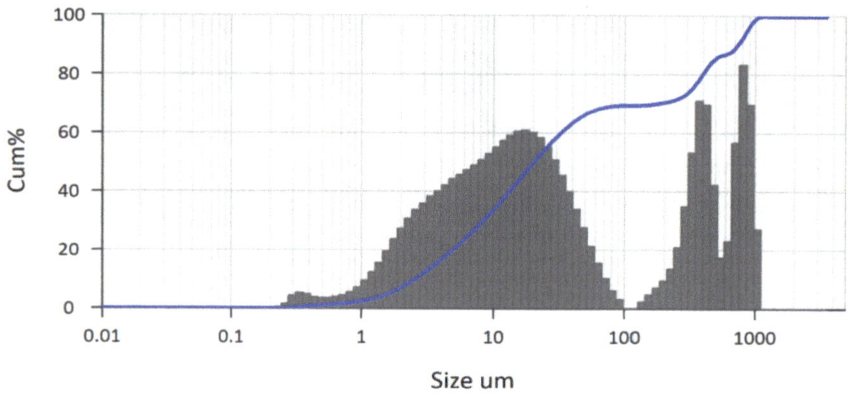

Fig. 1. Particle size distribution graph of LD slag.

After analyzing the untreated LD slag for alkaline activation and mechanical strength, the selected variants were identified based on the results, and the treated (leached) LD slag was then considered for further analysis. Almost identical experimental procedures were applied to both untreated and treated LD slag; the details of which can be found further in the subsections. Initially, the Blaine fineness, pycnometer measurements, and chemical composition of LD slag were considered. Based on these results, modified mixtures for alkaline activation were prepared to assess their mechanical properties. Additionally, microstructural analysis was performed using Scanning Electron Microscopy (SEM).

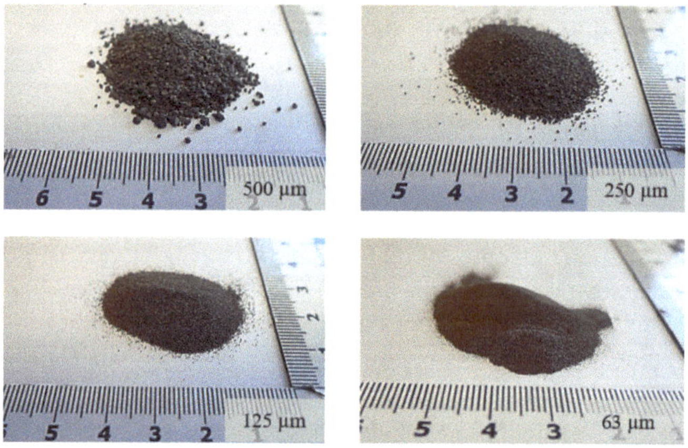

Fig. 2. LD slag with varying grain sizes.

To analyze the physical properties of LD slag, its density and Blaine specific surface area were determined. The fineness of the LD slag was measured using a Blaine apparatus in accordance with DIN EN 196-6 [5], while the specific density was measured using a pycnometer following the procedure specified in DIN EN 12350-6 [6]. The specific surface area, measured via the Blaine apparatus, and the specific density, determined using a pycnometer, of both untreated and treated LD slag are presented in Tables 1 and 2.

Table 1. Blaine surface area of untreated and treated LD slag.

	Untreated	Treated
LD 1	1763.3 cm^2/g	–
LD 2	2328.0 cm^2/g	5281.2 cm^2/g
LD 3	2894.1 cm^2/g	5380.6 cm^2/g

Table 2. Specific density measured values of untreated and treated LD slag.

	Untreated	Treated
LD 1	3.6 g/cm^3	–
LD 2	3.2 g/cm^3	2.2 g/cm^3
LD 3	3.0 g/cm^3	2.1 g/cm^3

On the other hand, Energy-Dispersive X-ray (EDX) spectroscopy was conducted to evaluate the chemical composition of the LD slag variants. The elemental compositions are presented in Table 3.

Table 3. Elemental Composition of LD slag variants [wt%].

Elemental IDs	LD 1	LD 2	LD 3
Si	5.3	5.1	5.6
Ti	1.3	1.1	0.8
Al	0.8	0.9	1.0
Fe	20.6	17.7	18.1
Mn	3.9	2.6	3.3
Mg	3.0	3.0	0.8
P	0.2	0.2	0.4
S	0.2	0.2	0.2
V	3.0	2.7	3.1
Ca	60.9	65.9	66.1
Cr	0.8	0.5	0.5

Elemental composition was analyzed over a 20 µm area. Calcium was found to be the most abundant element across all three LD slag variants, followed by iron, silicon, manganese, and magnesium. Small amounts of other elements were also detected.

2.2 Sample Preparation

The mixtures were formulated to assess the mechanical properties of LD slag. Granulated blast furnace slag (alkaline activated) and Stone ash/Saxo Carb 200 (being inert) were used as reference materials. The exchange between Saxo Carb 200 for the untreated mixture and Stone ash for the treated mixture recipe was conducted to evaluate different material types in this study. Saxo Carb 200 is generally recognized as a commercially produced mineral filler consisting primarily of calcium carbonate ($CaCO_3$) powder. By considering two reference materials, five different mixtures were prepared each time. These mixtures were prepared by varying the proportions of their constituents and incorporating alternative materials to examine their overall influence on the mixtures' properties. The quantity of LD slag variants was kept constant across various mixture recipes. Various alkaline solutions composed of alkali silicates and hydroxides play an important role in preparing mixture recipes. The most significant mixture compositions are summarized in Tables 4 and 5.

Table 4. Composition of the mixture for untreated LD slag and reference material.

Constituents (untreated)	kg/m^3
Considered grain sizes: 63 µm, $d_{80} < 63$ µm, and $d_{80} < 45$ µm	
LD slag 1,2, and 3/Granulated blast furnace slag/Saxo Carb 200	400
Alkaline activator based on sodium silicate	80
NaOH-solution (20 M.wt%)	165
Quartz sand 0–2 mm	950
Gravel 2–8 mm	700
w/b	0.44

In the treated material, several additional constituents were incorporated; these will be described in detail in Sect. 3. The sample preparation steps for conducting the bending, tensile strength, and compressive strength tests are illustrated in Fig. 3. The materials were weighed according to the designed mixture proportions. All dry constituents were first dry mixed for 60 s to ensure uniform distribution. Subsequently, the aqueous phase consisting primarily of alkaline solutions was added, and mixing was continued for an additional 2 min. The workability of the mixture was then evaluated. If adequate consistency was achieved, the mixture was cast into molds; otherwise, mixing was extended for a further 1 min. The prepared mixtures were monitored and tested over 112 days. Bending tensile strength was measured in accordance with DIN EN 196-1 (2016) [7, 8] after 28, 56, and 112 days. Compressive strength tests were also conducted following DIN EN 12390-3 (2019) [9, 10] at the same time intervals.

Table 5. Prepared mixture recipe for treated LD slag with reference material.

Constituents (treated)	kg/m^3
LD slag 1,2, and 3/ Granulated blast furnace slag/ Stone ash	400
Alkaline activator based on potassium silicate	90
KOH-solution (20 M.wt%)	250
Quartz sand 0–4 mm	700
Gravel 2–8 mm	460
w/b	0.50

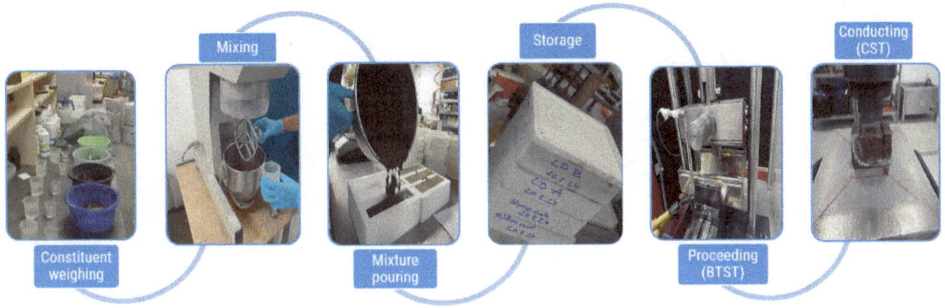

Fig. 3. Overview of mixture preparation till conducting the bending tensile strength test and compressive strength test of LD slag and reference material.

2.3 Scanning Electron Microscopy

To examine morphological variations in LD slag before and after leaching, Scanning Electron Microscopy (SEM) was performed using a ZEISS EVO 15 instrument. Prior to imaging, the samples were carefully prepared to enhance surface conductivity, as LD slag inherently exhibits low electrical conductivity. A thin conductive coating was deposited using a Quorum Q150T ES Plus sputter coater. Both untreated and leached LD slag samples were mounted on carbon adhesive tape attached to aluminum SEM stubs. A uniform gold coating with an approximate thickness of 7 nm was deposited on each specimen using a sputtering current of 20 mA for 1 min. Microstructural characterization was subsequently carried out primarily using a Secondary Electron (SE) detector to obtain detailed surface topography.

3 Results and Discussion

3.1 Investigation of Strength Characteristics of LD Slag

Following the evaluation of all primary test data and completion of the designated curing period, 4-point bending tensile and compressive strength tests were conducted. The graphical representation of the untreated LD slag strength is shown in Fig. 4. Among all variants, the LD 2 slag sample with $d_{80} < 63$ µm demonstrated the highest bending tensile strength of 2.1 MPa after 56 days, and LD 3 slag exhibited only slight variations in strength compared to LD 2. In comparison, the reference granulated blast-furnace slag achieved a maximum strength of 5.2 MPa, whereas Stone ash showed no measurable strength over the 112-day curing period due to its inert nature.

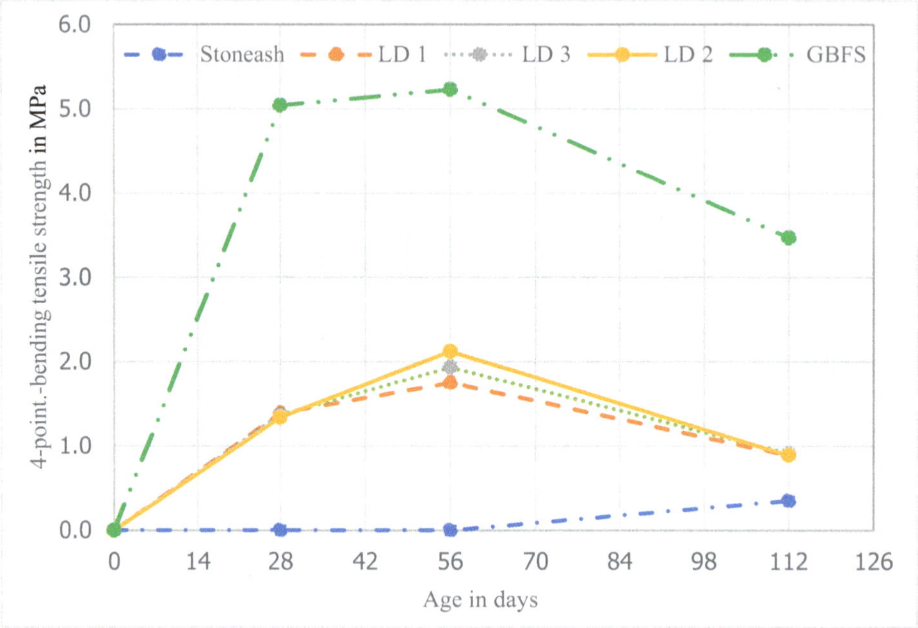

Fig. 4. Bending tensile strength of untreated LD slag samples with $d_{80} < 63$ µm grain size, GBFS, and Stone ash samples.

In contrast, different grain sizes were considered to compare the compressive strength of LD slag at various time intervals, as shown in Fig. 5. A stable trend was observed for the 63 µm grain size, with LD 2 slag achieving the maximum compressive strength of 5.6 MPa. More pronounced trends were observed for LD 2 and LD 3 slags. Among these, LD 3 exhibited the highest compressive strength of 12.8 MPa for the $d_{80} < 45$ µm grain size after 112 days, while LD 2 followed a nearly identical pattern. Similarly, LD 3 with $d_{80} < 63$ µm fraction demonstrated a comparable compressive strength of 12.5 MPa after 112 days, indicating consistent mechanical performance. Furthermore, Stone ash consistently exhibits low or negligible strength across the entire dataset. The analysis of untreated LD slag indicates that, in addition to the mixture constituents, grain size significantly influences both bending tensile strength and compressive strength. Larger grain sizes reduce specific surface area, limiting particle contact and inhibiting the formation of a dense, well-bonded matrix. This, in turn, leads to increased porosity, weaker interfacial bonding, and a higher susceptibility to crack initiation.

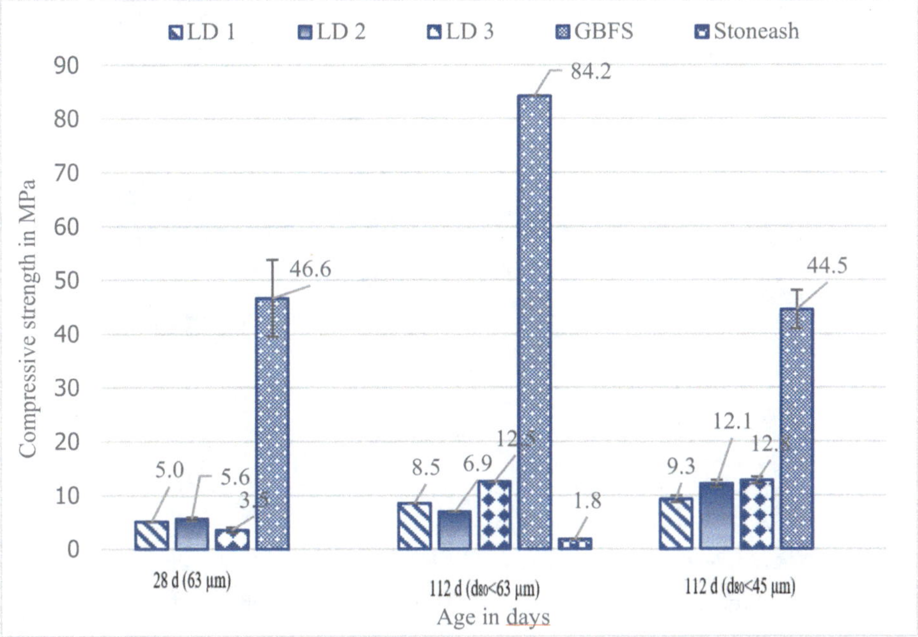

Fig. 5. Compressive strength of untreated LD slag samples with different grain sizes, GBFS, and Stone ash samples.

Based on the earlier outcomes, LD 2 and LD 3 slags were selected for further studies of treated LD slag. Treated LD slags exhibited several challenges during mixture preparation. The material exhibits minimal reactivity and extreme dryness, as shown in Fig. 6. Furthermore, an intense exothermic reaction occurred during preparation of the mixture for mechanical strength testing. As discussed earlier, LD slag primarily consists of free lime (CaO), iron oxides (FeO and Fe_2O_3), and calcium silicates, mainly dicalcium silicate (C_2S), with minor amounts of tricalcium silicate (C_3S), as well as minor oxides such as MgO, MnO, and Al_2O_3. Sulfuric acid reacts predominantly with free lime and other basic oxides present in the slag, leading to the formation of soluble metal sulfates and calcium sulfate, which precipitates as gypsum under aqueous conditions. The dissolution of these alkaline, moisture-retaining phases, together with the partial decalcification of calcium silicates, disrupts the hydrated, gel-like binding phases that retain water. As a result, the residual slag becomes porous, silica-rich, and non-hygroscopic, producing a dry material. The neutralization reactions between sulfuric acid and the basic oxides are highly exothermic. For example, the reaction of CaO with H_2SO_4 releases approximately 64 kJ/mol of heat, while reactions involving MgO release additional heat. Minor contributions from FeO further elevate the mixture's temperature. This exothermic behavior accelerates slag drying and complicates mixture preparation.

To address these issues, various measures were considered, including developing a new mixture recipe and using different mixing equipment to optimize the preparation process. After achieving the desired workability of the mixtures, the 4-point bending

Fig. 6. Pouring of the treated LD slag mixture into the mold.

tensile and compressive strengths were measured. Treated LD 2 and LD 3 slag were evaluated by increasing the amounts of potassium hydroxide (KOH) up to an optimal level and by adding superplasticizer. Increasing the KOH concentration can promote the dissolution of calcium silicates and other reactive phases, which may enhance early age strength development. Alternatively, introducing a superplasticizer into the mixture improves workability and particle dispersion, facilitating mixing and casting. This can lead to more uniform gel formation and potentially enhanced mechanical properties. The graphical representation of the four-point bending tensile strength is shown in Fig. 7. It has been observed that LD 2 slag exhibits a maximum strength of 1.1 MPa after 56 days, followed by a slight decline subsequently. In contrast, LD 3 slag revealed a maximum value of 1.4 MPa after 112 days of curing. Furthermore, over the entire curing periods of 28, 56, and 112 days, LD 3 slag exhibited the highest bending tensile strength. Compared with Fig. 4, there is no significant relative reduction in the strength of treated LD slag after 56 days, indicating reduced microcracking. Conversely, the untreated LD slag experiences a decline in strength by 112 days, likely due to sample expansion.

The compressive strength of LD 2 slag and LD 3 slag can be visualized in Fig. 8. After 28 days of curing, LD 3 slag recorded a higher strength than LD 2 slag. This trend continued at 56 days, with LD 3 slag reaching a maximum value of 7.4 MPa. This strength development is attributed to alkaline activation, which promotes the breakdown of silica and alumina phases and the subsequent formation of C-S-H-type gels. Later, a slight decline in strength was observed after 112 days of curing, possibly due to several factors. Internal stresses may develop due to shrinkage during curing. In addition, calcium-rich phases in LD slag could react with CO_2 over prolonged periods, leading to a minor

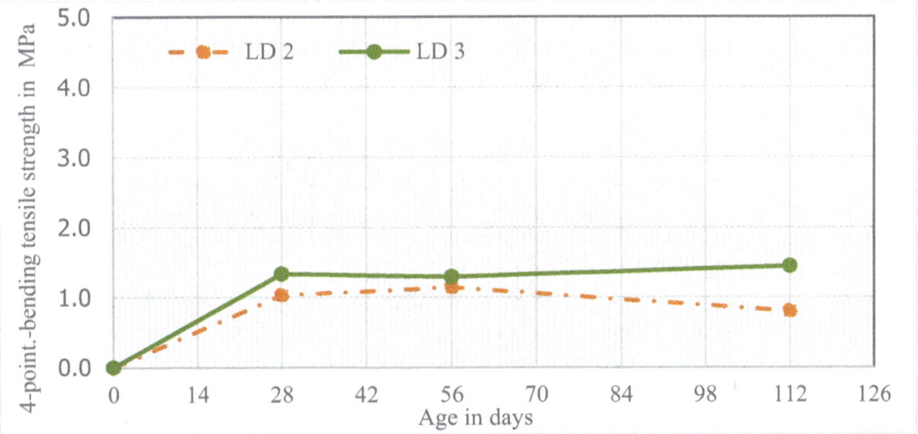

Fig. 7. Bending tensile strength of treated LD slag.

reduction in mechanical properties. Moreover, the C-S-H gel network may densify or undergo slight reorganization, occasionally resulting in a decrease in peak strength.

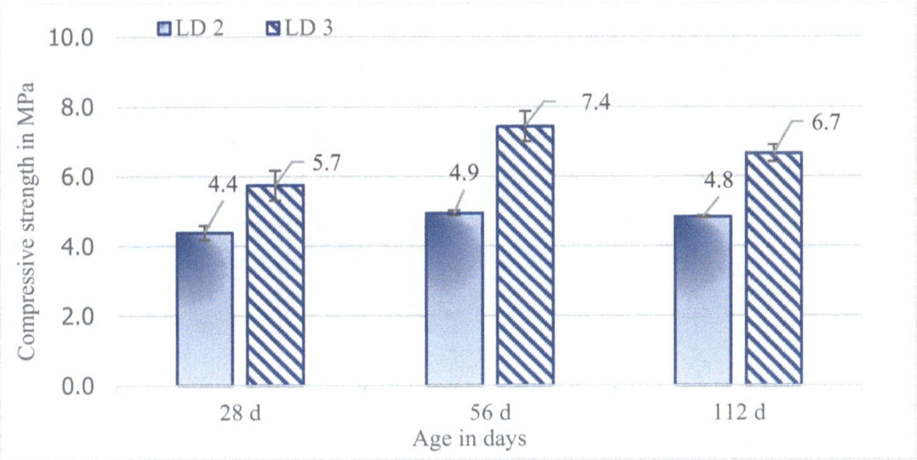

Fig. 8. Compressive strength of treated LD slag.

3.2 Scanning Electron Microscopy Study of Microstructural Changes

The SEM images of untreated and treated LD slags are shown in Figs. 9 and 10. Untreated LD slag is characterized by relatively large, irregular, and dense grains, which are typical of the raw material and suggest an intact matrix primarily composed of complex silicates and oxide phases. In contrast, leached LD slag displays a markedly different morphology, with smaller, rod-shaped, and longitudinal plate grains exhibiting a more porous and fragmented structure. This change in structure is likely attributable to the chemical dissolution of soluble phases, such as free CaO, MgO, and minor metal oxides, during leaching. Overall, these morphological changes reflect a chemical modification of the slag's microstructure.

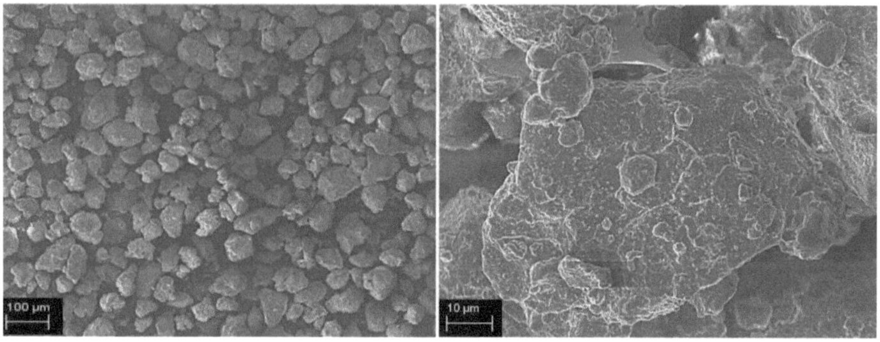

Fig. 9. SEM images of untreated LD slag at different magnifications.

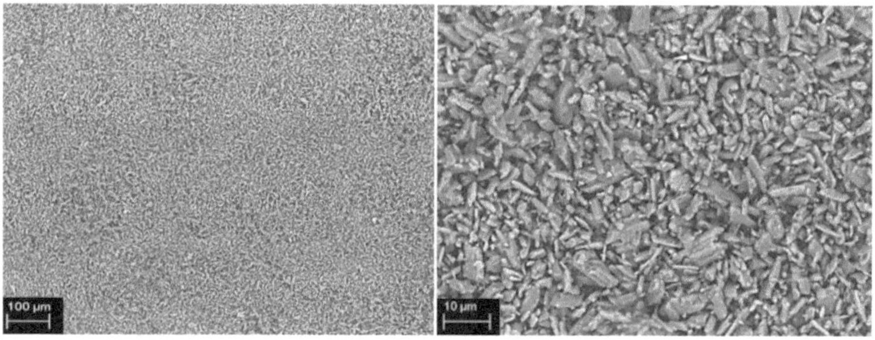

Fig. 10. SEM images of treated LD slag at different magnifications.

In addition, the elemental mapping of the leached LD slag is presented in Fig. 11. The maps reveal a heterogeneous distribution of silicon (Si), calcium (Ca), vanadium (V), and iron (Fe), reflecting the slag's complex chemical composition after the leaching treatment. Silicon is relatively uniformly distributed throughout the sample. In contrast, vanadium and iron exhibit localized enrichment in discrete regions. The non-uniform elemental distribution further highlights the phase heterogeneity of the leached slag and supports the morphological transformations observed by microscopic analysis. These changes in elemental distribution and microstructure are significant because they directly influence the reactivity, stability, and potential reuse of leached LD slag in applications such as metal recovery and environmental remediation. Overall, the elemental mapping confirms that the leaching process induces substantial chemical and structural modifications within the slag microstructure.

Fig. 11. Elemental mapping of treated LD slag.

3.3 Utilization of LD Slag as an Aggregate at Industrial Scale

Following the successful laboratory-scale trials with the optimized mixture recipe listed earlier in Table 4, the experiment was conducted at an industrial scale to assess its feasibility. Based on its previously observed better mechanical properties, untreated LD 3 was selected for further studies. The entire procedure required several days to complete. Multiple test areas were then prepared by incorporating recycled aggregates and varying the quantity of aggregates or other constituents. The overall process comprised several stages, including 7 to 8 steps of grinding, surface coating, and other preparatory treatments. A few of these steps are illustrated in Fig. 12.

After completing all the required steps, the prepared reference area, shown in Fig. 13, demonstrated potential for various structural applications. In addition, it exhibited resistance to both acidic and sulfuric environments.

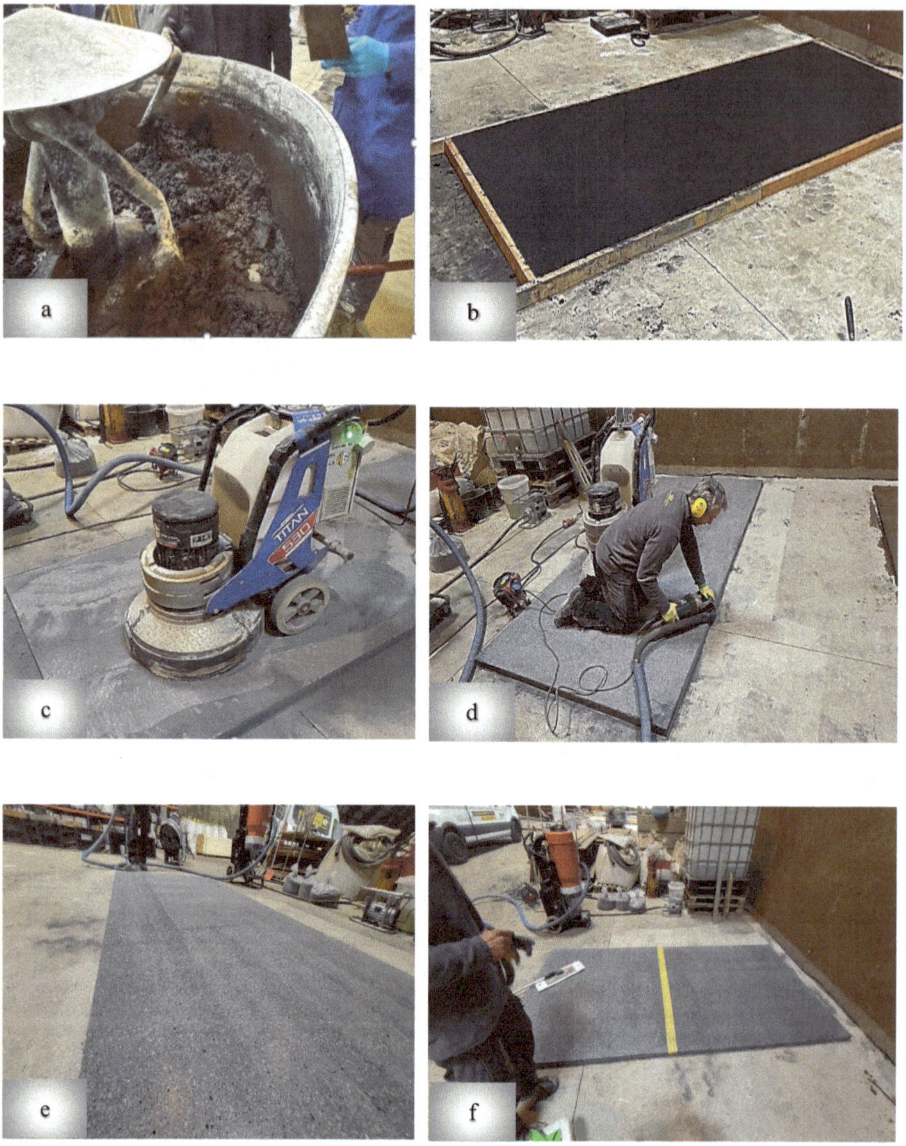

Fig. 12. Different steps involved in the preparation of the reference area: mixing of constituents (a), preparation of the area with the mixed material (b), grinding (c and d), and surface preparation for final polishing (e and f).

Fig. 13. Prepared reference surface area at industrial scale.

4 Conclusions

This study investigated the suitability of Linz-Donawitz (LD) slag with different compositional characteristics as a precursor for alkali-activated materials, with particular emphasis on the influence of leaching treatment. Mechanical testing showed that material produced from untreated LD slag exhibited the maximum bending tensile strength of 2.1 MPa and a compressive strength of 12.8 MPa. In contrast, material derived from treated slag demonstrated significantly lower mechanical performance, with maximum bending tensile and compressive strengths of 1.4 MPa and 7.4 MPa, respectively. The reduction in strength observed after leaching is due to the loss of soluble oxides (CaO and MgO). This process increases porosity, modifies mineral phases, and lowers the slag's alkalinity; as a result, the slag's overall structural integrity is diminished.

It was also observed that grain size significantly affects the mechanical strength of LD slag, particularly in untreated materials. In this study, samples with $d_{80} < 63$ µm, and even smaller grain sizes, exhibited the highest bending tensile strength and compressive strength. This indicates that finer grains enhance the reactivity of the slag, promoting more efficient alkali activation and contributing to improved mechanical performance. Additionally, the type and concentration of the alkali activator markedly influence the overall behavior of both untreated and leached LD slag.

Microscopic analysis revealed that the morphology of LD slag was altered following the leaching process, indicating changes in surface texture and grain structure. Based on the analysis, LD3 slag is used as an aggregate due to its better mechanical properties. Reference specimens were prepared using a combination of other aggregates at both laboratory and industrial scales. These reference specimens have potential applications in a variety of flooring systems.

Acknowledgments. The authors gratefully acknowledge the financial support provided by the "BMBF- Bundesministerium für Bildung und Forschung" (Federal Ministry of Education and Research).

Disclosure of Interests The authors declare no competing interests related to the content of this article.

References

1. Singh SK, Rekha P, Surya M (2020) Utilization of Linz–Donawitz slag from steel industry for waste minimization. J Mater Cycles Waste Manag 22:611–627
2. GMK Center. https://gmk.center/en/news/eu-recycled-92-of-steel-slag-in-2024-euroslag. Last accessed 21 June 2026
3. Chand S, Paul B, Kumar M (2016) Sustainable approaches for LD slag waste management in steel industries. Metallurgist 60:116–128
4. Chand S, Paul B, Kumar M (2015) An overview of use of Linz-Donawitz (LD) steel slag in agriculture. Curr. World Environ. 10(3):975–984
5. DIN EN 196-6:2019-03: Methods of testing cement—Part 6: determination of fineness. German version EN 196–6:2018
6. DIN EN 12350–6:2019–09: Testing fresh concrete—Part 6: density. German version EN 12350–6:2019
7. DIN EN 12390–2:2019–10: Testing hardened concrete—Part 2: making and curing specimens for strength tests. German version EN 12390–2:2019
8. DIN EN 196-1:2016-11: Methods of testing cement—Part 1: determination of strength. German version EN 196–1:2016
9. DIN EN 12390–3:2019–10: Testing hardened concrete—Part 3: compressive strength of test specimens. German version EN 12390–3:2019

Comparative Analysis of Two Aggregate Types in Terms of Their Influence on Concrete Properties and Environmental Impact

Amanda Akram(✉) and Marta Słowik

Lublin University of Technology, Nadbystrzycka 40, Lublin 20-618, Poland
a.akram@pollub.pl

Abstract. Concrete is one of the most widely used construction materials worldwide, but its production has a significant environmental impact, primarily from cement-related CO_2 emissions and the extraction and transport of aggregates. This study investigates the influence of two coarse aggregate types on the mechanical performance of concrete and its potential implications for sustainability. Two concretes were compared: one containing natural river gravel and the other crushed dolomite aggregate. Mixtures for both concretes were designed with the same maximum aggregate size of 16 mm. Aggregate properties, including porosity, crushing strength, and surface roughness, were evaluated. The results indicated that concrete with dolomite aggregate exhibited improved mechanical properties compared to concrete with natural gravel. Despite lower crushing resistance, the higher roughness of dolomite aggregate enhanced the bond between the grains and the cement matrix. This led to increased compressive strength, splitting tensile strength, elastic modulus, and fracture toughness of concrete. The findings suggest that crushed dolomite aggregate can improve concrete quality and potentially extend the service life of structural members. Consequently, selecting appropriate aggregate types may contribute to more durable and environmentally sustainable concrete structures, although detailed analyses are required for specific applications.

Keywords: Concrete · Type of Coarse Aggregate · Environmental Impact.

1 Introduction

Concrete is the most widely used construction material, playing a fundamental role in modern infrastructure. However, its large-scale production is associated with considerable environmental impact [1–4]. Cement manufacturing is the dominant contributor to concrete-related CO_2 emissions, accounting for the majority of concrete production's total footprint [5, 6]. At the same time, coarse aggregates constitute more than 70% of the concrete mixture by mass, making their processing a non-negligible source of environmental burden [7, 8]. In particular, the exploitation of natural river aggregates is recognized as environmentally harmful, as it disrupts river ecosystems, alters hydrological conditions, and contributes to landscape degradation and dust emissions [9, 10]. While natural aggregates may require less intensive processing than crushed materials,

Z. Perkowski et al. (Eds.): ECCE 2026, LNCE 929, pp. 19–31, 2027.
https://doi.org/10.1007/978-3-032-39515-3_2

long transport distances can offset this advantage. Consequently, the use of locally available crushed aggregates may represent a rational strategy for reducing transport-related emissions, provided that adequate mechanical performance of concrete is maintained.

Beyond environmental considerations, mechanical performance remains a decisive factor in material selection. The type and morphological characteristics of coarse aggregate significantly influence the behavior of concrete, especially within the interfacial transition zone (ITZ) between aggregate particles and hardened cement paste [11–17]. Aggregate surface roughness, angularity, mineral composition, and stiffness affect stress transfer mechanisms and crack propagation paths. Rough and angular particles tend to enhance mechanical interlocking and bond strength at the ITZ [18, 19], whereas rounded natural gravel often promotes smoother crack paths along weaker interfaces [20].

These mechanisms are reflected in fracture parameters such as tensile strength, fracture energy (G_f), or the critical stress intensity factor (K_{Ic}). The enhancement of fracture properties indicates greater resistance to unstable crack growth under tensile and flexural loads, thereby improving post-peak ductility.

From a sustainability perspective, improved strength and fracture parameters may enhance the durability-related properties of concrete structures by increasing their resistance to crack propagation. Consequently, selecting an appropriate aggregate type can be one of the strategies for enhancing both the mechanical efficiency and environmental performance of concrete. Therefore, the aim of this study is to assess the influence of coarse aggregate type on the strength properties of concrete, while accounting for transport-related environmental factors. Two ordinary concretes were examined, differing only in the type of coarse aggregate used: natural river gravel and crushed dolomite. The effects of aggregate characteristics on the tested concretes are investigated. Furthermore, the environmental assessment focuses specifically on the impact of transport logistics, as delivery distances can be a more decisive factor in the overall carbon footprint in aggregate sourcing than the energy consumed during aggregate production. This approach highlights how regional sourcing of crushed materials can be a primary strategy for reducing concrete-related greenhouse gas emissions.

2 Materials and Experimental Methods

2.1 Materials and Preparation of Testing Specimens

The experimental program involved two series of concrete mixtures designed to evaluate the influence of coarse aggregate type on the mechanical properties and fracture resistance of concrete:

- L16N – concrete containing natural river gravel,
- L16D – concrete containing crushed dolomite.

Both mixtures were designed with a maximum aggregate size (D_{max}) of 16 mm (Table 1). The gravel aggregate was obtained from the gravel pit "Żwirbet" in Suwałki, Poland, and the dolomite aggregate was sourced from the mine "Nowy Staw" in Łagów, Poland.

The cement used in both series was CEM II 42.5R, and a constant water-to-cement (w/c) ratio of 0.50 was kept. The fine aggregate consisted of natural sand (fraction 0.0–2.0 mm), while the coarse aggregate fractions were maintained in similar proportions to ensure comparable grading curves, as shown in Fig. 1.

Table 1. Composition of concrete mixtures for L16N and L16D series.

Series	Cement, kg/m^3	Water, kg/m^3	Sand 0.0–2.0 mm, kg/m^3	Coarse Aggregate 2.0–16.0 mm, kg/m^3
L16N	350	192	497	1,344
L16D	350	192	497	1,344

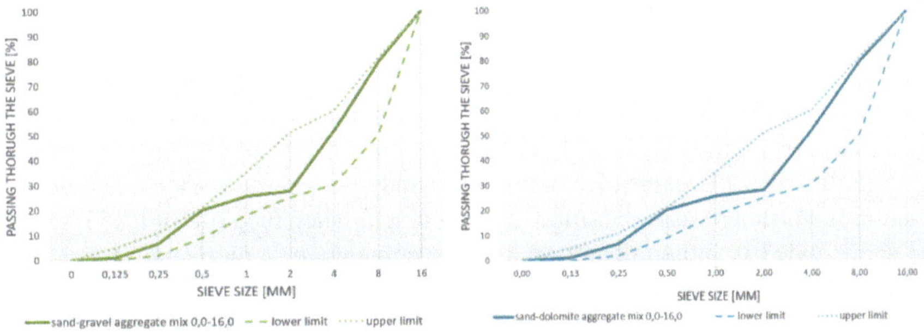

(a) **(b)**

Fig. 1. Grain size distribution curves for the aggregates: a) sand-gravel mix, b) sand-dolomite mix.

All prepared specimens were cast and cured in water for 28 days in accordance with Eurocode requirements [21]. For strength and modulus of elasticity tests, standardized cylinders (150 mm × 300 mm) were used [22]. For the critical stress intensity factor evaluation, beams with dimensions of 700 mm × 150 mm × 80 mm were prepared, with an initial central notch of 50 mm, cut with a diamond saw prior to testing in the middle of the beam span.

2.2 Testing of Aggregate Characteristics

The coarse aggregates used in this study consisted of natural river gravel and crushed dolomite, both within the 2.0–16.0 mm grain size range. To assess the influence of aggregate morphology on concrete characteristics, selected shape-related properties were measured. The physical and morphological properties of the aggregate were tested:

- flakiness index (FI),
- shape index (SI),
- porosity (P),
- crushing strength (ACV),
- surface roughness parameters (arithmetical mean height S_a, the maximum height from the highest peak to the lowest valley S_z, and the root mean square height S_q).

The flakiness index was evaluated in accordance with [23] using the bar sieves with a gap width between bars equaling $d_{max}/2$, where d_{max} was the upper limit of the tested fraction. Firstly, aggregate samples were divided into 4.0–8.0 mm and 8.0–16.0 mm fractions for testing; particles outside the 4.0–16.0 mm range were excluded. The aim of the test was to assess the mass percentage of coarse aggregate particles whose thickness was less than 0.6 times their nominal size. After sieving, the total percentage of flat grains in each sample was determined, and the flatness index (FI) was calculated using the following formula for each separated fraction:

$$\mathrm{FI} = \frac{\mathrm{M}_1 \times 100}{\mathrm{M}_2} \tag{1}$$

where: M_1 – mass of flaky particles after sieving [g], M_2 – total mass of a fraction sample [g].

The shape index of both coarse aggregates was determined in accordance with [24]. The test was performed on representative aggregate samples separated into size fractions in the same way as for the FI testing. For each fraction, individual particles were measured using a Schultz caliper to determine their largest and smallest dimensions: length (L) and thickness (E). Particles were classified as non-cubical when the ratio of L to E exceeded 3. The shape index was then determined as the percentage of elongated grains relative to the total mass of the sample for each separated fraction:

$$\mathrm{SI} = \frac{\mathrm{M}_1 \times 100}{\mathrm{M}_2} \tag{2}$$

where: M_1 – mass of the particles with the L/E ratio greater than 3 [g], M_2 – total mass of a fraction sample [g].

In addition to geometrical characterization, other properties relevant to aggregate-cement mortar interaction and the fracture behavior of concrete were tested, including porosity and crushing resistance. The aggregate porosity was assessed indirectly from density measurements performed in accordance with Eurocode [25]. The oven-dried particle density and saturated surface-dry density were determined separately for the 4.0–8.0 mm and 8.0–16.0 mm fractions for both gravel and dolomite aggregate. The open porosity of the aggregates was estimated from the difference between these density values, providing an indirect measure of the internal pore structure.

The resistance to crushing was evaluated using the Aggregate Crushing Value (ACV) test, conducted in accordance with the code [26]. Dried, standardized samples of fractions: 4.0–8.0 mm and 8.0–16.0 mm were prepared for both gravel and dolomite aggregate. The samples were compressed separately in a cylindrical mould and sieved to determine the mass loss after crushing. The ACV was then calculated as the percentage loss of the crushed material, obtained after applying a compressive load. The aggregate crushing value for each fraction was calculated using the presented formula:

$$\mathrm{ACV} = \frac{(m_1 - m_2) \times 100}{m_1} \tag{3}$$

where: m_1 – the mass of the dried sample prepared for testing [g], m_2 – the mass of the sample after testing and sieving [g].

The surface roughness of the coarse aggregates was characterized using non-contact optical profilometry. Measurements were performed with a Keyence optical microscope, enabling high-resolution three-dimensional surface scans of selected aggregate particles.

The surface topography was quantified using standard roughness and surface porosity parameters derived from the acquired 3D profiles. These measurements provided information on surface irregularities and local depressions, which are known to affect mechanical interlocking and bond strength at the aggregate-cement mortar interface. Optical profilometry enabled a quantitative comparison of surface characteristics between the rounded gravel and the crushed dolomite aggregates, complementing standard geometrical and mechanical tests of the aggregates. The results of conducted laboratory experiments were combined to establish a comprehensive characterization of the properties of coarse aggregates that could influence the strength and fracture toughness of the designed concretes.

2.3 Experimental Methods for Concrete Properties

The experimental program included the determination of selected concrete characteristics for both series L16N and L16D. The two concrete series differed only in the type of coarse aggregate used—natural river gravel (L16N) and crushed dolomite (L16D), while all other mixture design parameters were kept identical. This approach enabled a direct assessment of the influence of aggregate type on concrete mechanical performance and fracture behavior.

The following concrete properties were assessed for both tested series:

- compressive strength (f_c),
- splitting tensile strength ($f_{ct,sp}$),
- modulus of elasticity (E_c),
- critical stress intensity factor (K_{Ic}).

The compressive strength of concrete was determined in accordance with the regulations in [27]. The cylindrical specimens were tested under uniaxial compression using a calibrated testing machine. The tensile strength of concrete was tested by the splitting tensile test according to the Eurocode [28]. The modulus of elasticity of concrete was determined using Method A, in compliance with the standard [29]. The test was conducted under controlled cyclic loading conditions within the specified stress range calculated based on the compressive strength tests.

The critical stress intensity factor (K_{Ic}), representing the fracture toughness under mode I, was determined using the RILEM TC-89 recommendations [30]. The three-point bend test was conducted on notched beams using a closed-loop testing machine. The crack mouth opening displacement (CMOD) and applied load were recorded continuously. The K_{Ic} was calculated based on the maximum load (F_{max}) and the critical effective crack length (a_c), which accounts for stable crack growth prior to failure:

$$K_{Ic} = \frac{3 \cdot (F_{max} + 0.5 \cdot w) \cdot \left[s \cdot (\pi \cdot a_c)^{0.5} \cdot f(a_c/d)\right]}{2 \cdot d^2 \cdot b} \tag{4}$$

where: F_{max} – the peak load [N], $w = (w_o \bullet s)/l$ [N], w_o– self-weight of the tested beam [N], d – beam depth [mm], b – beam thickness [mm], s – loading span [mm], l – specimen length [mm], and $f(a_c/d)$ – shape function in accordance with [30].

The experimental setups for each test used for determining the chosen concrete properties are presented in Fig. 2.

(a) (b) (c) (d)

Fig. 2. Testing set-ups for: a) compressive strength, b) splitting tensile strength, c) modulus of elasticity, d) the critical stress intensity factor.

The combination of compressive, tensile, elastic, and fracture-mechanics-based tests provided a comprehensive characterization of the mechanical behavior of concretes in the L16N and L16D series. The results were used to correlate aggregate properties with concrete performance and to assess the role of aggregate type in controlling fracture resistance and service-life-related durability.

3 Test Results

3.1 Aggregate Properties

The results of the tests performed on the coarse aggregates showed clear differences between natural river gravel and crushed dolomite aggregates. The gravel exhibited lower shape and flatness indices, indicating more rounded, regular particle shapes. In contrast, the crushed dolomite showed higher values of both indices, indicating a greater proportion of angular and elongated particles. Density-based tests revealed differences in aggregate porosity. The dolomite aggregate exhibited a higher calculated porosity than the natural gravel. The results indicated a more developed internal pore structure for the dolomite aggregate. The crushing test results showed lower ACV values for the natural gravel and higher values for the crushed dolomite aggregate, indicating greater resistance to crushing for the gravel. Mean values of all investigated aggregate characteristics are presented in Table 2. Two samples were tested for each property, separately for each fraction, 4.0–8.0 mm and 8.0–16.0 mm. Firstly, a mean value was calculated for each fraction, and, as the final result, a mean value was determined for the whole aggregate mixture 4.0–16.0 mm.

Surface characterization using optical profilometry revealed distinct differences in the surface texture of the tested grains. The gravel aggregate exhibited relatively smooth surfaces with limited micro-scale roughness, whereas the dolomite aggregate showed significantly higher surface irregularity and a more pronounced texture. The examples of three-dimensional scans from the Keyence profilometer are shown in Fig. 3.

3.2 Concrete Properties

The results of mechanical testing of concretes demonstrated consistent behavior within each series, L16N and L16D. The compressive strength (f_c) obtained for L16D concrete

Table 2. The aggregate properties – test results.

Property	Gravel			Dolomite		
	4.0–8.0 mm	8.0–16.0 mm	Mean value for 4.0–16.0 mm mixture	4.0–8.0 mm	8.0–16.0 mm	Mean value for 4.0–16.0 mm mixture
FI, %	2.24 1.01	3.11 2.80	2.30	1.96 2.45	5.20 4.91	3.63
SI, %	1.71 2.18	3.80 4.32	3.03	1.54 2.62	5.51 4.94	3.65
P, %	2.23 2.36	2.63 2.59	2.45	2.84 2.98	3.48 3.62	3.23
ACV, %	6.98 5.49	9.20 11.21	8.20	7.01 7.45	12.24 13.24	9.99
S_a, μm	8.02 7.64	8.63 8.15	8.11	11.56 12.24	13.33 11.91	12.26
S_z, μm	76.87 87.86	92.16 82.41	84.83	121.04 104.99	115.20 106.54	111.94
S_q, μm	10.14 11.02	11.27 13.01	11.35	13.99 15.45	16.92 14.66	15.26

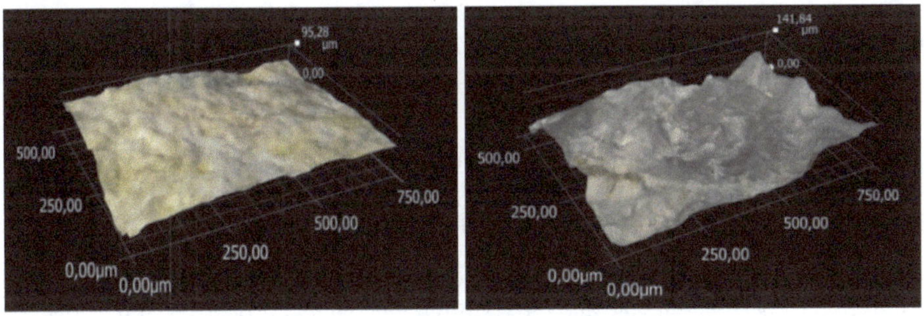

(a) (b)

Fig. 3. Examples of 3D surface morphology scans: a) gravel grain, b) dolomite grain.

was higher than that measured for the L16N series. The results showed low scatter, indicating good quality of the laboratory data collected. The splitting tensile strength ($f_{ct,sp}$) followed the same trend. The static modulus of elasticity also exhibited higher values for the L16D series than for the L16N series. The recorded stress-strain relationships during cyclic loading were stable and allowed reliable determination of the secant modulus for both concrete series.

The fracture toughness results, expressed by the critical stress intensity factor (K_{Ic}), showed distinct differences between the tested concretes. Higher K_{Ic} was obtained for concrete series L16D, while lower values were recorded for L16N. The load-displacement curves obtained from three-point bending tests showed a clear peak corresponding to crack initiation, followed by post-peak softening behavior. Stable crack propagation was observed prior to failure, enabling determination of K_{Ic} based on the unloading compliance, peak load (F_{max}), and the critical mouth-opening displacement (CMOD).

The mean values of all tested concrete strength and fracture properties are juxtaposed for both series in Table 3, where: N – number of tested specimens for a given property, σ – standard deviation for a given property, f_c – compressive strength, $f_{ct,sp}$ – splitting tensile strength, E_c – modulus of elasticity, K_{Ic} – mean critical stress intensity factor.

Table 3. Mechanical and fracture properties of the tested concrete series.

Property	L16N			L16D		
	mean value	σ	N	mean value	σ	N
f_c, MPa	35.1	2.73	6	37.5	3.43	6
$f_{ct,sp}$ MPa	2.68	0.261	6	2.76	0.246	6
E_c, GPa	29.3	0.42	6	30.0	1.19	6
K_{Ic}, MPa$\sqrt{m}$	1.12	0.101	3	1.18	0.043	3

The obtained results provide a comprehensive dataset describing both aggregate properties and the resulting mechanical and fracture-related behavior of concretes L16N and L16D, forming the basis for analysis and discussion presented below.

4 Analysis of Test Results and Discussion on Environmental Impact

4.1 Influence of Aggregate Properties on Concrete Characteristics

The experimental results demonstrate a correlation between the morphological characteristics of the coarse aggregate and the resulting mechanical properties of the concrete. The physical and morphological evaluation of dolomite and gravel revealed distinct differences between the two aggregates. The dolomite aggregate exhibited higher flatness and shape indices than the gravel. While the crushed dolomite exhibited a higher aggregate crushing value and higher porosity compared to natural gravel, suggesting a lower intrinsic resistance to fragmentation, these factors were effectively offset by its better surface attributes.

The 3D optical profilometry confirmed that dolomite particles exhibit a significantly more complex surface topography, as evidenced by surface roughness parameters. From a physical standpoint, this increased roughness, combined with the more angular and irregular grain shapes, facilitated improved mechanical interlocking between the aggregate and the hardened cement paste. In contrast, the rounded and smoother surfaces of the river gravel provided a more limited contact area, which typically results in a weaker bond. The obtained properties for both types of aggregate are summarized in Fig. 4.

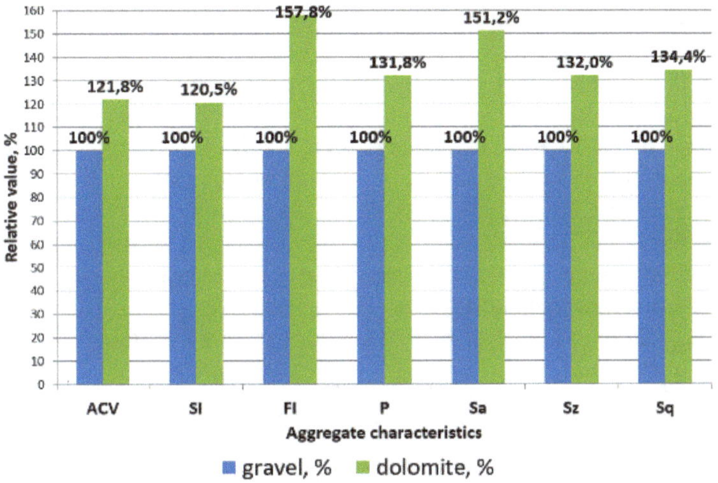

Fig. 4. Comparison of tested aggregate properties, where the gravel results were taken as the base for comparison and calculating the relative increase for dolomite properties.

These aggregate parameters translated into consistent improvements in concrete properties. The dolomite-based concrete in the L16D series generally exhibited higher strength than the gravel-based concrete in the L16N series. The compressive strength for the L16D series was 37.3 MPa, representing a 6.3% increase over the L16N series (35.1 MPa). The splitting tensile strength also showed a similar trend, increasing by 3.0% (from 2.68 MPa in L16N to 2.76 MPa in L16D). Furthermore, the modulus of elasticity reached 30.0 GPa for the dolomite-based concrete, which was 2.4% higher than the value obtained for the gravel-based concrete (29.3 GPa). These results suggested that the crushed aggregate's characteristics have contributed to a stronger internal structure of the tested concrete.

The evaluation of fracture toughness, as measured by the critical stress intensity factor, indicated that the type of coarse aggregate influences resistance to crack propagation. The L16D series achieved a K_{Ic} of 1.18 MPa$\sqrt{m}$, which was 5.4% higher than the 1.12 MPa$\sqrt{m}$ recorded for the L16N series. The juxtaposition of all tested concrete properties in both series is shown in Fig. 5.

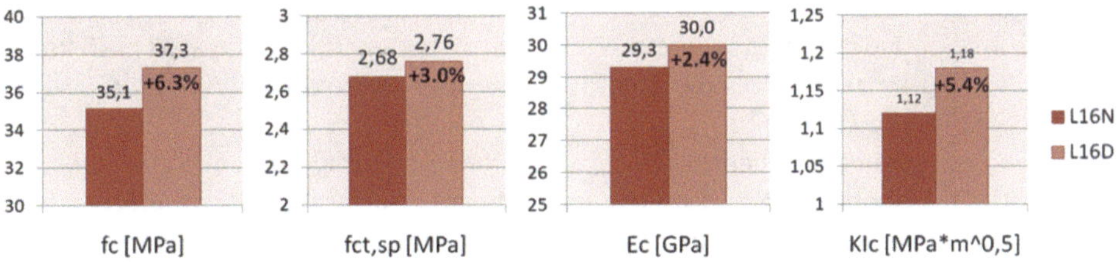

Fig. 5. The comparison of concrete properties.

The higher critical stress intensity factor for the L16D series could potentially be attributed to enhanced mechanical interlocking between the rough, angular aggregate particles and the hardened cement paste. The increased surface area and topographical complexity of the dolomite grains promoted a stronger bond within the so-called interfacial transition zone between the two main phases of concrete's internal structure: coarse aggregate and cement matrix. This enhanced bonding mechanism led to greater energy dissipation during fracture than the smoother paths typically observed around rounded gravel inclusions.

These findings suggest that crushed dolomite aggregate can be a technically viable and potentially more sustainable alternative to natural river gravel, with the potential to improve structural durability by enhancing resistance to crack initiation and propagation.

4.2 Environmental Impact of Tested Coarse Aggregates

While the previous section addressed the influence of aggregate properties on the mechanical and fracture-related performance of concrete, current engineering practice requires that material selection also is considered in relation to the environmental impact. Therefore, this study additionally examines the environmental implications of the tested coarse aggregates using the actual experimental framework. It should be noted that the environmental assessment presented in this study does not constitute a full Life Cycle Assessment (LCA). The adopted approach is limited to transport-related CO_2 emissions associated with aggregate delivery to the laboratory site, reflecting the actual sourcing conditions of the investigated materials. Production-stage emissions, extraction processes, equipment use, and end-of-life scenarios were not included within the system boundaries of this analysis.

The analysis is based on information provided by the local aggregate supplier that delivered the materials for the laboratory experiments (Lublin, Poland), taking into account the locations of aggregate extraction. In the considered case, natural river gravel was transported from a distance of approximately 440 km (the gravel pit "Żwirbet" in Suwałki, Poland), whereas crushed dolomite originated from a source located about 140 km away (the mine "Nowy Staw" in Łagów, Poland). This comparison enabled an assessment of transport-related environmental impact associated with aggregate selection.

Transport-related emissions from road transport by heavy-duty diesel trucks were estimated, assuming an emission factor of 0.15 kg CO_2 per ton-kilometer, consistent with values adopted in the National Life Cycle Assessment Indicators [31]. Under this assumption, the transport of gravel resulted in approximately 66 kg CO_2-eq/t (440 km $\times$ 0.150 kg CO_2/(t·km)), whereas the dolomite generated approximately 21 kg CO_2-eq/t (140 km $\times$ 0.150 kg CO_2/(t·km)).

Despite the higher production-related emissions of crushed dolomite, its substantially shorter transport distance results in a significantly lower overall carbon footprint. This comparison indicates that transport distance is an important parameter controlling aggregate-related environmental impact, exceeding the influence of aggregate processing technology itself. From a sustainability perspective, these findings indicate that aggregate selection based solely on material type may be misleading if transport logistics are not explicitly considered. The results further suggest that optimizing aggregate sourcing at a regional scale may yield greater reductions in concrete-related greenhouse gas emissions than marginal improvements in aggregate production efficiency.

5 Conclusions

In the present study, a comparative analysis was conducted on the influence of coarse aggregate type on concrete properties. Additionally, some environmental aspects related to transport were considered, as transport can be a major factor in the overall global warming potential associated with coarse aggregate use in concrete mixtures. Two aggregates were considered: natural river gravel and crushed dolomite. The following conclusions could be drawn:

- The type of coarse aggregate has a noticeable impact on the mechanical and fracture-related properties of concrete. The use of crushed dolomite aggregate resulted in systematically improved concrete properties (L16D series) compared to concrete produced with natural river gravel (L16N series). However, it should be noted that, although the observed improvements in fracture-related parameters for the dolomite-based concrete were systematic, a more in-depth statistical evaluation would be required to assess the significance of this tendency.
- The aggregate morphology and surface characteristics were identified as key factors influencing the behavior of tested concretes. The higher irregularity and surface roughness of crushed dolomite grains enhanced aggregate-cement matrix bonding, leading to increased compressive strength, splitting tensile strength, elastic modulus, and critical stress intensity factor. Even though the dolomite aggregate is intrinsically more porous and easier to crush than gravel, its ability to form a stronger bond with the cement mortar resulted in a concrete with improved fracture characteristics, which is one of the factors influencing concrete's durability.
- Improved fracture resistance directly contributes to better crack control and reduced susceptibility to durability-related degradation mechanisms. As a consequence, concrete with crushed dolomite aggregate could be expected to have a longer service life and reduced maintenance requirements than concrete containing natural gravel. However, since no direct durability tests were performed within the present study, the implications for service life should be interpreted as indicative rather than conclusive.
- When transportation distances were taken into account, the environmental advantages of crushed dolomite became more pronounced. This combination of enhanced durability and reduced environmental burden supports the use of locally sourced crushed aggregates as a strategy for more sustainable concrete production.

It can be recapitulated that a more detailed analysis is required to select the most appropriate aggregate type for each specific construction investment, balancing the need for higher concrete quality with the goal of reducing environmental impact. Further analyses are necessary to investigate other aggregate types and enhance various properties to a greater extent, thereby ensuring a more substantial improvement in concrete's potential durability. While the current findings demonstrate a consistent trend, it should be noted that the observed improvements in the mechanical properties of the tested concrete require further verification through a detailed analysis of statistical significance in future research. Although the implementation of advanced low-emission technologies is still in progress, improving the performance of concrete made with conventional materials remains important for extending the potential service life of structures.

References

1. Chen S, Teng Y, Zhang Y, Leung CKY, Pan W (2023) Reducing embodied carbon in concrete materials: a state-of-the-art review. Resour Conserv Recycl 188:106653. https://doi.org/10.1016/j.resconrec.2022.106653
2. Braga AM, Silvestre JD, de Brito J (2017) Compared environmental and economic impact from cradle to gate of concrete with natural and recycled coarse aggregates. J Clean Prod 162:529–543. https://doi.org/10.1016/j.jclepro.2017.06.057
3. Arbeláez O, Senior V, Gómez J, Herrera S, Rodríguez C, Santis A (2024) Carbon dioxide emissions from traditional and modified concrete: a review. Environ Dev 52:101036. https://doi.org/10.1016/j.envdev.2024.101036
4. Adesina A (2020) Recent advances in the concrete industry to reduce its carbon dioxide emissions. Environ Chall 1:100004. https://doi.org/10.1016/j.envc.2020.100004
5. Miller SA, Juenger MCG, Kurtis KE, Weiss J (2024) Cement and alternatives in the Anthropocene. Annu Rev Environ Resour 49:309–335
6. Scrivener KL, John VM, Gartner EM (2018) Eco-efficient cements: potential economically viable solutions for a low-CO_2 cement-based materials industry. Cem Concr Res 114:2–26
7. Linares R, López-Uceda A, Piccinali A, Martínez-Ruedas C, Galvín AP (2024) LCA applied to comparative environmental evaluation of aggregate production from recycled waste materials and virgin sources. Environ Sci Pollut Res Int 31:44023–44035
8. Los Santos-Ortega J, Fraile-García E, Ferreiro-Cabello J (2025) Environmental assessment of natural coarse aggregate production in gravel pits—assessing CO_2 offsets through vine cultivation. Appl Sci 15(4):1868. https://doi.org/10.3390/app15041868
9. Koehnken L, Rintoul MS, Goichot M, Tickner D, Loftus A-C, Acreman MC (2020) Impacts of riverine sand mining on freshwater ecosystems: a review of the scientific evidence and guidance for future research. River Res Appl 36(3):362–370
10. de Bortoli A (2023) Understanding the environmental impacts of virgin aggregates: critical literature review and primary comprehensive life cycle assessments. J Clean Prod 415:137629. https://doi.org/10.1016/j.jclepro.2023.137629
11. Darvishvand HR, Taghia HS, S.A., Ebrahimi, M. (2021) The effect of size and shape of aggregate on compressive strength and ductility of concrete. Cem Lime Concr 26(1):46–54
12. Mohammed B et al (2021) The effect of aggregate types on the properties of concrete. Traektoriâ Nauki = Path Sci 7(12):5001–5012
13. Zhao H, Zhang L, Wu Z, Liu A, Imran M (2023) Aggregate effect on the mechanical and fracture behaviours of concrete. Int J Mech Sci 243(1):108067. https://doi.org/10.1016/j.ijmecsci.2022.108067
14. Golewski GL, Xi X, Zheng Y, Zhuo J, Zhang P, Panek R (2025) Experimental evaluation of mode II fracture and microstructure of matrix-aggregate bond of concrete with crushed limestone. Case Stud Constr Mater 23:e05183. https://doi.org/10.1016/j.cscm.2025.e05183
15. Qudoos A, Atta-ur-Rehman, Kim HG, Ryou J-S (2018) Influence of the surface roughness of crushed natural aggregates on the microhardness of the interfacial transition zone of concrete with mineral admixtures and polymer latex. Constr Build Mater 168:946–957. https://doi.org/10.1016/j.conbuildmat.2018.02.205

16. Skarżyński Ł, Tejchman J (2010) Calculations of fracture process zones on meso-scale in notched concrete beams subjected to three-point bending. Eur J Mech A Solids 29(4):746–760
17. Akram A, Słowik M (2025) Comparative analysis of the critical stress intensity factor tested by the RILEM method for two different concretes. In: Kanavaris F, Jędrzejewska A, Benboudjema F, Azenha M (eds) Early-age and long-term cracking in RC structures CRC 2025, RILEM Bookseries, vol 62. Springer, Cham, pp 65–75
18. Jia J-Y, Gu X-L (2021) Effects of coarse aggregate surface morphology on aggregate-mortar interface strength and mechanical properties of concrete. Constr Build Mater 294:123515. https://doi.org/10.1016/j.conbuildmat.2021.123515
19. Teymen A, Nacar S, Arslan A (2023) Statistical investigation of the effects of different origin aggregates on the interfacial transition zone and concrete mechanical performance. Revista de la Construcción 22(2):482–497
20. Golewski GL (2024) Effect of coarse aggregate type on the fracture toughness of ordinary concrete. Infrastructures 9(10):185. https://doi.org/10.3390/infrastructures9100185
21. EN 12390–2:2019 (2019) Testing hardened concrete—Part 2: making and curing specimens for strength tests. European Committee for Standardization (CEN), Brussels, Belgium
22. EN 12390–1:2021 (2021) Testing hardened concrete—Part 1: shape, dimensions and other requirements for specimens and moulds. European Committee for Standardization (CEN), Brussels, Belgium
23. EN 933–3:2012 (2012) Tests for geometrical properties of aggregates—Part 3: determination of particle shape—Flakiness index. European Committee for Standardization (CEN), Brussels, Belgium
24. EN 933–4:2008 (2008) Tests for geometrical properties of aggregates—Part 4: determination of particle shape—Shape index. European Committee for Standardization (CEN), Brussels, Belgium
25. EN 1097–6:2022 (2022) Tests for mechanical and physical properties of aggregates—Part 6: determination of particle density and water absorption. European Committee for Standardization (CEN), Brussels, Belgium
26. PN-B-06714-40:1978 (1978) Kruszywa mineralne. Badania. Oznaczanie wytrzymałości na miażdżenie. Polski Komitet Normalizacyjny, Warsaw, Poland (in Polish)
27. EN 12390–3:2019 (2019) Testing hardened concrete—Part 3: compressive strength of test specimens. European Committee for Standardization (CEN), Brussels, Belgium
28. EN 12390–6:2024 (2024) Testing hardened concrete—Part 6: tensile splitting strength of test specimens. European Committee for Standardization (CEN), Brussels, Belgium
29. EN 12390–13:2021 (2021) Testing hardened concrete—Part 13: determination of secant modulus of elasticity in compression. European Committee for Standardization (CEN), Brussels, Belgium
30. RILEM TC-89 FMC (Draft Recommendation) (1991) Determination of fracture parameters (K_{Ic} and $CTOD_c$) of plain concrete using three – point bend tests. Mater Struct, RILEM Publications SARL 23:457–460
31. EPD Polska Deklaracje Środowiskowe Produktu EPD (Environmental Product Declaration). https://epd.org.pl. Last accessed 15 June 2026

The Influence of Animal Glue on Selected Mechanical and Moisture Parameters of Lightweight Clay Composites

Przemysław Brzyski(✉), Krystian Patyna, and Wojciech Zbyszyński

Faculty of Civil Engineering and Architecture, Lublin University of Technology, Nadbystrzycka 40, Lublin 20-618, Poland
p.brzyski@pollub.pl

Abstract. Clay is one of the unprocessed materials that can be used as a binder in composite building materials. Clay in this form is used in air-dried bricks and blocks, as well as in plasters and mortars. When combined with organic fillers, such as cereal straw or hemp shives, clay forms a lightweight insulating mixture called lightweight clay. One limitation of unfired clay-based materials is their lack of water resistance. Consequently, they are used in interior wall layers. Their low mechanical strength is also a limitation, particularly at higher moisture content. Exploring methods to stabilize such materials is important to improve their performance and expand their range of applications to include exterior wall layers and load-bearing applications. One sustainable way to modify clay is to use biopolymers derived from plants or animals. Some of these additives have a hydrophobic effect, while others improve the clay's strength due to their adhesive and hardening properties.

The work focused on developing a clay-hemp mixture for unfired insulating bricks. The developed building material was stabilized with animal glue obtained from bones. The impact of the modification was evaluated using flexural and compressive strength tests. The effect of the additive on the clay's resistance to water washout was also assessed by examining capillary action. It was demonstrated that the adhesive-modified samples exhibited lower strength in the dry state, whereas adhesive modification proved beneficial at higher relative humidity. The reference and adhesive-modified samples both exhibited low water resistance, whereas the samples without glue proved more durable.

Keywords: Hemp Shives · Animal Glue · Clay · Compressive Strength · Capillary Uptake.

1 Introduction

Clay, with its centuries-old tradition of use in construction [1], is currently being rediscovered. Researchers are examining the potential applications of clay materials in historic buildings for protective [2] and conservation purposes, particularly with regard to earthenware heritage [3]. They are also investigating the impact of clay materials on

Z. Perkowski et al. (Eds.): ECCE 2026, LNCE 929, pp. 32–43, 2027.
https://doi.org/10.1007/978-3-032-39515-3_3

humidity regulation [4] and improving the microclimate in the interiors of contemporary buildings [5], including through the use of clay as a finishing plaster. Researchers are also exploring the potential of clay in combination with other materials. In Streiff et al.'s study [6], a modified COB mixture was tested. The experiment involved adding hemp fibers and ground shives to the clay as a substitute for cereal straw. The tests showed that adding fibers retained the mixture's self-supporting properties without improving its thermal properties. Adding shives reduced the mixture's mechanical properties but improved its thermal parameters. There is a need for further research and the development of solutions that enhance clay's resistance to wetting, as excessive moisture can diminish its mechanical properties and stability [7]. The clay's hardening mechanism, drying, is advantageous. Therefore, blocks made from such mixtures can be artificially dried at elevated temperatures, accelerating their readiness for installation on walls. One limitation of unfired clay-based materials is their lack of water resistance [8, 9]. Their low mechanical strength is also a limitation, particularly at higher moisture content.

One potential solution to stabilize the material and mitigate the impact of excessive moisture on clay behavior is to modify it with natural biopolymers [8–12]. This group includes biopolymers of animal origin, such as protein biopolymers (e.g., casein, gelatin, and collagen adhesives) and chitosan. These biopolymers can form hydrogen, ionic, and electrostatic interactions with clay minerals. Research indicates that their presence can increase material cohesion, stabilize microstructures, and limit clay swelling and degradation under elevated humidity conditions. However, these solutions are only effective in the short term and require further research for construction applications. This group of protein biopolymers includes collagen glues, such as bone glue, which are traditionally used as organic binders in painting and wood technology, among other applications.

Further research is warranted on the modification of clay-hemp composites with natural biopolymers of animal origin to enhance comfort of use and assess the potential to obtain a material with greater resistance to mechanical and moisture-induced damage.

This study examined the mechanical and moisture properties of lightweight, eco-friendly clay and hemp shive composites intended for use as wall materials. The goal was to explore the potential of modifying the composite with a natural, animal-derived adhesive: bone glue. Since this glue is used for wood, among other materials, its adhesive properties may strengthen the bond between wooden organic fillers. The introduction of a protein glue may also contribute to the hydrophobic effect; therefore, this work investigates the composite's ability to transport water via capillary uptake. The study examined flexural and compressive strength under dry and humid conditions, as well as capillary action and resistance to water washout.

2 Materials and Methods

2.1 Materials

This study used raw red clay (manufacturer: Ceradbud, Poland) with dominant oxides, including SiO_2 (60.75%), Al_2O_3 (15.60%), and Fe_2O_3 (10.41%). Quartz sand with a grain size range of 0–2 mm was used as the mineral aggregate. The clay-to-sand weight ratio was assumed to be 2:1. Fine hemp shives up to 5 mm long (manufacturer: Podlaskie Konopie, Poland) were used. The clay-to-shive weight ratio was 3:1. Fine shives were used in this study to create a denser composite structure suitable for plaster mixtures. Shives have a porosity of approximately 78% [13], which means they require a large amount of mixing water to achieve the proper consistency during preparation. The key ingredient is bone glue, a protein-based biopolymer obtained from animal bones. It is supplied by the manufacturer Kremer as beads. According to traditional recipes, it is a strong, water-soluble glue suitable for bonding wood, paper, and fabrics. The study developed and tested two mixtures: the first, a hemp clay mixture without biopolymer (HC-REF); and the second, a mixture containing 5% bone glue by dry weight of the clay (HC-5BG). In the HC-REF sample, a water to clay ratio of 0.91 was assumed, while in the case of HC-5BG, a water to clay plus glue ratio of 0.98 was assumed to obtain a comparable consistency.

2.2 Sample Preparation

First, the hemp shives were mixed with some of the mixing water. Then, clay and sand were added, and the mixture was stirred. The bone glue was soaked in water at room temperature for approximately one hour (glue-to-water ratio 1:5). After heating the water to 70 °C for approximately 30 min, the glue was mixed and dissolved. This procedure was adapted from traditional bone glue recipes. Mixing was performed using a magnetic stirrer. The resulting glue solution was added to the remaining ingredients and mixed until a uniform mixture was obtained. A 1300-watt mortar mixer was used to mix all the ingredients. The mixture was compacted using a wooden tamper to apply a uniform force and avoid discrepancies in sample density. After 24 h, the mold walls were removed from the compacted mixture, and the mixture was left to dry in the laboratory for 3 days. After this time, the samples were placed in a dryer at 50 °C for 2 days to accelerate drying. Before each test, the samples were dried again at 70 °C until the mass stabilized.

2.3 Methods

Apparent Density, Total Porosity, Shrinkage. Apparent density is expressed as the ratio of the dry weight to the volume of a 100 mm × 100 mm × 100 mm sample. Specific density was measured and calculated using the pycnometric method with ethyl alcohol as the test liquid. Total porosity was calculated based on the apparent and specific density results. Shrinkage was calculated as the difference between the mold length and the samples' average dimension after complete drying. Three samples from each mixture were measured for density and shrinkage.

Mechanical Properties. The flexural strength was determined in accordance with EN 1015-11:2001 using an MTS 809 hydraulic press, in which the load can be controlled by adjusting the assumed increase in head displacement. Three standard prisms, each measuring 40 mm × 40 mm × 160 mm, were tested from each recipe. The compressive strength of the hardened mortar was determined using the same press on broken half-samples. In the flexural strength test, the head displacement was assumed to be 0.2 mm/min, whereas in the compression test, it was assumed to be 2 mm/min. The samples were loaded in a direction perpendicular to the compaction of the mixture. Samples were tested under the following conditions: a) in the dry state, b) seasoned in the RH =

57% conditions, to demonstrate the effect of standard relative air humidity on strength, and c) loaded in a direction parallel to the direction of mixture formation, to demonstrate the anisotropic properties of the material (only flexural strength).

Capillary Uptake and Water Resistance. Capillary uptake testing was performed on 100 mm × 100 mm × 100 mm samples. The side surfaces were coated with bitumen. The sample was immersed in water to a depth of approximately 10 mm. The weight gain of the samples was measured after 15 min, 30 min, 1 h, 3 h, 6 h, and 24 h of immersion. The direction of water absorption corresponded to the direction of compaction of the mixture in the molds. To determine resistance to water washout, the samples were immersed in water for 7 and 30 min. Changes, damage, or weight loss were then observed after this time.

3 Results and Discussion

3.1 Apparent Density, Total Porosity, and Shrinkage

Table 1 shows the basic physical parameters of the clay-shive composites. These are the average values of the three samples.

Table 1. Basic physical properties of the hemp-clay composite.

Type of the composite/ Parameter	Apparent density (kg/m^3)	Total porosity (%)	Length loss due to shrinkage (%)
HC-REF	818.2	63.4	1.7
HC-5BG	825.6	63.1	2.1

The adhesive-modified composite had a slightly higher density than the reference one. In these shive-based composites, mixture consistency affects compaction susceptibility; however, the developed mixtures achieved comparable consistencies. In the adhesive-modified mixture, however, slightly more water was added due to the additional binder in the composite, which may have contributed to increased compaction. This was not noticeable during sample preparation, though.

The high porosity of the composites is due to the use of porous hemp shives, and the formation of macropores between them after densification. The presence of the biopolymer did not directly affect total porosity. Other studies using corn starch in similar clay-hemp composites reported slightly higher density, but they had higher water content, resulting in greater susceptibility to densification [14]. As a protein-based substance, bone glue could influence pore distribution, as demonstrated in clay/lime mortars and lime pastes modified with biopolymers such as casein [15] or gum arabic [8]. Biopolymers containing proteins can create air bubbles during mixing or dissolving [15, 16]. Their introduction into a mixture may alter pore distribution, a topic that should be investigated in future studies.

Samples containing glue shrank more upon drying. This may be because the water content of the mix was increased to achieve proper workability and consistency. Furthermore, the HC-5BG mix contained a lower overall percentage of shives in the binder, since the glue was an additional component rather than a partial replacement for clay. Typically, increasing the binder weight in these composites leads to greater shrinkage.

3.2 Mechanical Properties

The graphs (Figs. 1, 2, 3, 4 and 5) illustrate how the samples behave under static loading (displacement versus force). The behavior of each sample is demonstrated to show repeatability or variability.

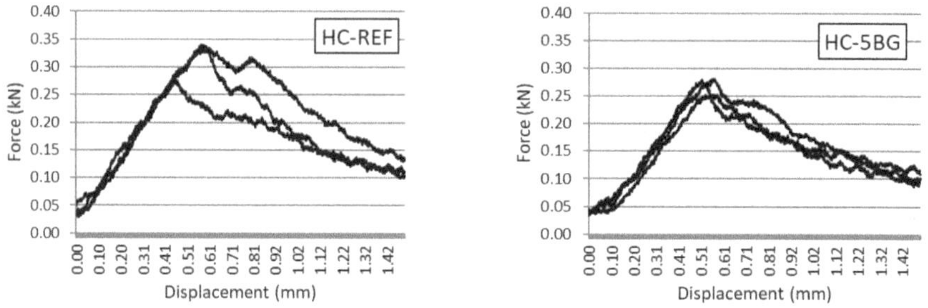

Fig. 1. Displacement-force relationship in the flexural strength test of samples in the dry state, loaded in a direction perpendicular to the direction of sample formation.

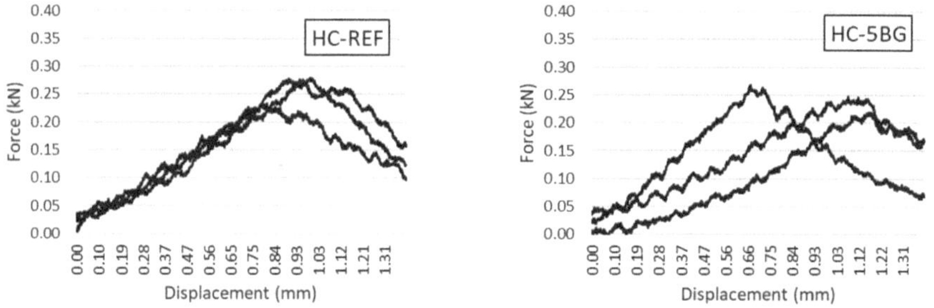

Fig. 2. Displacement-force relationship in the flexural strength test of samples in the dry state, loaded in a direction parallel to the direction of sample formation.

The graphs in Figs. 1 and 2 show the behavior of dried samples during bending tests in two different directions. Changing the direction of the applied force resulted in a noticeable change in the samples' behavior. The shives tend to be arranged with their fibers perpendicular to the direction of compaction. Horizontally oriented shives (perpendicular to the direction of force) increased the flexibility and deflection tolerance of the samples due to fiber orientation. Failure occurred at a displacement of approximately

1 mm (Fig. 2) for samples loaded parallel to the force, whereas for samples loaded perpendicular to the bending direction, failure occurred at around 0.6 mm (Fig. 1), indicating greater stiffness.

Comparing the behavior of the reference and adhesive-modified samples in the dry state, it can be seen that, in both bending (Fig. 1) and compression (Fig. 3), the curves' slopes and the failure displacements were similar for both composites. The adhesive-modified samples showed less variability in their behavior under loading and failed at a lower force than HC-REF.

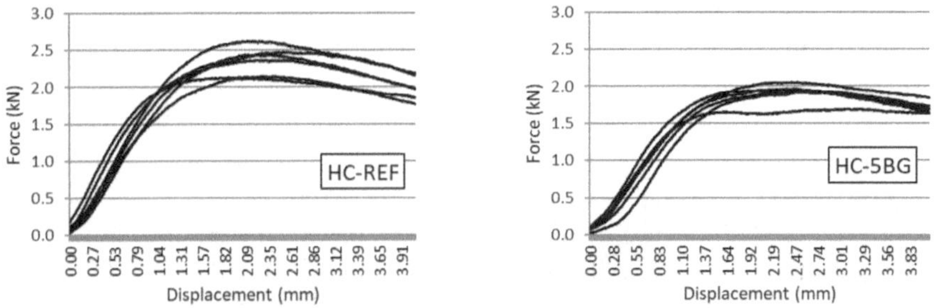

Fig. 3. Displacement-force relationship in the compressive strength test of samples in the dry state, loaded in a direction perpendicular to the direction of sample formation.

Although the samples contained only around 3% moisture by weight, they exhibited different behavior under load when aged at 57% humidity. They failed at a lower force than dry samples, with the difference more pronounced in the HC-REF. The bending test curves (Fig. 4) are disturbed, which may be due to the lower sensitivity of the sample surface contact with the compression head, possibly caused by loosening of the matrix structure due to moisture content. The compression test also showed that samples with adhesive exhibited smaller differences in behavior (Fig. 5), but the curves were more disturbed than those of dry samples shown in Fig. 3.

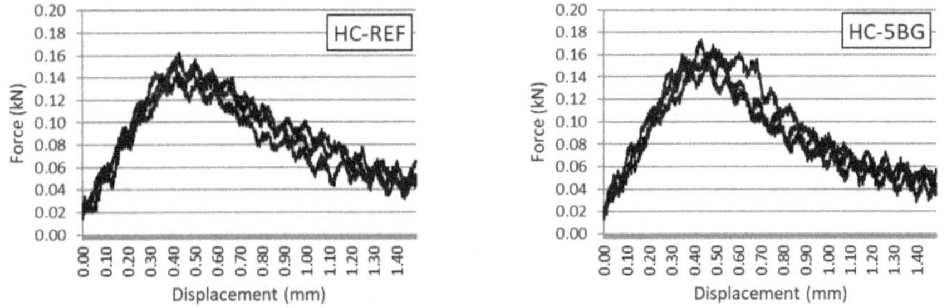

Fig. 4. Displacement-force relationship in the flexural strength test of samples matured in RH = 57% conditions (loaded in a direction perpendicular to the direction of sample formation).

The results of the flexural strength and the compressive strength test of the composites are shown in Fig. 6. Error bars represent standard deviation.

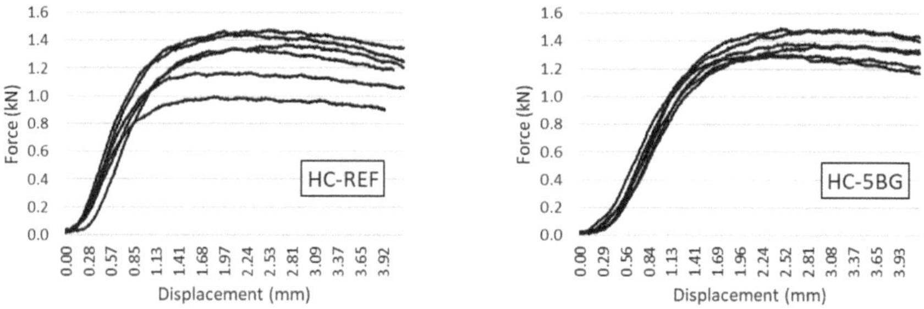

Fig. 5. Displacement-force relationship in the compressive strength test of samples matured in RH = 57% conditions (loaded in a direction perpendicular to the direction of sample formation).

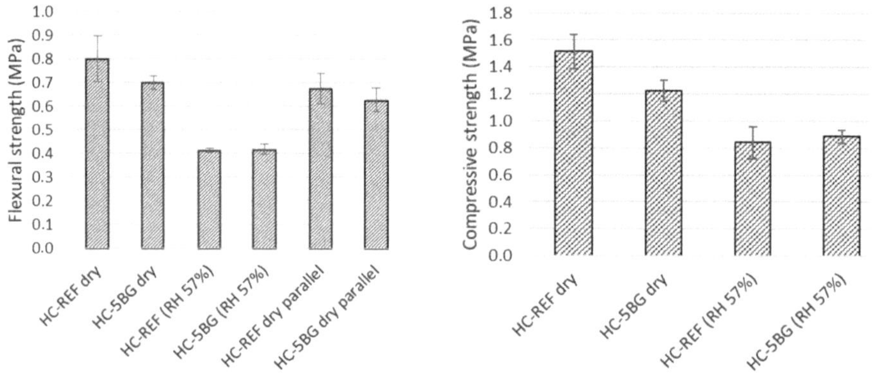

Fig. 6. Average values of flexural and compressive strength of the developed composites.

For dry samples, adding glue reduced strength. Due to the sensitivity of clay to moisture, the strength of the composites in the dry state was significantly higher. However, the strength results for the samples aged at 57% RH are more relevant from a practical perspective. In this case, the flexural and compressive strengths of the composite containing animal glue were slightly higher than those of the reference samples. This suggests that unmodified samples are more susceptible to moisture in the air and that adsorption has a negative impact on strength compared to samples in which the clay and shive particles are surrounded by a bone glue solution. As shown in the SEM images (Fig. 7), the unmodified clay matrix exhibits microcracks, possibly resulting from drying. Separated agglomerates of clay particles are also visible. In contrast, the clay matrix modified with bone glue presents a continuous, uniform layer that covers the shive surfaces.

Therefore, water vapor adsorption may be hindered by limited access to the hydrophilic surface of the hemp shives. The hydrophobic effect of the proteins contained in the glue could also hinder water vapor adsorption. The moisture content of the reference samples at RH = 57% was 3%, whereas that of the samples containing bone glue was 2.6%. The samples were seasoned for only 7 days, and equilibrium was not verified. This should be verified in further, more extensive studies. Another aspect of the study examined the effect of the direction of force application on flexural strength.

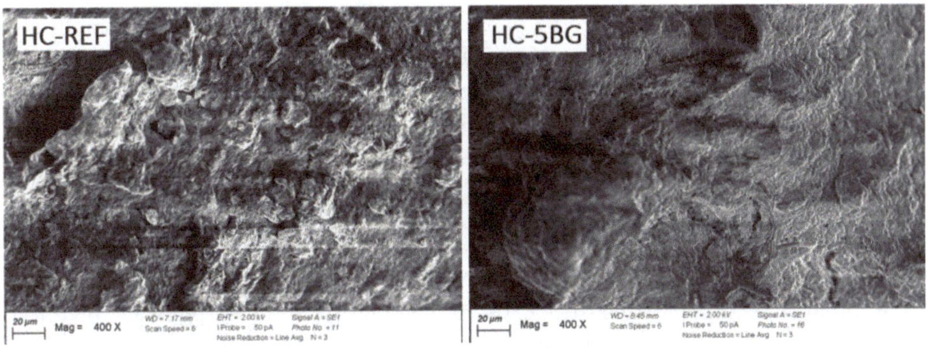

Fig. 7. Surface of composite samples under SEM observation.

Samples loaded in the direction of the mixture compaction in the mold were found to be weaker than those loaded perpendicular to the molding direction. The difference in strength values between the reference samples and the samples modified with glue was smaller when loaded in the same direction as the molding. Hemp shives tend to align their fibers perpendicular to the direction of the compaction force. Other works, [17] and [18], also demonstrated differences in the behavior of samples based on hemp shives under load in two mutually perpendicular directions.

3.3 Capillary Uptake and Water Resistance

The photos (Fig. 8) show cut samples after the capillary uptake test, which was carried out for 24 h (HC-REF) and 1 h (HC-5BG). The height to which the water was pulled can be seen.

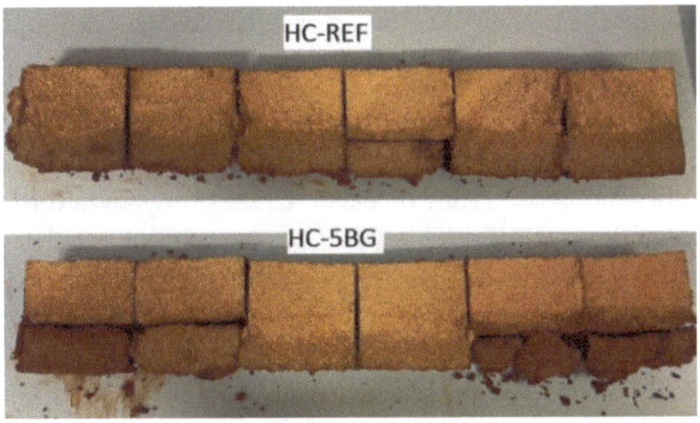

Fig. 8. Sample fractures after 24 h (HC-REF) and 1 h (HC-5BG) of capillary uptake, indicating the height of the water column.

In the reference samples, damage to the lower part was visible as cavities and water absorption-related swelling. One sample was damaged during cutting. While the samples appeared to retain their shape, the bottom zone was noticeably damaged as water had loosened the bonds between the clay particles. This area of the sample was soft, and even minimal pressure caused it to deform. The sample walls were coated in bitumen to protect them from excessive leaching and destruction. The capillary rise in the reference samples was approximately 4–5 cm over 24 h. In the samples modified with glue, the matrix structure was found to be more susceptible to water damage. After just one hour of the capillary uptake test, the lower part that had become saturated detached when the samples were removed from the water, as shown in the photo (Fig. 8). This demonstrates the detrimental impact of animal glue on the binder's adhesion to the shive particles. Only one sample was undamaged; this may be due to stronger compaction of the sample or a different arrangement of the shives, resulting in lower water absorption and therefore lower stress on the lower part of the sample. Due to significantly lower water absorption, this one sample was excluded from the calculation of the average capillary action value. Within one hour, the modified samples pulled water up to a height of 3.5–4 cm. Due to sample damage, the test was terminated after 1 h of saturation. The graph (Fig. 9) shows the mass increase of samples that were subjected to water saturation via capillary action.

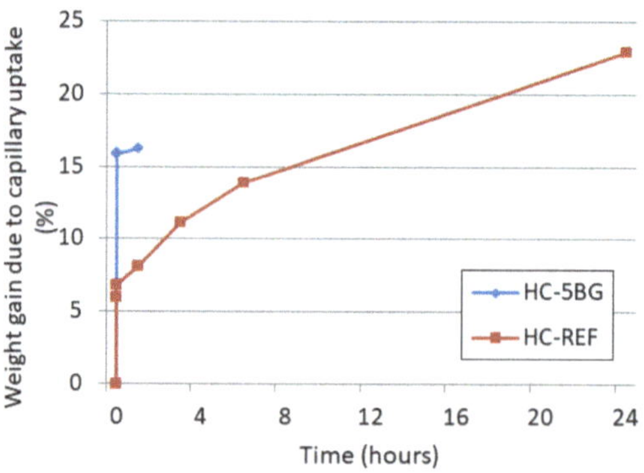

Fig. 9. Percentage of capillary water absorbed by the tested samples.

As shown, water absorption in the first 15 min was significantly higher in the samples modified with bone glue than in the HC-REF. This suggests that the structure was loosened considerably by the presence of water. Cracks may have formed in the sample due to destruction, creating additional spaces that were quickly filled with water. This resulted in a significant load, leading to part of the sample detaching. In the reference sample, saturation was also most intense in the first few minutes, albeit to a lesser extent than in the HC-5BG. This indicates greater resistance to cracking and destruction and allows for water penetration. The initial high water saturation capacity is related to the presence of porous shives and the lower part of the sample being in direct contact with water. Later, the upper portions of the sample that were not immersed in water became saturated. This saturation was related to the distribution of capillary pores within and between the shives coated with the binder. Access to higher levels by water may have been hindered by the sealing of pores by water-saturated clay particles. In other studies where analogous samples containing different biopolymers were tested, the test duration was 48 h, and the capillary lift height was comparable [9, 14]. However, when the samples were positioned so that capillary transport proceeded perpendicular to their formation direction, the lift height increased to 9 cm [9, 14]. In this configuration, some of the shives were oriented with their capillary pores aligned with the direction of water transport, thereby facilitating greater water uptake. This demonstrates the composite's anisotropic properties due to the presence of shives with oriented capillary pores.

Different observations were made during the water immersion test. Figure 10 shows the damage to the sample after immersion in water for 7 and 30 min. While the glue-modified samples remained intact, they were very soft and deformed under light pressure. However, the glue prevented the binder from being washed away. Mass loss was negligible. The samples' shape, mass, and uniform saturation with water (at approximately 56% by mass) ensured that no part detached, unlike in the capillary action test (Fig. 8). The reference samples, on the other hand, were washed away and destroyed. After 30 min, the mass loss was approximately 7%, but as can be seen in the photograph, the samples were completely destroyed. This small mass loss is due to the binding of the matrix with shives, as has also been demonstrated in other studies on clay composites modified with plant and animal glues.

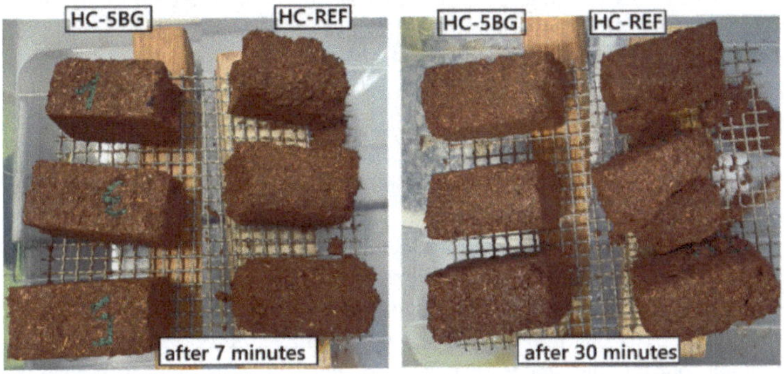

Fig. 10. The form of the samples after 7 and 30 min of immersion in water.

4 Conclusions

This paper presents selected research results on the modification of clay-hemp composites. It was demonstrated that in the dry state, samples modified with glue exhibited lower strength (flexural by about 12%, compressive by about 19%), whereas under higher relative humidity (57%), the glue modification proved beneficial. In this case, the modified composite was slightly stronger than the reference composite. In practical wall applications, the material will operate under various RH conditions, including around 50%, which is why this is a positive aspect of using bone glue in these composites. However, the strength of both decreased due to increased RH. Both the reference samples and the samples with added glue showed low water resistance, whereas the samples without glue were more durable. Further research should consider modifying the animal glue, for example, with an alkaline substance such as lime or thermal treatment at a different temperature range. It would also be advisable to examine the effect of higher glue contents on selected parameters to ultimately assess the feasibility of using this preparation in clay-based composites.

References

1. Fernandes FM, Lourenço PB, Castro F (2010) Ancient clay bricks: manufacture and properties. In: Dan MB, Přikryl R, Török Á (eds) Materials, technologies and practice in historic heritage structures. Springer. https://doi.org/10.1007/978-90-481-2684-2_3
2. Michette M, Lorenz R, Ziegert C (2017) Clay barriers for protecting historic buildings from ground moisture intrusion. Herit Sci 5:31. https://doi.org/10.1186/s40494-017-0144-3
3. Camões A, Eires R, Jalali S (2012) Old materials and techniques to improve the durability of earth buildings. In: CIAV-ICOMOS 2012
4. Jiang Y et al (2020) Moisture buffer, fire resistance and insulation potential of novel bio-clay plaster. Constr Build Mater 244:118353. https://doi.org/10.1016/j.conbuildmat.2020.118353
5. Peterková J, Zach J, Novák V, Korjenic A, Sulejmanovski A, Sesto E (2024) Optimizing indoor microclimate and thermal comfort through sorptive active elements: stabilizing humidity for healthier living spaces. Buildings 14(12):3836. https://doi.org/10.3390/buildings14123836
6. Streiff F et al (2020) CobBauge – A hybrid walling technique combining mechanical and thermal performance. In: Tagungsbeiträge Der 8. Internationalen Fachtagung Für Lehmbau 8th International Conference on Building with Earth
7. Affan H, Arairo W, Barraj F, Khatib M, Saba M, El Mendili Y (2025) Mechanical characteristics of clay-based masonry walls. Eng 6(10):260. https://doi.org/10.3390/eng6100260
8. Trigueiro P, Pereira JP d L, Ferreira MG, Silva LB, Neves L, Peña-Garcia RR (2024) Clay minerals and biopolymers in film design: overview of properties and applications. Minerals 14(6):613. https://doi.org/10.3390/min14060613
9. Brzyski P (2024) Influence of gum arabic admixture on selected properties of clay mortar. Constr. Build. Mater. 451:138865. https://doi.org/10.1016/j.conbuildmat.2024.138865
10. Aguilar R et al (2016) The potential use of chitosan as a biopolymer additive for enhanced mechanical properties and water resistance of earthen construction. Constr Build Mater 114:625–637. https://doi.org/10.1016/j.conbuildmat.2016.03.218
11. Brzyski P, Wankiewicz J, Pavlík Z, Sędzielewska E, Łagód G (2025) Selected properties of clay-shives wall material modified with protein-based polymer. AIP Conf Proc 3355:020002. https://doi.org/10.1063/5.0291565
12. Minke G (2022) Building with earth: design and technology of a sustainable architecture, 4th and revised edn. Birkhäuser, Berlin, Boston https://doi.org/10.1515/9783035622553
13. Kosiński P, Brzyski P, Tunkiewicz M, Nagy B, Gotkowska-Płachta A (2024) Moisture properties, microstructural and microbiological characteristics of flax and hemp shives used in construction industry. J Nat Fibers 21(1):2356695. https://doi.org/10.1080/15440478.2024.2356695
14. Patyna K, Zbyszyński W, Raczkiewicz A, Brzyski P (2025) The effect of modifying the clay-hemp composite with corn starch on its ability to transport water by capillary action. J Phys Conf Ser 3146:012023. https://doi.org/10.1088/1742-6596/3146/1/012023
15. Brzyski P, Suchorab Z, Łagód G (2021) The influence of casein protein admixture on pore size distribution and mechanical properties of lime-metakaolin paste. Buildings 11(11):530. https://doi.org/10.3390/buildings11110530
16. Dinç-Şengönül B, Yüzer N, Boylu S, Erdil K, Güdendede YA (2023) Utilization of waste animal blood as an additive in hydraulic lime-based mortars. Constr Build Mater 400:132909. https://doi.org/10.1016/j.conbuildmat.2023.132909

17. Williams J, Lawrence M, Walker P (2018) The influence of constituents on the properties of the bio-aggregate composite hemp-lime. Constr Build Mater 159:9–17. https://doi.org/10.1016/j.conbuildmat.2017.10.109
18. Williams J, Lawrence M, Walker P (2017) The influence of the casting process on the internal structure and physical properties of hemp-lime. Mater Struct 50:108. https://doi.org/10.1617/s11527-016-0976-4

Shrinkage of Cement Composites Containing Waste Polystyrene

Elżbieta Kuźmińska(✉), Mateusz Zakrzewski, and Jacek Domski

Koszalin University of Technology, 2 Śniadeckich Street, Koszalin 75-453, Poland
elzbieta.kuzminska@tu.koszalin.pl

Abstract. The use of waste polystyrene as aggregate in cement composites is a solution that supports plastic waste management and reduces the consumption of natural aggregates. One of the key issues affecting the practical application of such materials is their shrinkage.

This paper presents the results of linear shrinkage tests of cement composites modified with two types of polystyrene aggregate: a mixture of waste expanded and extruded polystyrene (EPS + XPS) and waste extruded polystyrene (XPS). Two mix formulations were developed for each type of aggregate: the first, in which the aggregate consisted exclusively of waste polystyrene, and the second, additionally containing fine aggregate in the form of sand.

The tests were conducted on cylindrical specimens, and shrinkage was determined from length changes measured between gauge points placed on their lateral surface. The results showed high and similar shrinkage values for composites containing only polystyrene aggregate, regardless of the aggregate type. The addition of fine sand led to a significant reduction in shrinkage values and modified its development over time. In addition, the basic properties of the composites were assessed by determining their compressive strength and bulk density.

Keywords: Concrete Shrinkage · Cement Composites · Waste Polystyrene · EPS/XPS Aggregates.

1 Introduction

Lightweight concretes are widely used in construction in situations where reducing the self-weight of elements, improving thermal insulation, and increasing energy absorption capacity are important. One effective way to obtain materials with reduced density is to use lightweight polymer aggregates, in particular waste expanded polystyrene (EPS) and extruded polystyrene (XPS) [1–5]. This solution is mainly applied in non-load-bearing elements and layers, where strength requirements are secondary to functional properties.

The incorporation of polystyrene aggregates into the cement matrix allows for a significant reduction in composite density; however, it also leads to a deterioration in their mechanical properties and an increased susceptibility to long-term deformations, including shrinkage and creep [6, 7]. These effects result primarily from the very low stiffness of polystyrene aggregate and its limited ability to restrain deformations of the

Z. Perkowski et al. (Eds.): ECCE 2026, LNCE 929, pp. 44–54, 2027.
https://doi.org/10.1007/978-3-032-39515-3_4

cement matrix [8, 9], which distinguishes these composites from lightweight concretes with mineral aggregates.

Shrinkage of lightweight concrete containing EPS is one of the main factors limiting its application. Studies have shown that an increase in the volume fraction of polystyrene results in greater linear shrinkage than in normal-weight and lightweight concretes with mineral aggregates [6, 10]. This effect is further intensified by the high cement paste content and the lack of effective restraint of deformation by the polystyrene aggregate [6, 8]. Therefore, shrinkage is a key practical issue in the design of this type of material.

To reduce unfavorable shrinkage deformations, numerous attempts have been made to modify the composition of cement mixtures, including the use of mineral additions, chemical admixtures, and dispersed reinforcement [11–14]. In particular, it has been shown that the introduction of mineral phases with higher stiffness may reduce composite shrinkage and result in stiffer behavior under load [15, 16].

A characteristic feature of lightweight concretes with polystyrene aggregates is the nonlinear stress-strain relationship and the ability to undergo significant deformations before failure [13, 17, 18]. For this reason, the mechanical properties of these materials are increasingly determined with reference to a specified level of axial strain, instead of the conventional compressive strength [17, 19]. This approach enables a more accurate assessment of the deformability of cement composites containing polystyrene aggregates.

Despite numerous studies on EPS-based concretes, issues related to the shrinkage of composites containing XPS and EPS-XPS mixtures, as well as the addition of sand, remain insufficiently recognized. In the available literature, the influence of both the type of polystyrene and the presence of sand on the material's shrinkage and mechanical properties is rarely analyzed simultaneously [5, 6, 10].

The aim of this study is to analyze the linear shrinkage of cement composites containing waste polystyrene aggregates (EPS + XPS and XPS), taking into account the effect of sand addition on the development of deformations over time. In addition, the basic properties of the tested composites were determined by measuring their compressive strength and bulk density.

2 Materials and Methods

2.1 Materials

A cement binder with the trade name iX CPP20, intended for floor and leveling layers, was used. Its main constituents include portland cement clinker, limestone, ground granulated blast furnace slag, siliceous fly ash (V), and gypsum; the detailed composition is provided in the technical data sheet [20]. Within the research program, no mineral additions or chemical admixtures were applied that could affect the development of shrinkage deformations.

The lightweight aggregate consisted of crushed polystyrene granulate obtained from recycled insulation materials: a mixture of EPS + XPS (Fig. 1) and XPS (Fig. 2), with a particle size of 0–8 mm. The aggregate was not surface-modified or pre-treated.

Fig. 1. Recycled EPS + XPS granulate.

Fig. 2. Recycled XPS granulate.

In selected mixtures, fine mineral aggregate in the form of hydroclassified sand with a particle size of 0–2 mm was used [21]. Tap water complying with the requirements for concrete mixing water was used [22].

2.2 Mix Composition

The experimental program included four series of cement mixtures, differing in:

- the type of polystyrene aggregate (EPS + XPS or XPS),
- the presence of fine quartz aggregate (with sand/without sand).

For each type of polystyrene aggregate, one mixture containing only polystyrene aggregate and one mixture modified with sand were designed.

The water-to-binder ratio (w/b) was kept constant within the compared pairs of mixtures. For mixtures designated as MAI and MCI, the w/b ratio was 0.32, while for mixtures MAIII and MCIII it was 0.38. The detailed mix compositions, including the w/b values and the quantities of individual components, are presented in Table 1.

Table 1. Cement mix compositions.

Name	MAI [kg/m^3]	MCI [kg/m^3]	MAIII [kg/m^3]	MCIII [kg/m^3]
iX CPP20 binder	420.00	420.00	335.00	335.00
Water	135.00	135.00	128.00	128.00
Recycled polystyrene EPS + XPS	18.30	–	16.10	–
Recycled polystyrene XPS	–	22.50	–	18.30
Hydroclassified sand	–	–	320.00	400.00
w/b	0.32	0.32	0.38	0.38

All mixture variants were prepared under laboratory conditions. The constituent materials were weighed using precision laboratory scales. The mixing process was carried out mechanically in a BP-135 mixer, with the components added in a fixed sequence. First, the polystyrene granulate was mixed with the binder, then water was added, and finally sand, if applicable. No segregation of components was observed during mixing. The proportions of individual components were determined experimentally, with the mixture's consistency as the main criterion.

3 Research Methodology and Results

3.1 Density of Composites

The bulk density of hardened composites was determined on cylindrical specimens with dimensions of 150 mm × 300 mm after 28 days of curing under laboratory conditions (temperature 20 ± 2 °C, relative humidity 50 ± 5%) [23]. For each mix design, nine measurements were performed, and the results were statistically verified using the Dixon test, which did not indicate any outliers.

The density results for mixtures with similar compositions show comparable relative differences between the MAI–MCI and MAIII–MCIII series. In the case of mixtures without sand, the average density of mixture MCI was 13.2% higher than that of MAI. In the group of mixtures containing sand, the density of mixture MCIII was 13.4% higher than that of MAIII.

The statistical analysis confirmed the consistency of the results within each measurement series. The density results together with the basic statistical parameters are presented in Table 2.

Table 2. Summary of the density of cylindrical specimens.

Parameter	Unit	MAI	MCI	MAIII	MCIII
Density	[kg/m^3]	491	556	717	813
Standard deviation σ	[kg/m^3]	21.9	28.2	41.3	29.3
Coefficient of variation υ	[%]	4.5	5.1	5.8	3.6

3.2 Compressive Strength of Composites

Compressive strength tests were carried out on cylindrical specimens with dimensions of 150 mm × 300 mm after 28 days of curing under laboratory conditions (temperature 20 ± 2 °C, relative humidity 50 ± 5%) [23]. Prior to testing, the top surfaces of the specimens were trimmed using a concrete cutting saw to ensure planarity. For each tested composite, the stress-strain relationship was determined. Based on this, the maximum force recorded during the test was identified. The tests were performed on six specimens per series, and the Dixon test did not indicate any outliers. The results of the compressive tests are presented in Table 3.

Table 3. Compressive strength and corresponding maximum forces.

Material	F_{max} [kN]	Compressive strength [MPa]
MAI	9.770	0.553
MCI	15.31	0.866
MAIII	15.80	0.894
MCIII	23.12	1.308

Higher compressive strength values were recorded for series MAIII and MCIII containing sand compared to the series without the fine natural aggregate fraction. This trend was observed for both mixtures with EPS + XPS-based aggregate and those with XPS aggregate.

3.3 Linear Shrinkage of Composites

Linear shrinkage tests were carried out on cylindrical specimens with dimensions of 150 mm × 300 mm, in accordance with the requirements of the standard [24]. Three specimens were prepared for each mix design. Shrinkage measurements began after demolding, and the specimens were then placed in a climatic chamber under controlled environmental conditions: temperature 20 ± 2 °C and relative humidity 50 ± 5%.

On the lateral surface of each specimen, three lines were marked, and two measuring studs were attached to each line at an axial spacing of 250 mm (Fig. 3). Shrinkage strain was determined as the average of the measurements taken on the individual gauge lengths. Readings were performed at equal time intervals after 3, 7, 14, 28, 56, 90, 180, and 360 days from the start of the test. A digital Tensotast measuring device (Huggenberger) with a reading accuracy of 0.001 mm was used, as shown in Fig. 4.

Table 4 presents the average linear shrinkage values, measured in ‰, for each composite series at successive measurement times.

Table 4. Linear shrinkage of the tested composites as a function of time [days].

Test time [day]	MAI [‰]	MCI [‰]	MAIII [‰]	MCIII [‰]
3	0.663	0.532	0.184	0.263
7	1.297	1.235	0.543	0.535
14	2.011	1.811	0.876	0.984
28	2.631	2.503	1.285	1.405
56	3.067	2.944	1.541	1.720
90	3.325	3.189	1.696	1.901
180	3.475	3.403	1.763	1.984
360	3.932	3.861	1.881	2.228

Fig. 3. Shrinkage test setup.

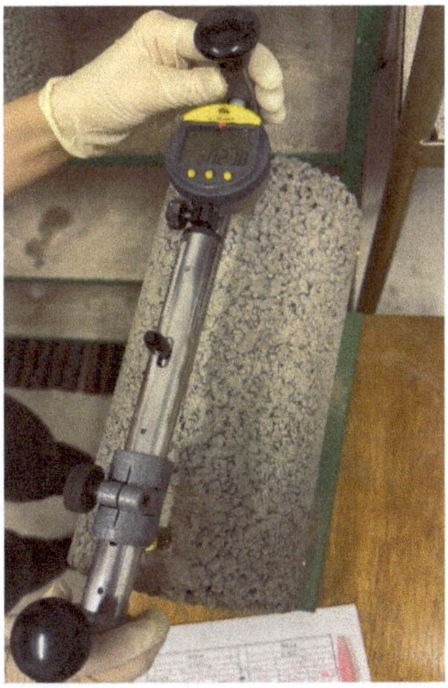

Fig. 4. Shrinkage measurement.

The linear shrinkage results show differences among the analyzed mix groups throughout the observation period. Composites without sand (MAI and MCI) reached higher shrinkage strain values compared to mixtures containing mineral aggregate (MAIII and MCIII). The differences between the series were visible at an early stage of curing and persisted until the end of the observation period.

The development of shrinkage over time had a similar pattern for all tested materials. During the initial curing period, a rapid increase in strain was observed, followed by a gradual slowdown. In order to provide a quantitative description of the linear shrinkage development over time, an approximation function of the following form was determined for each composite:

$$\varepsilon_{cs}(t) = E_1\left(1e^{-k_1 t}\right) + E_2\left(1e^{-k_2 t}\right) \quad (1)$$

where $E_1$and E_2represent the components of shrinkage strain, while $k_1$and $k_2$are coefficients describing the rate of its development.

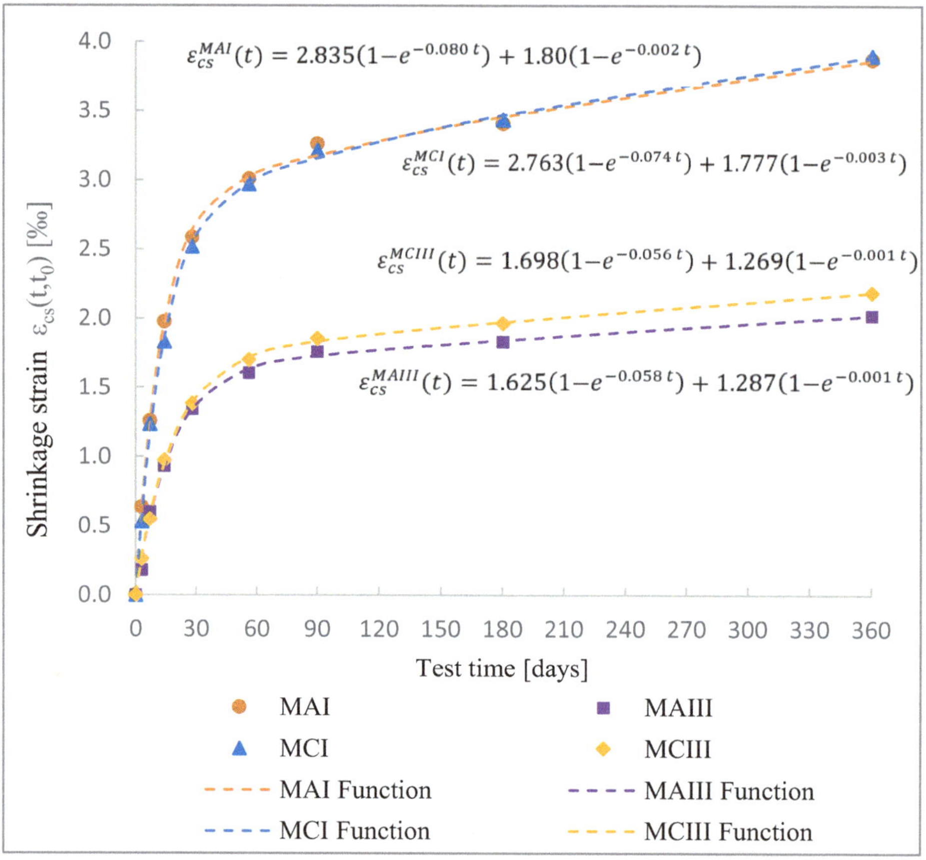

Fig. 5. Development of linear shrinkage as a function of time.

Table 5. Parameters of the approximation function describing the development of linear shrinkage and the goodness-of-fit measures.

Parameter	Unit	MAI	MAIII	MCI	MCIII
E_1	[‰]	2.835	1.625	2.763	1.698
k_1	[1/d]	0.08	0.058	0.074	0.056
E_2	[‰]	1.804	1.287	1.777	1.269
k_2	[1/d]	0.002	0.001	0.003	0.001
MEA	[‰]	0.042	0.038	0.041	0.016
δ	[%]	0.72	0.90	0.77	0.45

The function parameters were determined by least squares, minimizing the sum of squared differences between the measured values and those calculated from the model, over the full set of measurement points from 0 to 360 days. The mean absolute error (MAE) and the relative error δwere adopted as measures of the goodness of fit (Table 5). The obtained relative error values, ranging from 0.45% to 0.90%, indicate very good agreement between the adopted approximation function and the experimental results.

Figure 5 presents the development of linear shrinkage over time for the tested composites, along with the approximation curves obtained using the proposed approximation function.

In addition, shrinkage calculations were performed in accordance with EN 1992-1-1 [25]. In the calculations, the experimentally determined mean compressive strength of the composites, f_{cm}, obtained from the maximum failure force, F_{max}, was used as an input parameter for the EC2 model (Table 3). The calculated shrinkage values, as per [25], for the individual test days are presented in Table 6.

Table 6. Shrinkage calculated according to EN 1992-1-1.

Test time [day]	MAI [‰]	MCI [‰]	MAIII [‰]	MCIII [‰]
3	0.108	0.107	0.107	0.107
7	0.163	0.162	0.162	0.162
14	0.227	0.226	0.226	0.225
28	0.31	0.309	0.309	0.308
56	0.414	0.412	0.412	0.41
90	0.493	0.491	0.491	0.488
180	0.608	0.605	0.605	0.602
360	0.707	0.704	0.704	0.701

Over the entire analyzed time range, the shrinkage values calculated according to [25] were significantly lower than the experimentally obtained values for the tested composites. These differences were already visible at the early stage of curing and persisted until the end of the observation period.

4 Discussion

The comparison between mixtures containing the EPS + XPS blend and those based solely on XPS aggregate showed no significant differences in shrinkage values or in the character of their development over time. This indicates that, within the study's scope, the overall elastic deformability of polystyrene aggregate as a material group is of primary importance, rather than its technological form. Similar conclusions have been reported in studies analyzing the effects of EPS type and particle size, showing that shrinkage changes depend more strongly on the volume fraction of aggregate and the continuity of the cement matrix than on the specific characteristics of the aggregate itself [3, 9, 26].

Significant differences in shrinkage values were observed in sand-modified mixtures. The presence of a fine natural aggregate fraction reduced shrinkage throughout the test period, regardless of the type of polystyrene aggregate used. This effect can be interpreted as a result of the increased proportion of mineral phases with higher stiffness and improved particle packing within the cement matrix, which promotes more effective restraint of shrinkage deformations. Similar trends have been observed in studies on EPS concretes modified with mineral additions, where improved dimensional stability was associated with a denser composite structure [12, 15, 27].

It should be emphasized, however, that the scope of the research was limited to a small number of mix designs and a single set of curing conditions. The influence of environmental variables or other mix modifications, such as chemical admixtures or dispersed reinforcement, was not analyzed, although these factors have been shown in the literature to significantly affect shrinkage and mechanical behavior of concretes with polystyrene aggregates [13, 14]. Despite these limitations, the results provide insight into the influence of polystyrene aggregate type and mix grading on the shrinkage and deformability of cement composites.

5 Conclusions

1. Cement composites containing waste polystyrene aggregates show significant susceptibility to linear shrinkage, resulting from the limited ability of polystyrene aggregate to restrain deformations of the cement matrix.
2. The type of polystyrene aggregate used (EPS + XPS blend or XPS) is not a decisive factor for the level of shrinkage or the character of its development over time, provided that comparable mix compositions are maintained.
3. The use of a fine aggregate fraction in the form of sand leads to a clear reduction in the linear shrinkage of cement composites, regardless of the type of polystyrene aggregate applied.
4. The development of shrinkage over time was similar for all tested composites and included an intensive increase in strain at the early stage of curing, followed by gradual stabilization at later ages.
5. The adopted function describing shrinkage development over time enabled an accurate representation of linear shrinkage, as confirmed by the low fitting error values for all mix designs.
6. The shrinkage calculated according to EN 1992-1-1 was significantly lower than the experimentally obtained values, indicating the limited applicability of standard models for the quantitative description of shrinkage in composites with a high content of polystyrene aggregates.

References

1. Sabaa B, Rasiah S (1997) Engineering properties of lightweight concrete containing crushed expanded polystyrene waste. https://www.researchgate.net/publication/265916085. Last accessed 21 June 2026
2. Rucińska T, Kiernożycki W (2010) Mechanical properties of concrete with foamed styrene aggregate. Cem Wapno Beton 15:289–295

3. Xu Y, Jiang L, Xu J, Li Y (2012) Mechanical properties of expanded polystyrene lightweight aggregate concrete and brick. Constr Build Mater 27(1):32–38. https://doi.org/10.1016/j.conbuildmat.2011.08.030
4. Al-lami MS, Al-saadi EA-M (2021) The relationships between compressive strength and density of polystyrene lightweight concrete and their component ratios. J Appl Eng Sci 19(1):175–185. https://doi.org/10.5937/jaes0-27471
5. Prasittisopin L, Termkhajornkit P, Kim YH (2022) Review of concrete with expanded polystyrene (EPS): performance and environmental aspects. J Clean Prod 366:132919. https://doi.org/10.1016/j.jclepro.2022.132919
6. Tang WC, Lo Y, Nadeem A (2008) Mechanical and drying shrinkage properties of structural-graded polystyrene aggregate concrete. Cem Concr Compos 30(5):403–409. https://doi.org/10.1016/j.cemconcomp.2008.01.002
7. Tang WC, Cui HZ, Wu M (2014) Creep and creep recovery properties of polystyrene aggregate concrete. Constr Build Mater 51:338–343. https://doi.org/10.1016/j.conbuildmat.2013.10.093
8. Major M, Halbiniak J (2019) Effect of adhesion between EPS granules and cement matrix on the characteristics of lightweight concretes. IOP Conf Ser Mater Sci Eng 603(3):032054. https://doi.org/10.1088/1757-899X/603/3/032054
9. Babu DS, Babu KG, Tiong-Huan W (2006) Effect of polystyrene aggregate size on strength and moisture migration characteristics of lightweight concrete. Cem Concr Compos 28(6):520–527. https://doi.org/10.1016/j.cemconcomp.2006.02.018
10. Demirboga R, Kan A (2012) Thermal conductivity and shrinkage properties of modified waste polystyrene aggregate concretes. Constr Build Mater 35:730–734. https://doi.org/10.1016/j.conbuildmat.2012.04.105
11. Sun Y et al (2022) An investigation of the properties of expanded polystyrene concrete with fibers based on an orthogonal experimental design. Materials 15(3):1228. https://doi.org/10.3390/ma15031228
12. Babu KG, Babu DS (2003) Behaviour of lightweight expanded polystyrene concrete containing silica fume. Cem Concr Res 33(5):755–762. https://doi.org/10.1016/S0008-8846(02)01055-4
13. Babavalian A, Ranjbaran AH, Shahbeyk S (2020) Uniaxial and triaxial failure strength of fiber reinforced EPS concrete. Constr Build Mater 247:118617. https://doi.org/10.1016/j.conbuildmat.2020.118617
14. Chen B, Liu J (2004) Properties of lightweight expanded polystyrene concrete reinforced with steel fiber. Cem Concr Res 34(7):1259–1263. https://doi.org/10.1016/j.cemconres.2003.12.014
15. Bouvard D et al (2007) Characterization and simulation of microstructure and properties of EPS lightweight concrete. Cem Concr Res 37(12):1666–1673. https://doi.org/10.1016/j.cemconres.2007.08.028
16. Babu DS, Babu KG, Wee TH (2005) Properties of lightweight expanded polystyrene aggregate concretes containing fly ash. Cem Concr Res 35(6):1218–1223. https://doi.org/10.1016/j.cemconres.2004.11.015
17. Cui C, Huang Q, Li D, Quan C, Li H (2016) Stress-strain relationship in axial compression for EPS concrete. Constr Build Mater 105:377–383. https://doi.org/10.1016/j.conbuildmat.2015.12.159
18. Liu H et al (2023) Experimental scrutiny of uniaxial compressive stress–strain relationship for expanded polystyrene concrete. Iran J Sci Technol Trans Civ Eng 47:2931–2948. https://doi.org/10.1007/s40996-023-01112-y
19. Rucińska T, Kiernożycki W (2014) Strain of axially compressed concrete with styrofoam aggregate. Cem Wapno Beton 1:33–39

20. Technical data sheet, safety data sheet for iX CPP20, www.holcim.pl/kontakt/. Last accessed 21 June 2026
21. PN-EN 12620:2004 + A1:2008 (2008) Aggregates for concrete. Polish Committee for Standardization (PKN), Warsaw, Poland
22. PN-EN 1008:2004 (2004) Mixing water for concrete—Specification for sampling, testing and assessing the suitability of water, including water recovered from processes in the concrete industry, as mixing water for concrete. Polish Committee for Standardization (PKN), Warsaw, Poland
23. PN-EN 12390-3:2019-07 (2019) Testing hardened concrete—Part 3: compressive strength of test specimens. Polish Committee for Standardization (PKN), Warsaw, Poland
24. PN-EN 12390-16:2019 (2019) Testing hardened concrete—Part 16: determination of the shrinkage of concrete. Polish Committee for Standardization (PKN), Warsaw, Poland
25. EN 1992-1-1:2004 + A1:2014 (2014) Design of concrete structures—Part 1–1: general rules and rules for buildings. European Committee for Standardization (CEN), Brussels, Belgium
26. Chung SY, Abd Elrahman M, Stephan D (2018) Effects of expanded polystyrene (EPS) sizes and arrangements on the properties of lightweight concrete. Mater Struct 51:57. https://doi.org/10.1617/s11527-018-1182-3
27. He H et al (2023) A new type of mineral admixture and its impact on the carbonation resistance of EPS concrete. Sustainability 15(9):7233. https://doi.org/10.3390/su15097233

Technology Development for the Production of Fibre-Based Curved Reinforcement for Concrete

Marco Lindner[1(✉)], Peter Schneider[2], and Sandra Gelbrich[1]

[1] University of Technology Chemnitz, Institute of Lightweight Structures, Reichenhainer Str. 31, Chemnitz 09126, Germany
marco.lindner@mb.tu-chemnitz.de

[2] thoenes® Dichtungstechnik GmbH, Zeppelinstr. 1, Klipphausen 01665, Germany

Abstract. Fibre-reinforced composites have been identified as a viable alternative to traditional concrete reinforcement methods. A new steel-free stiffening system with reduced curvature radii has been developed to expand the range of applications for reinforcement elements in fibre composite construction. Extensive investigations, underpinned by international test methods and bolstered by empirical evidence, have substantiated an increase in load-bearing capacity. Furthermore, a technology has been developed for economical production. The high potential for lightweight construction and resource efficiency is demonstrated by newly reinforced precast concrete elements from the water and transport infrastructure sector.

Keywords: Bending Rebars · FRP Stirrup · Concrete Reinforcement · Reduced Radii of Curvature

1 Introduction

Concrete is a high-performance material and the most widely used building material worldwide, thanks to its durability and excellent mechanical properties when reinforced with steel. The metal reinforcement elements are available in a wide range of diameters and designs. However, due to steel's tendency to corrode, concrete structures are very heavy, as minimum concrete coverings must be maintained depending on environmental conditions. Furthermore, these steel-reinforced elements affect magnetic and electric fields, necessitating specialised, complex decoupling measures in rail-bound traffic-area systems (e.g., traffic control systems with induction loops). New steel-free reinforcement systems made of innovative fibre-reinforced plastic (FRP) composites offer significant potential for durability and lightweight designs (e.g., reducing concrete consumption for angle support elements by 52% [1–5]).

Increasingly, FRP components are being recognised as a key alternative to steel-based reinforcement in concrete. FRP typically comprises a plastic matrix reinforced with fibres to enhance tensile strength. The selection of suitable fibre materials is crucial for achieving high specific strengths and stiffness. According to current research, FRP

Z. Perkowski et al. (Eds.): ECCE 2026, LNCE 929, pp. 55–68, 2027.
https://doi.org/10.1007/978-3-032-39515-3_5

reinforcements are primarily manufactured in straight form by pultrusion (an extrusion process) of thermosetting polymer systems combined with glass, basalt, or carbon fibres [5, 6]. Deformation is no longer possible once the thermosetting matrix has hardened because of the strong cross-linking of the polymer chains. This means that, for example, it is impossible to bend the reinforcement again using energy.

The properties of FRP reinforcements in stirrup shape clearly differ from those of straight bars due to their modified shape (see Fig. 1). The fibre architecture and surface profile of curved FRP reinforcements are significantly different, resulting in altered component properties. The manufacturing process limits the individual curvature radii (D_{min}) of the FRP reinforcement stirrups. The limit is a minimum of 7 times the bar diameter ($Ø_{nm}$). The D_{min} must be no less than $7Ø_{nm}$ (see steel reinforcement $D_{min} = 4Ø_{steel}$). The D_{min} must also be within the service temperature range of up to 60 °C. The glass transition temperature (T_g) must be greater than 100 °C. Furthermore, curved FRP stirrups are available only with curvature in one plane, which greatly limits their range of applications and makes it difficult to establish them on the market [7–21].

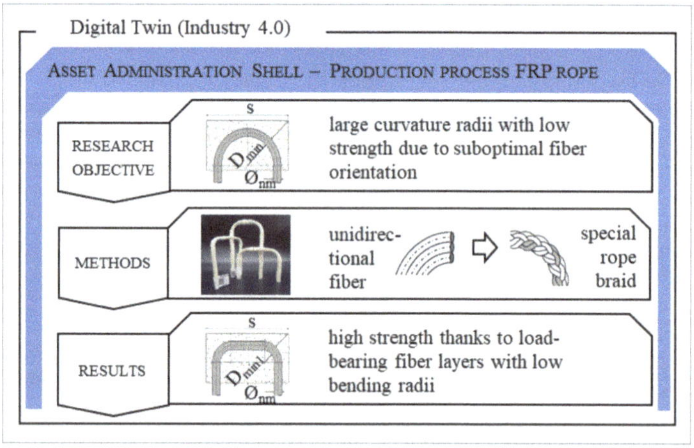

Fig. 1. State-of-the-art and the project's goal.

The research project at Chemnitz University of Technology, in collaboration with *thoenes*® Dichtungstechnik GmbH, focused on developing FRP stirrup reinforcements with high load-bearing capacity and reduced curvature radii, as well as establishing a practical production facility. These developments have been validated in accordance with American Concrete Institute (ACI), American Society for Testing and Materials (ASTM), International Organization for Standardization (DIN ISO), and German Committee for Structural Concrete (DAfStb) test specifications.

The present study concludes that the recently developed stirrups have considerably greater potential for lightweight buildings than existing FRP stirrups, aligning with the current state of the art. The most important results are the significant reduction in the radius of curvature (to $D_{min1} = 4Ø_{nm}$) and the increase in the tensile strength of the stirrups.

A technology for the mass production of curved rods has been developed, enabling curved sections with stretched fibre layers across the entire cross-section. During bending, the fibre layers are stretched significantly. In the context of Industry 4.0, the Asset Administration Shell (AAS) is considered a global open-source standard. The digital twin serves as the basis for simulating, evaluating, and directly controlling real processes. The implementation of FRP stirrup reinforcement will therefore open up new areas of application in the construction industry, characterised by reduced weight (i.e., reduced cement) and increased efficiency.

2 Materials and Methods

2.1 Material of Reference

The test series comprised two reinforcement profiles with different curvatures, made of FRP and with a bar diameter of 12 mm. The investigation revealed that the geometry of the FRP stirrup reinforcements, as shown in Fig. 2, differed in design, achievable radius of curvature, fibre content, and manufacturing technology.

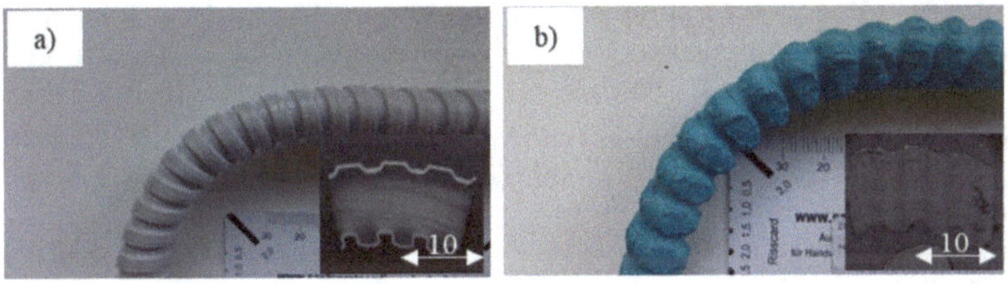

Fig. 2. Reference material: a) FRP tube, b) FRP winding.

The first stirrup variant was produced using a production technology that involves pulling an impregnated E-CR glass fibre bundle through a corrugated plastic cladding tube, which is then curved and subsequently hardened (referred to as an FRP tube). This technology was developed by Schoeck GmbH (Combar®; Bent Bars [20]). The minimum radii of curvature in this production technology must be no less than seven times the bar diameter. The second variant utilises a divergent technological approach (FiReP® Group; Rebar [22]). The formation is achieved by entwining a resin-impregnated glass fibre bundle (vinyl ester resin) over pulleys (FRP winding) in this technological process. The minimum bending roll diameter is eight times the bar diameter ($D_{min} \geq 8Ø_{nm}$).

Another carbon fibre manufacturer (SGL Carbon GmbH; CFRTP [23]) offers a 4.5 mm diameter reinforcement bar fabricated from carbon fibre (CFRP). As indicated by the extant literature, the profile is embossed and utilises a proprietary PA66 (thermoplastic) as a matrix. As indicated by the supplied data, the manufacturing facility allows for subsequent bending. The glass transition temperature of at least 100 °C specified by the ACI was not attained in the tests conducted; the maximum recorded was 80 °C [22, 23].

2.2 Analysis of Current Stirrup Reinforcements

It is imperative to ascertain the fibre content when evaluating the mechanical properties of FRP profiles, as it is of the greatest significance in the assessment process. Moreover, the fibre volume content (FVC) provides an essential characteristic value and the basis for calculating the dimensions of concrete elements. The FVC was reliably determined using the combustion method (oxygen-reduced atmosphere). For the purpose of this study, specimens were exposed to a temperature range (500 °C (CFRP) and 625 °C for glass fibre-reinforced composite and basalt fibre-reinforced composite) at which the pyrolytic decomposition of the plastic matrix takes place. The resulting residues and the undecomposed fibres were weighed using a high-precision balance. The ratio of the reinforcement fibre mass (residues) to the total mass of the fibre-reinforced material provides the fibre content, which is subsequently converted into the fibre content, considering the different densities of the composite components. Moreover, a microphotograph was obtained for each profile employing a high-resolution light microscope. This procedure was undertaken to evaluate the FRP reinforcement bars with regard to fibre impregnation within the matrix system and to illustrate fibre orientation.

A research study was initiated to evaluate the performance of steel-free bars in concrete and to ascertain the long-term behaviour of the FRP-concrete composite. The present study accelerated the ageing of FRP reinforcement bars in an alkaline environment. To assess the alkali resistance of the FRP reinforcement bars, a series of experiments was conducted utilising a 1 M sodium hydroxide solution with a pH greater than 13. In the present experiment, the specimens were heated at 60 °C for 28 days. The assessment of alkali resistance was conducted by measuring the mass loss during the course of the tests [25–27].

2.3 Suitability as Reinforcement Stirrups

The average nominal cross-section (A_{nm}) and the derived equivalent nominal diameter ($Ø_{nm}$) are generally considered as the most significant parameters for reinforcement calculations and classifications. The measurement is conducted in accordance with ISO 10406-1 (see Fig. 3a), which stipulates that the nominal diameter is determined indirectly by measuring the bar's volume ($V_{FRP\ sample}$) and length (l_0) [27]:

$$V_{FRP\ sample} = A_{nm} \cdot l_0 = V_1 - V_0 \tag{1}$$

$$Ø_{nm} = \sqrt{\frac{4 \cdot V_{FRP\ sample}}{\pi \cdot l_0}} \tag{2}$$

The modulus of elasticity and tensile strength of the fibre composite material are particularly significant for determining the material's properties and performance under different conditions. The test setup has been meticulously designed in accordance with the prevailing guidelines, namely those established by ACI, ASTM, and DAfStb (see Fig. 3b). To ensure meaningful measurement results, sleeves were attached to the clamping points of the test specimen. The primary function of these sleeves is to prevent potential damage to the fibre from the tension exerted during the tensile test. Furthermore,

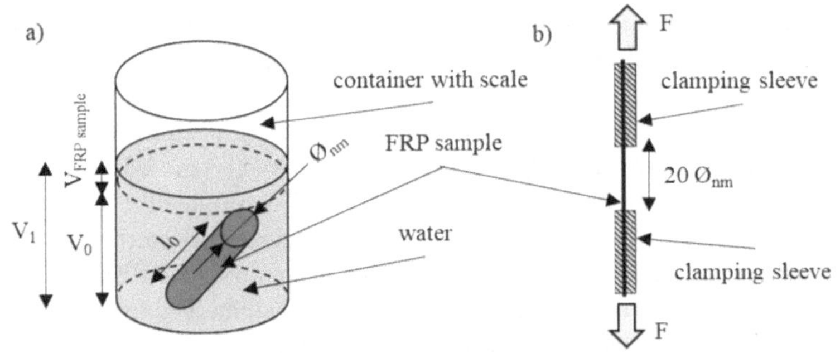

Fig. 3. Determination of basic parameters. a) Measuring setup for diameter determination. b) Test setup for tensile strength.

they ensure uniform force transmission to the tensile specimen up to the point of failure. The tensile test was conducted at a rate of 3 mm/min.

It has been established that, in general, curved FRP stirrup reinforcements possess tensile strengths that are significantly lower than those of their straight counterparts. The high correlation between tensile strength and radius of curvature is attributable to stirrup reinforcement in concrete components being subject to shear forces from the concrete matrix at their points of deflection. At the onset of 2024, the DAfStb guideline, entitled 'Concrete components, non-metallic reinforcement: 2024-01', was published. This guideline demonstrated a high degree of congruence with both DIN EN 1992-1-1 (metallic reinforcement) and the ACI standard. To ensure practical development for national and international markets, suitability tests for FRP stirrups in concrete and analyses of existing stirrups are conducted in accordance with the ACI guidelines [25–27].

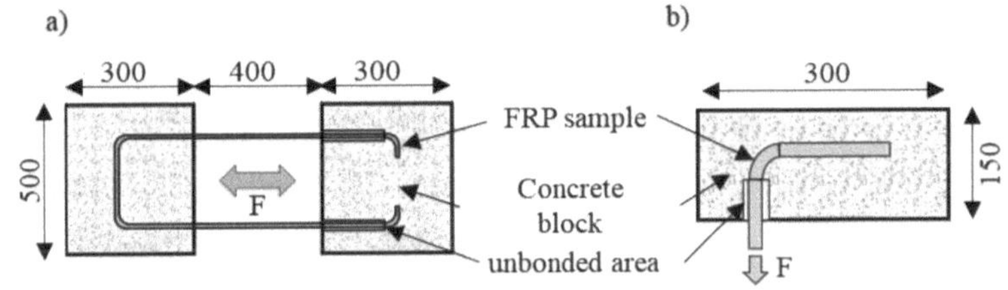

Fig. 4. Determination of basic parameters for stirrups. a) Test setup according to ACI 440.3R-04 B.5 b) Test setup after DAfStb.

In order to ascertain the tensile strength of stirrups in accordance with the provisions of ACI 4403R-04 B.5 (see Fig. 4a), the stirrup ends are embedded in two concrete cubes. Tensile stresses are applied using a hydraulic crosshead. In reinforced concrete design, the influence of lateral loads on the stirrup reinforcement is a critical consideration. The utilisation of sleeves, utilised to define a bond length, is a method of mitigating the undesirable effects of these loads. Furthermore, strain gauges are attached to the test specimen in the free areas between the concrete blocks. Therefore, it can be hypothesised that there is consistency in the distribution of bond stress along the bond length. The test is concluded by the reduction of force, which is defined as the failure criterion in the bending area of the stirrup.

As illustrated in Fig. 4b, the DAfStb stirrup tensile test procedure entails embedding a stirrup end in a concrete block and applying defined tensile forces using a tensile testing machine, such as those manufactured by Zwick/Roell. To mitigate the adverse effects of transverse pressure on the reinforcing stirrup, a sleeve with a predetermined bond length up to the onset of the bend was specified. This indicates that the distribution of bond stress along the bond length can be approximated as constant. The test is terminated by a drop in force, which serves as the failure criterion in the stirrup's curvature region.

In the present study, the pull-out test (see ACI 440.3R04 B.3) was selected to assess the bond behaviour between reinforcing bars and the concrete matrix. This test is highly analogous to the test described in the DAfStb guidelines (see Fig. 5a) and other publications [25–27]. This specific test setup has been developed to minimise the undesirable influence of transverse pressure on the reinforcing bar. Simultaneously, it allows for the assumption of an almost constant local distribution of the bond stress along the bond length. In accordance with the DAfStb guideline, the test configuration enables the production of test specimens with reduced geometric dimensions, thereby significantly simplifying the handling of a large number of tests. As illustrated in the provided figure, the test specimen consists of a cube-shaped concrete matrix with a reinforcing bar of defined bond length, embedded on one side.

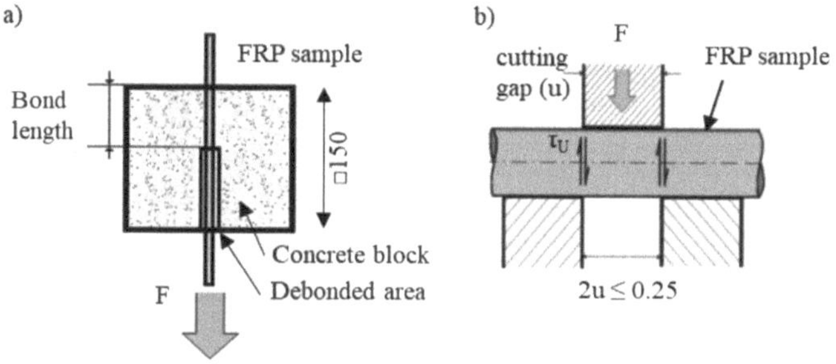

Fig. 5. Test set-up according to: a) pullout in the DAfStb-guideline, b) shear in ACI 440.3R04 B.4.

The determination of shear strength was achieved using a device in accordance with ACI guidelines (ACI 440.3R04 B.4; see Fig. 5b), as the DAfStb does not provide a description of such a device [26, 27]. This apparatus facilitates the adjustment of crack widths through an adjustable mechanism. It is imperative to acknowledge that the tensile stress component was not documented in this specific shear test. During the test, the FRP test bar was positioned within the die prism and secured with a hold-down device. The cutting device was then clamped in the tensile-compression testing machine to determine the force-displacement curves. Therefore, in accordance with the established guidelines, the initial gap size (u) was set at 0.1 mm. The measurement was then continued until the specimen broke.

2.4 Process Development

In pursuit of the aforementioned objective, a circular fibre architecture was conceptualised, as theoretically and numerically validated models indicated the optimal configuration. This architecture entails the rotational positioning of incoming and outgoing fibres within the fibre's bending range. It is theoretically postulated that this fibre arrangement provides a constant flat fibre stress in the profile cross-section. The prevention of wrinkling of the fibre layers in the inner bending range is thus achieved. Consequently, a bending radius that is considerably smaller than the stipulated sevenfold value of the bar diameter (Dmin <$7\varnothing_{nm}$) can be achieved. This development is predicated on conducting intricate experiments on resin impregnation and wrapping. The objective of these experiments is to generate a yarn that exhibits adequate capillary action for the resin impregnation process and, concomitantly, increases its compressive strength during the yarn bending process. The procedure ensures the yarn's suitability for the subsequent beating and braiding processes. It is imperative to note that, throughout the braiding process, the twine is meticulously positioned in accordance with the established parameters (see Fig. 6a and b).

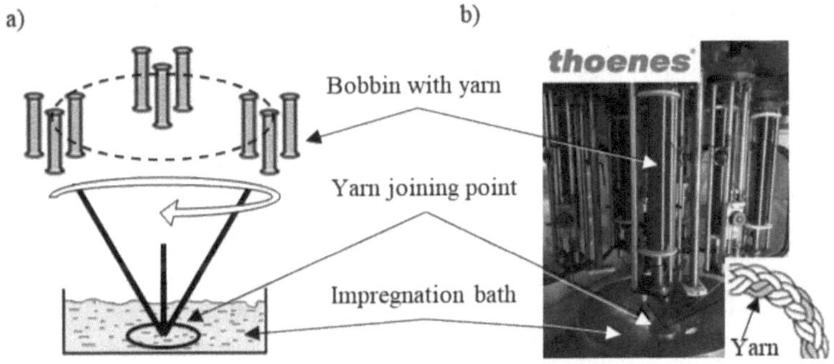

Fig. 6. Production Braiding FRP rope braid: a) scheme, b) production.

As evidenced by extensive research in the relevant literature, carbon fibres are typically distinguished by their remarkably high tensile strength and elastic modulus, accompanied by minimal elongation at break in the fibre direction [1–18]. The increased costs and significantly reduced strength perpendicular to the longitudinal direction have a negative impact. Furthermore, processing and testing electrically conductive carbon fibres required significant technical effort to ensure the protection of all electrical equipment (computers, test machines, sensors, etc.). The mechanical strength values of isotropic basalt and glass fibres are lower than those of anisotropic carbon fibres. A direct comparison of the two inorganic fibres reveals that the basalt fibres demonstrate a slightly steeper increase in the elastic modulus than the glass fibres. A fundamental requirement of the project is the cost-effective production of the new concrete components. Consequently, basalt and AR glass fibres, which are significantly less expensive, are considered the optimal choice for this study. The current research focus is also on carbon fibres. The construction industry trend shows that carbon fibres are evolving from a niche application to a key technology for sustainable, durable, and intelligent construction [29].

In the context of developing a production line, it is essential to consider the mechanical properties of braided ropes, such as elongation and breaking strength, to accurately determine the clamping force. The test was conducted in accordance with DIN EN ISO 2307 [28]. The use of this method entails clamping the sample material solely at the terminal point, thereby making it applicable even to highly sensitive fibre surfaces. The tensile stress on the test specimen is reduced by frictional locking on the force-reduction roller. In accordance with the test specification, the test cycles are continued until the force decreases due to fibre breakage between the two markings (see Fig. 7a). The Zwick/Roell Z150 tensile testing machine is shown in Fig. 7b.

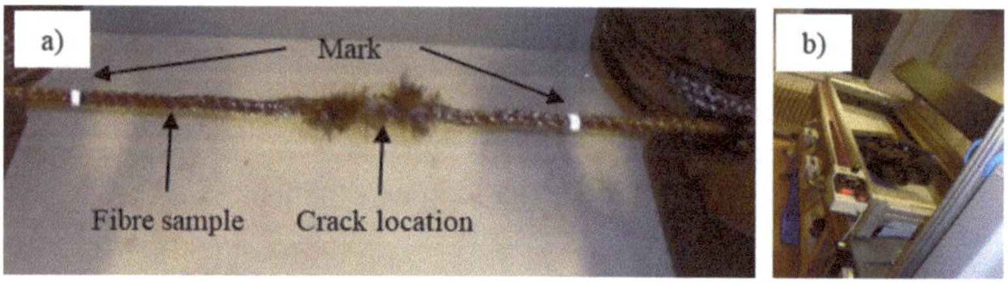

Fig. 7. Test set-up: a) sample after test, b) testing machine Zwick/Roell Z150.

The forming of FRPs that have not yet polymerised always poses a particular challenge. On the one hand, the forming force must be applied, while on the other hand, the free ends of the resin-impregnated fibre strand must be stretched in the elastic range. To ensure reproducibility and consistent results at all deformation points within the mould, a uniform preload force must be applied. The tension before and after the deflection point (see Fig. 8a) is measured using a rope tension measuring device (Honigmann GmbH; CableBull® TEXter® S-R16/600; see Fig. 8b). For this purpose, the rope is clamped with sensors or rollers that exert pressure on the rope. The resulting force is usually converted indirectly into tension through cable deformation. The determined tension is then immediately shown on the measuring device's display.

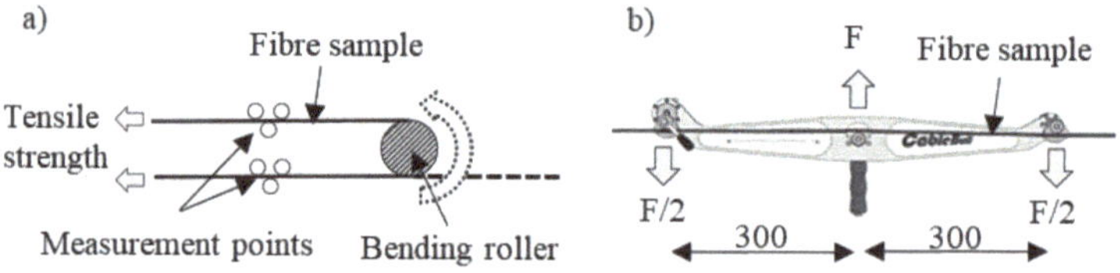

Fig. 8. Scheme: a) test set-up, b) cable tension meter [30].

The central topic of this study is the fabrication of high-strength fibre-reinforced composite stirrups with reduced curvature radii using an automated manufacturing process. During the investigation, several technology variants were developed and subsequently compared. The resulting manufacturing technology is essentially based on producing steel reinforcement stirrups, except that an additional tensile force is required in the fibre direction. The principle is illustrated in Fig. 9.

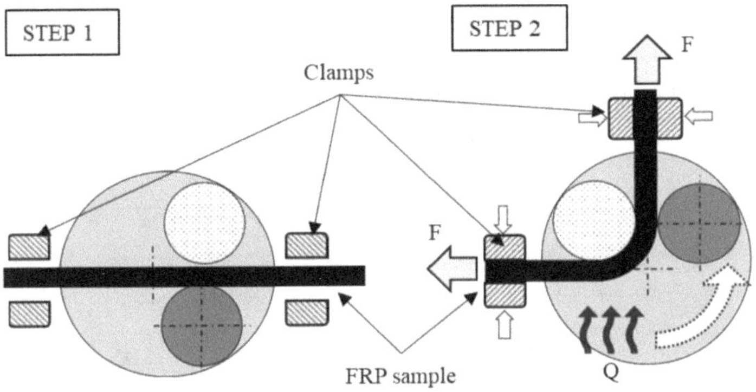

Fig. 9. Production of FRP stirrup.

The manufacturing process for FRP stirrups begins in step 1 by feeding unpolymerised FRP. Once the desired bend position on the mould segment has been reached, the feeding process is stopped. The clamping jaws close, and the FRP is formed by the angle-dependent movement of the mould unit (step 2). The clamping mechanism is activated. The polymerisation of the bend is supported by a radiant heat source. The clamping jaws open, the mould segments are retracted, and the braid is transported further. To ensure that production processes are equipped for the challenges of the future, it is essential to combine automation with the creation of a digital representation of the product, known as a 'digital twin'. Implementing this process requires entering results and parameters into the AAS Explorer software, thereby initiating the creation of a digital twin.

Construction of the production line is currently underway. *thoenes*® Dichtungstechnik GmbH has announced that it will commence prototype testing on the developed demonstration plant in April 2026.

3 Results

In order to facilitate a direct comparison with existing reinforcement stirrups (FRP tube and FRP winding), stirrups with a basalt fibre-epoxy resin matrix and a diameter of $Ø_{nm}$ = 12 mm were produced in a laboratory test (sample name: FRP rope). The radius of curvature meets the target specifications and is therefore four times the bar diameter (a comparable value is also obtained for steel reinforcement). The results are based on the test specifications of the ACI and the DAfStb. To ensure reproducibility, 5 individual samples were included in each batch.

Thermogravimetric analysis of the FRP rope revealed a high fibre content exceeding 55 Mass-Vol.-%. It was established that the fibre volume content of the samples of FRP winding was equivalent to that of the FRP rope. Conversely, the FRP tube demonstrated a significantly lower value of 37 Mass-Vol.-%. In addition, the impregnation of the fibre-resin samples was deemed satisfactory when evaluated using optical microscopy on at least five samples from each batch.

The uniaxial tensile strength and modulus of elasticity are of particular importance in the design of FRP reinforcement bars (see Fig. 10a). As illustrated in Fig. 10b, the values thus obtained are then used to assess the suitability of the aforementioned materials for use as reinforcement stirrups.

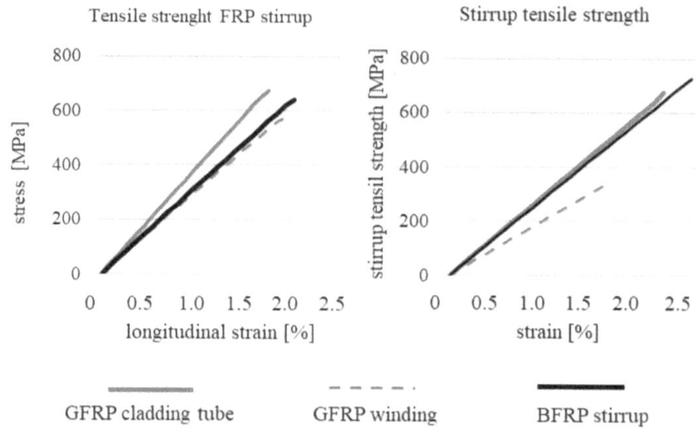

Fig. 10. Results: a) tensile strength, b) stirrup tensile strength.

Both diagrams provide a direct comparison of tensile strength and bow tensile strength, leading to the following conclusions. A direct comparison was made between 'FRP winding' (Young's modulus 26.3 GPa; $\sigma = 573$ MPa) and 'FRP tube' (Young's modulus 30 GPa; $\sigma = 634$ MPa). The comparison indicated that defined fibre layers made of basalt (FRP rope) resulted in higher tensile strength and a higher modulus of elasticity for an equivalent fibre volume. The fibres in the sample carrier tube with FRP winding exhibited a significantly increased Young's modulus (42 GPa) compared to the samples with angular deviation, while the tensile strength ($\sigma = 671$ MPa) of the developed basalt fibre-based reinforcement stirrups remained comparable. The tensile strengths of the stirrups in the FRP rope sample were found to be approximately 725 MPa ($D_{min} = 4$ $Ø_{nm}$; conventional FRP rope up to $\sigma = 675$ MPa) at a significantly smaller bending radius.

The pull-out test showed a maximum adhesive tensile strength of 19 MPa for the 'FRP tube' specimen (see Fig. 11a). The thermoplastic profile tube was subjected to shear when the glass fibre-reinforced thermoset core was loaded. This was evidenced by the exposed composite surface.

The hypothesis is that the shear strength of the thermoplastic pipe has a significant effect on the composite of the FRP tube and concrete (see Fig. 11a). The FRP winding sample (see Fig. 11b) batch demonstrated a higher adhesive tensile strength (22 MPa) in comparison to the FRP tube sample. As shown in Fig. 11c, the FRP rope exhibited the highest adhesive tensile strength, at 30 MPa.

The determined shear strength of the sample batch with inserted fibre layers in 'FRP tube' ($\tau = 224$ MPa) and the angle deviation 'FRP winding' ($\tau = 214$ MPa) of a matrix of vinyl ester resin were found to be within the same range of values. Conversely, a lower shear strength (FRP rope with $\tau = 160$ MPa) was observed in sample batches produced with epoxy resin for hand laminates. It can thus be demonstrated that the shear behaviour is decisively dependent on the resin matrix of the test specimen.

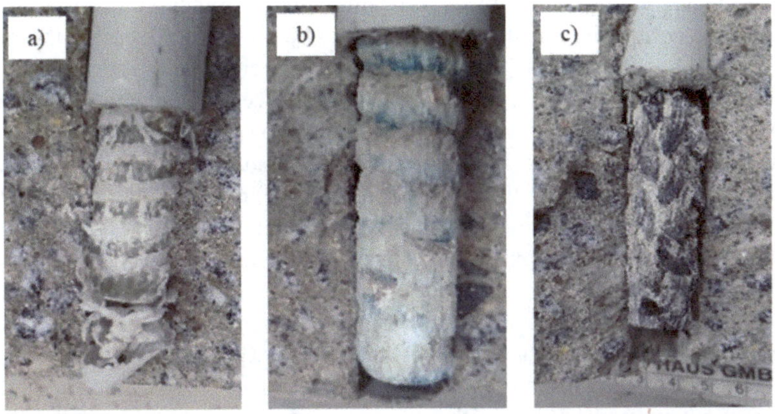

Fig. 11. Sample after pull-out test: a) FRP tube, b) FRP winding, c) FRP rope.

4 Reference

Chemnitz University of Technology is a pioneer in developing innovative bridges, including a steel-free pedestrian and bicycle bridge made of carbon concrete. Reference structures have been built in Wesseling, Germany, to demonstrate the advantages of lightweight construction and durability (see Fig. 12). In cooperation with *thoenes®* Dichtungstechnik GmbH, the developed FRP reinforcement stirrups (FRP rope) were used and tested in practice for the first time. To simplify the application and approval process, brackets made of approved alkali-resistant glass fibre were used.

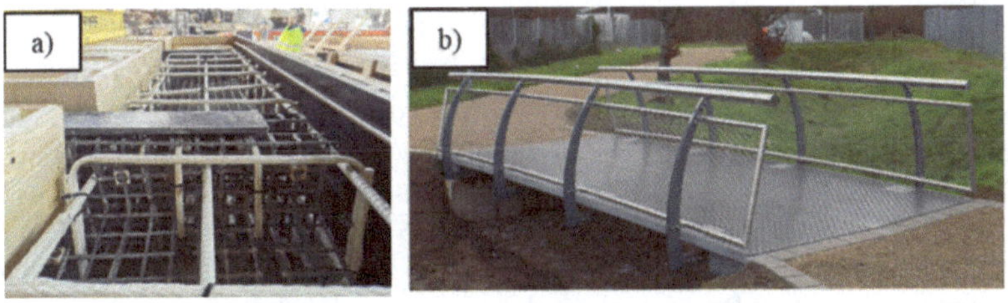

Fig. 12. Reference bridge: a) FRP rope (with glass fibre), b) carbon-concrete-bridge: Photo © Prof. Sandra Gelbrich [31].

In conclusion, it has been demonstrated that the recently developed reinforcement stirrups possess a highly intricate component structure and exhibit high load-bearing capacity.

5 Conclusions

The present study has demonstrated that the new stirrup reinforcements—made of fibre-reinforced plastic with reduced bending radii achieved through a specially developed manufacturing technology—exhibit significantly greater potential than conventional fibre-reinforced composite stirrups. The primary objective is to decrease the radius of curvature whilst concurrently augmenting the tensile strength of the stirrups. This leads to improved durability and, thus, greater cost and resource efficiency in precast concrete elements. Furthermore, this study presented the technological development for manufacturing curved FRP elements. In addition to the curvature, this process also achieves tensile stress at the free ends of the stirrup. Finally, a reference bridge incorporating the novel reinforcement stirrups was impressively implemented.

This pioneering contribution to the field of lightweight construction is predicated on a substantial reduction in material and weight consumption, as well as application-specific methodologies for functional integration (e.g., monitoring). The utilisation of lighter construction materials has been shown to limit transport costs and cement consumption, thereby leading to a more favourable CO_2 balance. The practice of lightweight construction is commonly linked with economic and ecological advantages, including reduced CO_2 emissions.

Acknowledgments. The authors gratefully acknowledge the financial support provided by the European Regional Development Fund (ERDF) through the Saxon Development Bank (SAB), as well as by the ZIM – Central Innovation Programme for SMEs of the Federal Ministry for Economic Affairs and Climate Action (BMWK). Their support was instrumental in enabling the successful implementation of this project.

Disclosure of Interests. The authors have no competing interests to declare that are relevant to the content of this article.

References

1. Lindner M et al (2025) Tool development for precast concrete elements in robot-assisted flow production. In: Perkowski Z, Beben D, Zembaty Z, Massimino MR, Oliveira MJ (eds) Environmental challenges in civil engineering III. ECCE 2024, Lecture Notes in Civil Engineering, vol 615. Springer. https://doi.org/10.1007/978-3-031-73776-3_8
2. Abstoss M, Senf F, Gliniorz R, Funke H, Gelbrich S (in press) Sustainable precast concrete construction using hybrid addivite manufacturing: 3D concrete printing and shotcrete with low-carbon concrete. In: Proceedings of the 2nd international conference on Net-zero built environment: innovations in materials, structures, and management practices
3. Senf F, Javed R, Funke H, Gelbrich S (2025) Enhanced environmental sustainability of 3D-printed concrete mixtures based on calcium sulfoaluminate (CSA) and blast furnace cement (CEM III/B). In: MATEC web of conferences, vol 409, p 06004
4. Senf F, Abstoss M, Funke H, Gelbrich S (in press) Advanced 3D concrete printing (3DCP) with low-carbon calcium sulfoaluminate (CSA) cement blends. In: Proceedings of the 2nd international conference on net-zero built environment: innovations in materials, structures, and management practices

5. Lindner M, Vanselow K, Gelbrich S, Kroll L (2019) Fibre-reinforced polymer stirrup for reinforcing concrete structures. Technol Lightweight Struct 3(1). https://doi.org/10.21935/tls.v3i1.117
6. Nanni A, De Luca A, Jawaheri Zadeh H (2014) Reinforced concrete with FRP Bars: mechanics and design. CRC Press, London. https://doi.org/10.1201/b16669
7. Xin W et al (2021) Mechanical and bonding behavior of a bendable fiber-reinforced thermoplastic rebar. Constr Build Mater 302:124222. https://doi.org/10.1016/j.conbuildmat.2021.124222
8. Qureshi J (2022) A review of fibre reinforced polymer structures. Fibers 10(3):27. https://doi.org/10.3390/fib10030027
9. Daniyan IA, Mpofu K, Adeodu AO, Adesina O (2021) Development of carbon fibre reinforced polymer matrix composites and optimization of the process parameters for railcar applications. Mater Today Proc 38(2):628–634. https://doi.org/10.1016/j.matpr.2020.03.480
10. Koumoulos EP et al (2019) Research and development in carbon fibers and advanced high-performance composites supply chain in Europe: a roadmap for challenges and the industrial uptake. J Compos Sci 3(3):86. https://doi.org/10.3390/jcs3030086
11. Davalos JF, Chen Y, Ray I (2012) Long-term durability prediction models for GFRP bars in concrete environment. J Compos Mater 46(16). https://doi.org/10.1177/0021998311427777
12. Askar MK, Hassan AF, Al-Kamaki YSS (2022) Flexural and shear strengthening of reinforced concrete beams using FRP composites: a state of the art. Case Stud Constr Mater 17:e01189. https://doi.org/10.1016/j.cscm.2022.e01189
13. Elmahdy A, Feldman E, Verleysen P (2024) Tensile behavior of basalt fiber reinforced composites at high strain rates. In: ECCM18 – 18th European conference on composite materials, pp 24–28
14. Li H, Xian G, Xiao B, Wu J (2011) Comprehensive characterization of BFRP applied in civil engineering. In: Ye L, Feng P, Yue Q (eds) Advances in FRP composites in civil engineering. Springer. https://doi.org/10.1007/978-3-642-17487-2_11
15. Feng G, Zhu D, Guo S, Rahman MZ, Jin Z, Shi C (2022) A review on mechanical properties and deterioration mechanisms of FRP bars under severe environmental and loading conditions. Cem Concr Compos 134:104758. https://doi.org/10.1016/j.cemconcomp.2022.104758
16. Guadagnini M, Pilakoutas K, Waldron P (2006) Shear resistance of FRP RC beams: experimental study. J Compos Constr 10(6). https://doi.org/10.1061/(ASCE)1090-0268(2006)10:6(464)
17. Liu H et al (2022) A review on basalt fiber composites and their applications in clean energy sector and power grids. Polymers 14(12):2376. https://doi.org/10.3390/polym14122376
18. Protchenko K, Zayoud F, Urbański M, Szmigiera E (2020) Tensile and shear testing of basalt fiber reinforced polymer (BFRP) and hybrid basalt/Barbon fiber reinforced polymer (HFRP) bars. Materials 13(24):5839. https://doi.org/10.3390/ma13245839
19. ACI 440 (2015) 1R-15: guide for the design and construction of structural concrete reinforced with Fiber-Reinforced Polymer (FRP) bars. American Concrete Institute, Farmington Hills, MI, USA
20. Schöck Bauteile GmbH. https://www.schoeck.com/de. Last accessed 15 June 2026
21. FiReP Rebar Technology GmbH. https://www.firepgroup.com/products/firep-rebar/. Last accessed 15 June 2026
22. Tsukamoto K (2014) WO2014/032846A1, FIREP Rebar Technology GmbH
23. Apitz A, Schmitz J, Hückler A, Schlaich M (2022) New thermoplastic carbon fiber reinforced polymer rebars and stirrups. Struct Concr 23(2):923–938. https://doi.org/10.1002/suco.202100434
24. El-Tahan M, Galal K, Hoa SV (2013) New thermoplastic CFRP bendable rebars for reinforcing structural concrete elements. Compos Part B 45(1):1207–1215. https://doi.org/10.1016/j.compositesb.2012.09.025

25. Ahmed EA, El-Sayed AK, El-Salakawy E, Benmokrane B (2009) Bend strength of FRP stirrups: comparison and evaluation of testing methods. J Compos Constr 14(1). https://doi.org/10.1061/(ASCE)CC.1943-5614.0000050
26. ACI 440 (2004) 3R-04: guide test methods for Fiber-Reinforced Polymers (FRPs) for reinforcing or strengthening concrete structures. American Concrete Institute, Farmington Hills, MI, USA
27. DAfStb (2024) Heft 660: Richtline Erläuterungen und Beispiele zur DAfStb-Richtlinie Betonbauteile mit nichtmetallischer Bewehrung. Print ISBN 978-3-410-65941-9
28. DIN EN ISO (2019) 2307:2019-12: Fibre ropes—Determination of certain physical and mechanical properties (ISO 2307:2019). German version EN ISO 2307:2019
29. REPORT 2026: carbon fiber construction repair market – global forecast 2026-2032: 360iResearch™, ID 5924791, https://www.researchandmarkets.com/report/carbon-fiber-construction-repair#product%2D%2Drelated-products. Last accessed 01 Jul 2026
30. Honigmann GmbH. https://honigmann.com/i533/Seilkraftmesser-Seilspannungsmesser-Seilspannungsmessger%C3%A4t-CableBull-TEXter.html&lang=1. Last accessed 15 June 2026
31. Steinebach M Premiere für filigrane Carbonbeton-Plattenbalkenbrücke "VariBridge", TUCaktuell., https://www.tu-chemnitz.de/tu/pressestelle/aktuell/11704. Last accessed 15 June 2026

Enhancing the Durability of Timber Elements by Applying a Natural Fiber-Reinforced Polymer (NFRP) Encapsulation

Fabian Nendel(✉), Denise Lenk, and Sandra Gelbrich

Chemnitz University of Technology, Chemnitz 09126, Germany
fabian.nendel@mb.tu-chemnitz.de

Abstract. Timber frame structures, particularly elements near the ground, are subject to increased moisture exposure, which significantly accelerates biological degradation by wood-destroying fungi and insects. Conventional chemical wood preservatives require frequent renewal and pose sustainability challenges, as treated timber must be disposed of as hazardous waste at the end of its service life. This research focuses on extending the service life of wooden components while enabling material reuse after their initial application. To achieve this goal, an alternative protective system to maintenance-intensive coatings was developed. The system consists of a diffusion-open matrix combined with a textile support structure made from natural fibers, forming a protective layer that diverts water and prevents moisture accumulation within the wood. Using flat knitting technology, a form-fitting, three-dimensional knitted semi-finished product was designed, offering high drapability and form stability through its mesh structure. In addition, RFID-based sensor technology was integrated into the coating to monitor temperature and wood moisture in critical areas. A key outcome of this research is the development of a removable interface between a wood and a textile coating that provides sufficient adhesive strength during the product's service life while allowing near-residue-free separation at the end of the product's life cycle. Extensive mechanical, chemical, and hydrodynamic testing confirmed the system's performance. The innovative layer structure enables a service life of up to 20 years in near-ground applications, while preserving over 96% of the wood cross-section for subsequent reuse and recycling.

Keywords: Timber Construction · Wood Protection · Durability · Textile Encapsulation

1 Introduction

Wood is an integral part of construction and architecture. As a renewable, sustainable raw material, it offers high stability and low weight [1]. Wood is easy to process and, with appropriate protection and professional treatment, is very durable. In the construction industry, timber frame structures, especially support feet close to the ground, are exposed to particularly high levels of moisture. Constant water absorption promotes the

Z. Perkowski et al. (Eds.): ECCE 2026, LNCE 929, pp. 69–82, 2027.
https://doi.org/10.1007/978-3-032-39515-3_6

colonization of wood-destroying fungi and insects, significantly reducing the service life of components [2]. Therefore, wood protection is very important, which can be achieved chemically, physically, or structurally. Chemical wood preservatives must be continuously renewed at intervals of 2–5 years and are considered unsustainable, as the treated wood must be disposed of separately as hazardous waste in waste wood class IV at the end of its service life [3].

The research work carried out focused on extending the service life of timber components, creating alternatives to conventional, maintenance-intensive coatings, and recycling, i.e., reusing timber components at the end of their first product life cycle.

In addition to increasing resource and material efficiency in construction through the use of renewable raw materials, the research aims to expand the range of applications for wood-composite construction methods. The innovative, biological, natural fiber-reinforced polymer (NFRP) coating for extending service life is designed to preserve the natural wood appearance by using natural fibers. The additional integration of sensor technology enables components to be monitored, wood moisture content to be measured and documented, and the long-term effectiveness of the wood protection system to be tested.

In recent years, significant research has focused on enhancing the durability of fiber-reinforced timber structures, particularly under conditions of increased moisture exposure. Current studies highlight that fiber-reinforced polymer (FRP) systems can effectively improve both mechanical performance and resistance to environmental degradation [4, 5]. However, most existing approaches rely on synthetic fiber systems, such as glass or carbon fibers, combined with diffusion-tight resin matrices, which may lead to moisture accumulation and local defects. Therefore, emphasis should be put on research on vapor-permeable or hybrid systems that enable controlled moisture transport.

2 Development of Contour-Adapted 3D Natural Fiber Coating

To increase the durability of wooden components, a coating has been developed that drains most water that comes into contact with it, preventing the wood from being damaged by moisture. This system consists of a textile support structure made of natural fibers and a diffusion-open matrix. The possible integration of load transfer elements is simulated using a column base with ground anchors.

2.1 Development of the Support Structure

Using flat-knitting technology, a form-fitting, three-dimensional semi-finished product tailored to the component was developed. Thanks to its mesh structure, this product offers good drapability while retaining its shape.

During the research, relevant wood profiles were determined as the basic form for the knitted fabrics to be developed. Programming was carried out using the M1Plus program from Stoll. Different types of weave were tested and implemented in a 3D knitted fabric. The types of weave play a decisive role in lamination with the resin system.

The resin must be able to penetrate the knitted structure evenly without forming air bubbles, cavities, or gaps. Due to the flexibility of the knit, parts of the fabric can be easily stretched and fitted precisely to convex areas or the ground anchor. To ensure consistent lamination, the mesh sizes and the behavior of the knitted fabric under tension were analyzed microscopically (see Fig. 1, left). By adjusting the knit architecture (e.g., adapted mesh sizes in the corner areas), overstretched meshes that previously led to cavities or other defects were successfully avoided, and uniform lamination was achieved (see Fig. 1, right).

The knitted support structure can be prefabricated to the desired length and cross-section, avoiding waste or rejects in the manufacturing process.

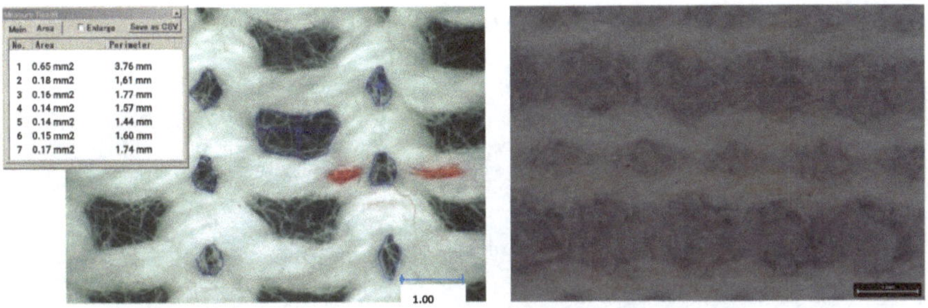

Fig. 1. Microscopic examination of mesh sizes and impregnation quality.

To enable practical application, various geometric adjustments must be made to the 3D knitted fabric. In addition to corners and edges, the development focuses primarily on penetrations and partially closed structures (similar to a stocking). Due to varying degrees of mesh expansion in various directions, a differentiated adjustment of the mesh architecture is required. The complex mesh construction for a circular opening for the threaded rod of the ground anchor is shown in Fig. 2 (left). Figure 2 on the right shows the entire mesh construction of the semi-finished textile product for a support foot. The side surfaces and the front side with the opening are connected fully automatically by the machine and processed into a closed mesh structure.

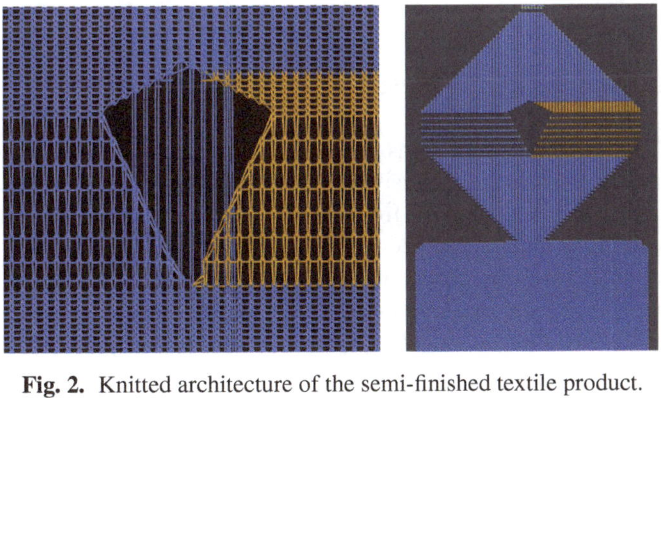

Fig. 2. Knitted architecture of the semi-finished textile product.

2.2 Fixing the Knitted Fabric to the Wooden Component

Various end designs were developed to secure the knitted fabric to the wooden component and evaluated in a series of practical tests. One solution proved particularly effective: a pull cord integrated into the fabric, combined with a specially shaped groove in the wooden component. This system ensures that the covering is securely fixed in place and retains its shape.

In addition, the targeted constriction at the end forms a defined drip edge on the upper wooden component, which diverts rainwater or condensation running down the uncoated wooden support to the outer covering. This effectively prevents moisture from penetrating the covered area from above (see Fig. 3).

Fig. 3. Transition between wooden component and NFRP-coating.

3 Material Testing of the Resin System

Research into available resin systems focuses on material properties relevant to the coating of timber components. A decisive factor is the resin system's diffusion capacity, which is necessary for the transfer of free water and for maintaining a defined moisture balance between the wooden component and its environment.

In addition, high breaking elongation of the resin is essential. Wood is subject to continuous swelling and shrinkage as moisture content changes. To absorb the resulting deformations without damage, a highly elastic resin system is required to prevent cracks in the coating.

Another important evaluation criterion is the processability of the resin system in combination with the knitted semi-finished product. The resin must reliably penetrate the textile structure and impregnate it completely and evenly to ensure a sufficient mechanical and functional bond. Finally, the UV stability of the selected resin systems is examined to assess their suitability for long-term outdoor use under real weather conditions.

3.1 Tensile Tests

A tensile test was performed to determine the modulus of elasticity, tensile strength, and elongation at break for each resin type in accordance with DIN EN ISO 527-2. The test samples were cut to size as 'shoulder bars' in accordance with DIN 527 Type 1B and glued to the clamps with fiberglass adhesive [6]. The test was carried out until component failure. Figure 4 illustrates the breaking forces achieved. The breaking force must be considered in conjunction with the elongation at break (see Fig. 5). The epoxy resin-based systems exhibited higher breaking forces with low elongation at break. They are therefore stable but less adaptable systems. The PUR system, on the other hand, exhibited significantly lower tensile strengths but outstanding elongation at break.

With an elongation at break of approx. 35%, it clearly stood out from the other epoxy resin-based resin systems (approx. 1.5% elongation at break). This high elongation at break is considered very positive, as it allows the wood to swell and shrink without damaging the coating in later applications. Furthermore, cracks that occur can be bridged to a certain extent, thereby preventing damage to the composite component.

The following test series, each with 10 test samples, were conducted:

- **P**oly**u**rethane-based **r**esin system (PUR)
- **B**io-based **e**poxy **r**esin system (BER)
- **P**etrochemical **e**poxy **r**esin system (PER)
- Epoxy **r**esin system for **n**autical applications (NER)

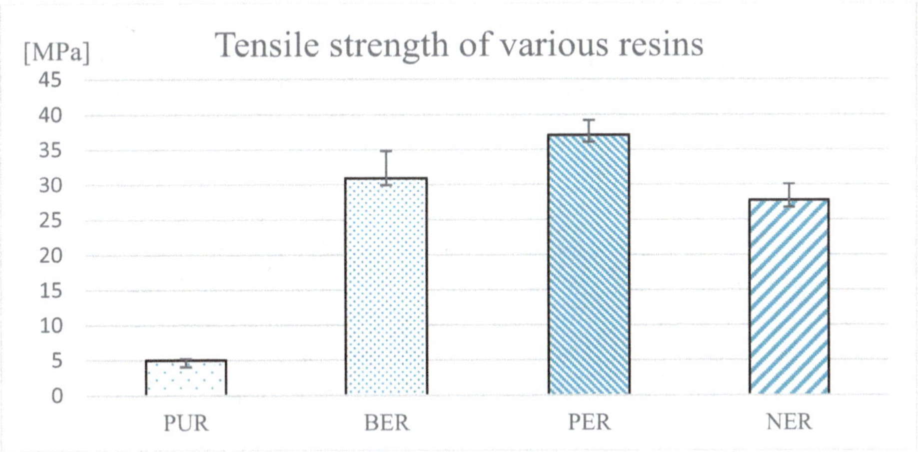

Fig. 4. Tensile strength of various resins.

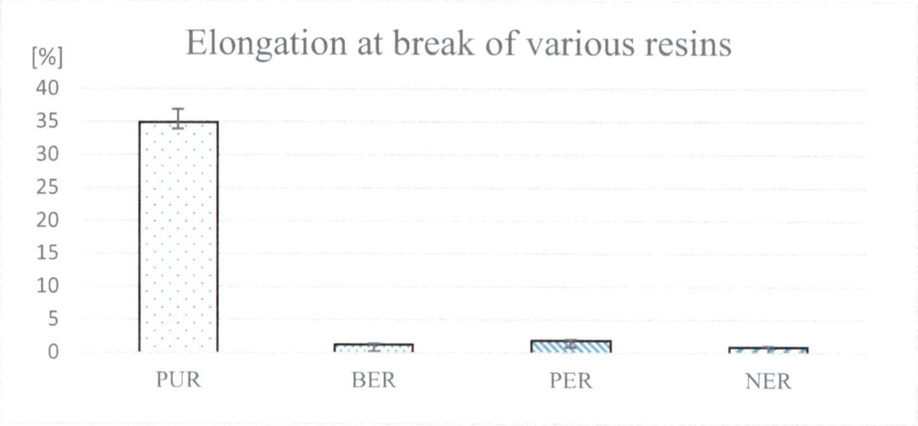

Fig. 5. Elongation at break of various resins.

3.2 Water Vapor Permeability

Water vapor diffusion capacity is an important criterion for evaluating resin systems. Based on a practical usability assessment, a completely diffusion-tight coating is considered unlikely. Moisture could penetrate the composite component through cracks or defects, but would be unable to escape again due to the diffusion-tight resin. This would cause significant damage to the composite component. The use of a diffusion-open resin enables limited moisture exchange with the environment, while simultaneously, the majority of the moisture that comes into contact with the coating (e.g., rainfall) is deflected.

To test the diffusion capacity in accordance with DIN EN ISO 12572, glass bowls are filled with dried salt ($CaCl_2$), covered with the NFRP laminate, and then sealed with sealing tape (see Fig. 6) [7]. The PU resin system exhibited an SD value of 60, which corresponds to a vapor barrier or a partially permeable laminate. The laminates with the bio-based resin systems Greenpoxy 56 and Super SAP showed SD values of 4,100 and 3,400, respectively, indicating they are diffusion-tight and rather unsuitable for this application.

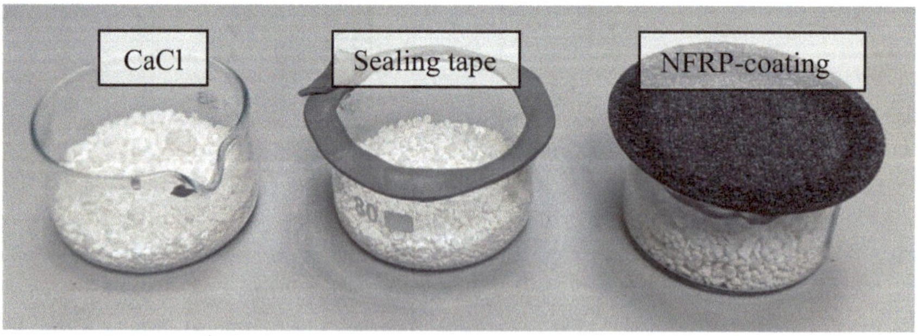

Fig. 6. Test samples for water vapor diffusion capacity.

3.3 Testing UV Stability

UV resistance was tested in accordance with DIN EN ISO 4892-2 using a xenon arc weathering tester [8]. For this purpose, different knitted fabric variants were combined with different resin systems. Selected samples were additionally given UV protection in the form of a coating varnish or a UV-stabilizing additive mixed into the resin. The test specimens were produced using a hand-lamination process, then cut to the required geometry using laser cutting, and fixed in specially designed sample holders (see Fig. 7).

Artificial weathering was conducted for a total duration of 1000 h in an Atlas Xenotest 440 testing machine. This involved combining 102 min of UV irradiation with 18 min of sprinkling in cycles to simulate realistic outdoor weathering conditions. After 500 h, the test was interrupted for an interim evaluation, and a visual inspection was carried out. Initial color changes were already visible at this point, but these could also be attributed in part to the dark cut edges of the laser cutout being washed out during the sprinkling phases.

The final evaluation was carried out after completion of the 1000 h of weathering. In addition to a visual assessment, mechanical tests were performed to determine tensile strength and elongation at break. The results indicated significant embrittlement of the PU-based resin under UV exposure. In contrast, the epoxy resin-based system exhibited greater UV stability, with mechanical characteristics remaining largely unchanged. In individual test series, an increase in strength was even observed, which can be attributed to post-curing of the resin under UV exposure. The added UV stabilizers only reduced embrittlement slightly.

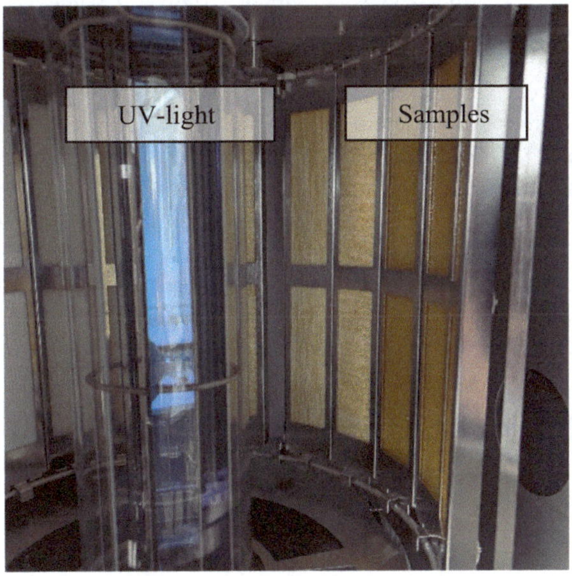

Fig. 7. Testing UV Stability

3.4 Mid-term Evaluation

Based on the evaluation of the previous test series, a differentiated interim conclusion can be drawn for further material development. First, it was found that bio-based yarns can be processed using knitting techniques without restrictions. The yarns can be reliably processed into the intended knitted structures and thereby fulfill the basic requirements for the production of semi-finished textile products.

Regarding the used resin systems, significant material-dependent differences can be observed. Diffusion-open resin systems enable controlled moisture balance between the wooden components and the environment, helping to regulate moisture within the components. In contrast, diffusion-tight resin systems provide almost complete protection of wooden components against external moisture when processed correctly. However, these systems require particularly high-quality processing, as even the smallest imperfections can lead to local moisture ingress.

In terms of mechanical properties, PUR-based systems have a significantly higher elongation at break. This property is particularly advantageous with respect to dimensional changes in wood caused by swelling and shrinkage, as the coating is better able to accommodate these deformations. In contrast, epoxy resin-based systems exhibit significantly higher UV resistance and are therefore better suited for long-term outdoor use under weathering conditions.

The results reveal a central conflict in objectives regarding the further development of material technology. On one side, there are diffusion-tight, UV-resistant resin systems; on the other, diffusion-open but less UV-stable systems. Similarly, there is a conflict between high elongation at break and flexibility, and between high stiffness and long-term stability. Therefore, further material development should aim to balance these conflicting requirements as much as possible, or to address them specifically through hybrid system approaches.

Based on the experimental investigations, a polyurethane (PUR)-based resin system was selected for further development and application within the NFRP coating. This decision is primarily justified by its high elongation at break (approx. 35%), which allows the system to accommodate dimensional changes in timber due to moisture-induced swelling and shrinkage without cracking the coating. In addition, the PUR system exhibits a moderate water vapor diffusion resistance (SD $\approx$ 60), enabling a controlled moisture exchange between the timber component and the environment. This behavior is considered essential to prevent moisture accumulation in the wood in the event of a local defect.

4 Functionalization

4.1 Sensor Integration

Sensors were integrated into the system to monitor the condition of the coated wooden components and to continuously or periodically record relevant climatic conditions within the structure. The applied sensor systems measure the temperature and relative humidity in the environment surrounding the components. Based on the recorded measurement data, the wood's balance (equilibrium) moisture content can be determined using a calculation matrix as described by Keylwerth (1949). This enables an indirect assessment of the moisture-related stress state of the wooden components [9].

Wired and wireless sensor systems were compared in order to select a suitable measurement concept. Flat, wired measurement systems offer high measurement accuracy and continuous data logging, even in hard-to-reach areas. The sensors are positioned directly on the wood surface, i.e., in the transition zone between the wooden component and the coating. However, these systems have certain disadvantages, including relatively high space requirements and the need for a complex cable layout. In particular, the required coating penetration for cable management poses a risk of leaks and can impair the system's protective function.

Wireless RFID-based sensor systems were considered as an alternative. These sensors only record measurements when actively read out by an external transponder and do not require a continuous power supply. They are integrated in a space-efficient manner via a special hole drilled directly into the wooden component (see Fig. 8). After the sensor has been inserted, the hole is sealed with a precisely fitting wooden plug, which permanently protects the sensor from mechanical damage and external influences. Depending on the sensor type, an installation depth of up to 15 mm into the wood can be achieved (see Fig. 9).

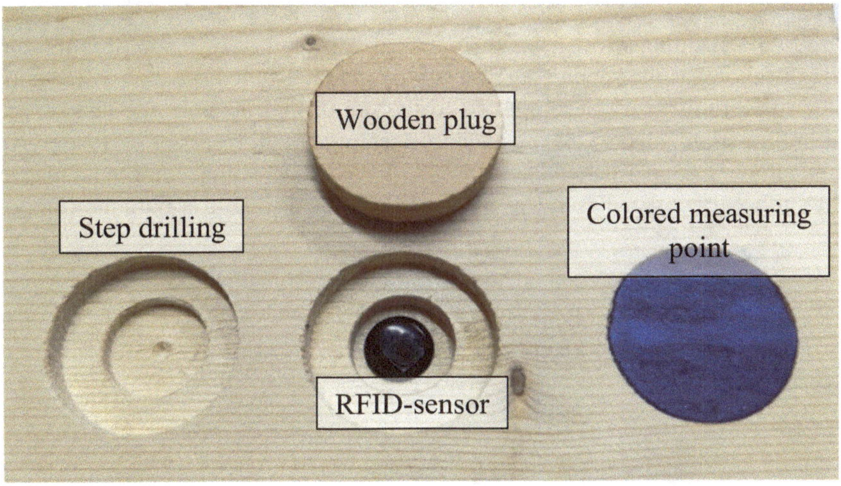

Fig. 8. Installation of the RFID sensor.

For the specific application of the column base, RFID sensors were identified as the most practical and cost-effective measurement system. Individual sensors can be installed with minimal space requirements and at a reasonable cost. In addition, the system enables periodic monitoring of wood moisture, for example, at six-monthly or annual intervals, without the need for complex measurement technology. The readings can be taken directly by the building owner using a transponder, which records, evaluates, and documents the measured values.

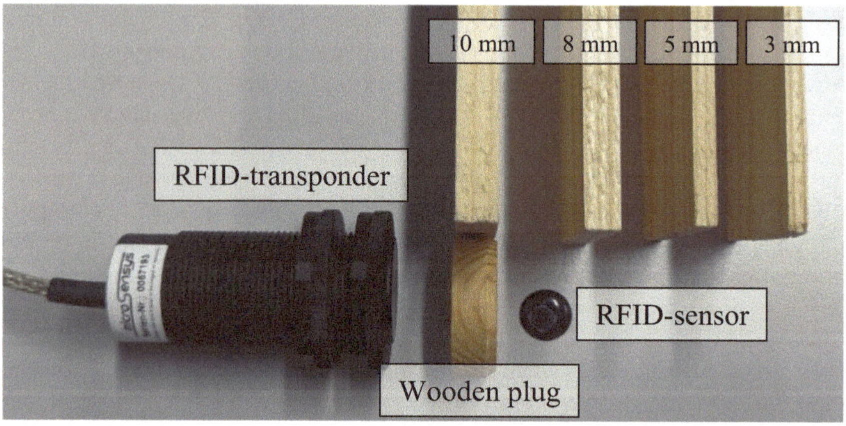

Fig. 9. Test setup for determining the measuring range of the RFID sensor.

4.2 Introduced Fasteners

Not all fasteners can be considered during the planning phase and incorporated into the knitted fabric for the coating. For subsequent fastener installations, such as threaded rods, screws, or similar elements, it is essential that they are sealed to the rest of the coating.

Due to the cladding's low layer thickness of approximately 1 mm, it is difficult to seal the drill hole, as the sealing surface is very small. To increase the sealing surface, chamfers of various sizes were applied to the pre-drilled holes in the test series. As shown in Fig. 10, this increased the available sealing height from approximately 1 mm to 5 mm, allowing the permanently elastic SIGA PRIMUR sealing compound to reliably seal the surface. Excess material was displaced into the drill hole or onto the component surface.

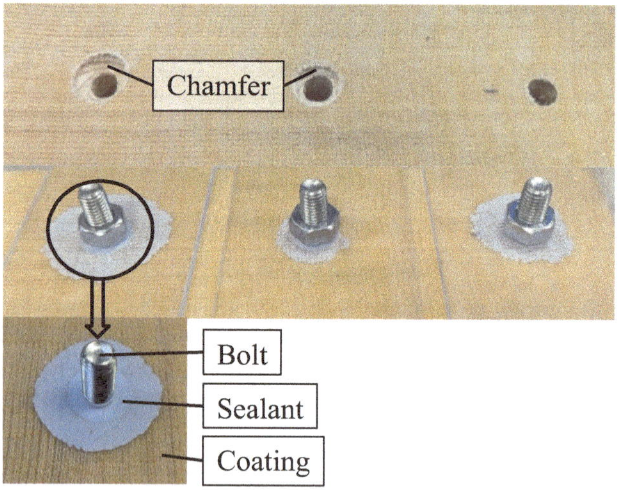

Fig. 10. Sealing of fasteners with permanently elastic sealant.

5 Recycling

To recycle natural fiber composites professionally, it is necessary to separate the material groups as completely as possible. This requirement is consistent with recent research, which highlights that mechanical and material separation is a key challenge in recycling natural fiber-reinforced polymer composites (NFRPCs), particularly due to the strong fiber-matrix interaction and material heterogeneity [10]. In this research, a method for separating and recycling the NFRP coatings of wooden beams was developed and evaluated.

To create a defined separation surface, the timber components were coated with a biological release agent. This specifically reduced the adhesive tensile strength between the timber and the cladding. The required release force was adjusted based on the type and amount of the applied release agent. Adhesive tensile tests yielded an optimum value of around 1.5 MPa, corresponding to around 50% of the transverse tensile strength of spruce timber (see Fig. 11).

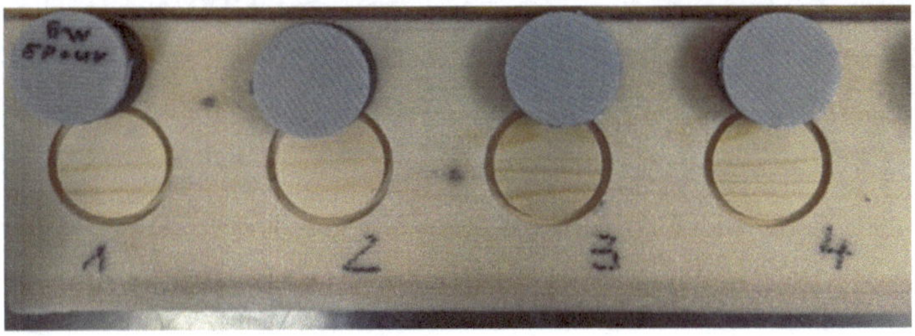

Fig. 11. Adhesion test of the coating.

Manual removal of the coating proved to be impractical. Instead, the coating was removed mechanically using a planing bench. The feed was approximately 2 mm in the thickness direction. An average of 1–1.5 mm of the NFRP coating and an additional 0.5–1 mm of wood was removed per planing pass. With a beam cross-section of 120 × 200 mm^2, this corresponded to a cross-section retention of around 97%.

The removed material was chipped, extracted, and then sieved for classification. The fractions included <1 mm, 1–2 mm, 2–4 mm, 4–8 mm, and > 8 mm. The wood content in the fractions decreased with increasing grain size. Fractions between 2 and 4 mm had the highest mass fraction. Large fractions (> 8 mm) showed almost pure separation of the coating with only minor wood adhesion. The fine fraction <1 mm contained a higher proportion of wood, but at around 15% by mass, this accounted for only a small share of poorly recyclable residues (see Fig. 12).

Given the small quantities of coating material to be recycled (< 1 t/a), the use of NFRP residues as a filler in concrete was investigated. For this purpose, 2% by mass of the residues were added to a concrete mix. After 28 days of storage, the flexural and compressive strengths were evaluated. The flexural strength decreased by about 10% to 8.42 MPa while the compressive strength remained at around 80 MPa. The modified concrete mixture is therefore suitable for use in foundation areas or support zones of precast bridges that are subject to low stress (see Fig. 13).

Fig. 12. Different chip fractions of the planed NFRP coating.

Fig. 13. Fracture pattern of the concrete test sample with 2% by mass NFRP filler content.

Overall, the developed process represents a resource-saving and cost-efficient recycling strategy. Additional CO_2 emissions are avoided while NFRP residues are returned to the building materials cycle. Larger fractions can also be processed in a cutting mill to further increase the amount of recyclable waste and reduce the proportion of materials that need to be thermally recycled.

6 Conclusions

As part of the study, a holistic system for coating wooden components based on knitted natural fiber semi-finished products was developed and evaluated. The knitting process for the bio-based yarns proved to be reliable, enabling the production of form-fitting, three-dimensional knitted structures with good drapability. This allowed the achievement of a load-independent, component-conforming coating. The developed system enables a service life extension of up to **20 years** in near-ground applications.

Various resin systems were investigated for impregnating and fixing the knitted fabrics. Diffusion-open systems allow defined moisture equalization in the timber component, while diffusion-tight systems offer complete moisture protection when processed correctly. PU-based resins exhibit a high elongation at break of approximately 35%, enabling effective compensation for swelling and shrinkage in timber. In contrast, epoxy resin systems exhibit significantly lower elongation at break (approx. 1.5%) but higher tensile strengths of up to ~40 MPa and better UV resistance. Water vapor diffusion tests showed SD values of approx. 60 for PU systems, indicating partial permeability, whereas epoxy-based systems reached SD values between 3400 and 4100, classifying them as diffusion-tight. UV aging tests over 1000 h revealed embrittlement of PU systems, while epoxy systems largely retained their mechanical properties. However, both material groups require additional UV protection for long-term outdoor use. The choice of material is therefore characterized by a conflict of objectives between diffusion behavior, flexibility, and durability.

To functionalize the system, temperature and relative humidity sensors were integrated into the design. RFID-based sensor systems, in particular, proved to be a space-saving, economical, and practical solution for periodical condition monitoring of wooden components. The sensors can be integrated into the timber to a depth of up to 15 mm and allow non-destructive evaluation of moisture conditions. Based on the measurement data, the equilibrium moisture content of the wood can be calculated, enabling early detection of potential moisture-related damage.

Finally, a recycling concept was developed for the natural fiber composite coatings. The use of a biological release agent enabled the creation of a defined separation plane between the wood and the coating. Mechanical separation using a planing bench preserved approximately 96–97% of the timber cross-section while allowing efficient removal of the coating. The resulting residues were successfully reused as fillers in concrete formulations at a proportion of 2 wt.%. This led to a reduction in flexural strength of approximately 10% (to around 8.4 MPa), while compressive strength remained largely unchanged at approximately 80 MPa, enabling reuse in low-load structural applications.

References

1. Kossakowski PG (2024) New advances in strengthening of structural timber. Materials 17(11):2545. https://doi.org/10.3390/ma17112545
2. Bühler J (2008) Holz als konstruktiver Baustoff, Holzbau Handbuch, vol Bd. 4, Teil 1, Folge 1. Holzabsatzfonds -Absatzförderungsfonds der deutschen Forst-und Holzwirtschaft, Godesberger Allee 142–148, D-53175 Bonn. ISSN 0466-2114
3. Wagenführ R, Scholz F (2021) Holzatlas, 7th edn. Carl Hanser Verlag, München
4. Bazli M, Heitzmann M, Hernandez BV (2022) Durability of fibre-reinforced polymer-wood composite members: an overview. Compos Struct 295:115827. https://doi.org/10.1016/j.compstruct.2022.115827
5. Huang S, Yan L, Bachtiar EV, Kasal B (2022) Durability of epoxy and polyurethane bonded timber-hybrid FRP joints under hygrothermal and weathering conditions. J Build Eng 58:104989. https://doi.org/10.1016/j.jobe.2022.104989
6. DIN EN ISO (2025) 527-2:2025–09: plastics—determination of tensile properties—part 2: test conditions for moulding and extrusion plastics (ISO 527-2:2025). German version EN ISO 527-2:2025
7. DIN EN ISO (2025) 12572:2025–03: hygrothermal performance of building materials and products—determination of water vapour transmission properties—cup method (ISO 12572:2016 + Amd 1:2024). German version EN ISO 12572:2016 + A1:2024
8. DIN EN ISO (2021) 4892-2:2021–11: plastics—methods of exposure to laboratory light sources—part 2: Xenon-arc lamps (ISO 4892-2:2013 + Amd 1:2021). German version EN ISO 4892-2:2013 + A1:2021

9. Keylwerth R (1949) Grundlagen der Hochtemperaturtrocknung des Holzes, Holz-Zentralblatt Jg. 75 Nr. 76, 1949, DRW-Verlag, Leinfelden-Echterdingen
10. Zhao X et al (2022) Recycling of natural fiber composites: challenges and opportunities. Resour Conserv Recycl 177:105962. https://doi.org/10.1016/j.resconrec.2021.105962

Experimental Study of Lead and Copper Immobilization in Soil Using Priestia sp. J4

Vaida Vitkauskaite[1(✉)], Saulius Vasarevicius[1], and Renata Gudiukaite[2]

[1] Department of Environmental Protection and Water Engineering/Faculty of Environment Engineering, Vilnius Gediminas Technical University, Sauletekio al. 11, Vilnius LT-10223, Lithuania
vaida.vitkauskaite@vilniustech.lt

[2] Institute of Biosciences, Life Sciences Center, Vilnius University, Sauletekis av. 7, Vilnius LT-10257, Lithuania

Abstract. Heavy metal contamination of soils, particularly with lead (Pb^{2+}) and copper (Cu^{2+}), remains a significant environmental and engineering challenge due to their persistence, toxicity, and potential for bioaccumulation. Immobilization-based remediation strategies are increasingly preferred as sustainable solutions for reducing metal mobility and environmental risk without disturbing soil structure. Among biological approaches, microbial remediation has gained considerable attention due to microorganisms' adaptive capacity and their ability to promote long-term contaminant stabilization. The aim of this study was to evaluate the efficiency of lead and copper immobilization in artificially contaminated soil using the bacterial strain *Priestia sp. J4* under controlled laboratory conditions. Soil samples were artificially contaminated with Pb and Cu salts and treated with different concentrations of a Priestia sp. J4 suspension. The experiment was conducted over a thirteen-week period, during which changes in soil pH and the mobility of heavy metals were monitored. The bioavailable fractions of Pb and Cu were assessed using $CaCl_2$ extraction, followed by quantitative analysis using inductively coupled plasma optical emission spectrometry (ICP-OES). The results demonstrated that inoculation with *Priestia sp. J4* significantly reduced the mobility of both lead and copper in soil, with immobilization efficiency strongly dependent on incubation time and microbial inoculum concentration. Lead exhibited more stable immobilization behavior, while copper showed greater variability during the early stages of the experiment. The most pronounced reduction in metal mobility was observed after prolonged incubation, indicating the importance of long-term microbial activity. These findings confirm that *Priestia sp. J4* has strong potential as a sustainable biological agent for immobilizing lead and copper in contaminated soils, contributing to long-term risk reduction and environmentally friendly soil remediation strategies.

Keywords: Bioremediation · Heavy Metals · Immobilization · Microorganism

Z. Perkowski et al. (Eds.): ECCE 2026, LNCE 929, pp. 83–98, 2027.
https://doi.org/10.1007/978-3-032-39515-3_7

1 Introduction

Soil contamination with heavy metals, particularly lead (Pb^{2+}) and copper (Cu^{2+}), represents a persistent global environmental problem associated with industrial activities, urban development, and intensive agriculture [1–3]. Due to their chemical stability and non-biodegradable nature, Pb^{2+} and Cu^{2+} accumulate in soils, posing long-term risks to ecosystems and human health through food-chain transfer and groundwater contamination [2, 4]. Numerous studies have demonstrated that even moderate concentrations of these metals can disrupt soil microbial activity, reduce soil fertility, and negatively affect essential biochemical processes in soil systems [3, 5].

From an engineering perspective, heavy metal-contaminated soils present additional challenges, as metal accumulation may alter soil physicochemical properties and compromise land usability and structural safety [6, 7]. As a result, bioremediation of contaminated soils has become a priority within sustainable environmental management and civil engineering research [6]. Conventional bioremediation approaches, including excavation, soil washing, and physicochemical stabilization, although effective in certain cases, are often costly, energy-intensive, and disruptive to soil structure, while potentially generating secondary pollutants [6–8].

In recent years, immobilization-based bioremediation strategies have gained increasing attention as environmentally friendly alternatives for managing heavy metal-contaminated soils [9, 10]. Immobilization aims to reduce metal mobility and bioavailability rather than remove contaminants entirely, thereby minimizing ecological risk while preserving soil structure and functionality [10, 11]. These processes involve adsorption, ion exchange, complexation, and precipitation reactions that transform metals into more stable and less bioavailable forms [9, 12].

Biological immobilization using microorganisms has emerged as a particularly promising approach due to the adaptive capacity of microbial systems and their ability to promote long-term stabilization of contaminants [3, 13]. Microorganisms immobilize heavy metals through biosorption onto cell surfaces, biofilm formation, bioprecipitation, biomineralization, and transformation of metals into less toxic chemical forms [13–15]. Compared with purely physicochemical methods, microbial remediation offers advantages in terms of sustainability, cost efficiency, and reduced environmental disturbance [2, 3].

Among microbial candidates, bacteria of the genus *Bacillus* have attracted considerable interest for soil remediation applications due to their spore-forming ability, resistance to environmental stress, and metabolic versatility [14, 16]. Numerous studies have demonstrated that *Bacillus spp.* can immobilize Pb and Cu by binding metal ions to functional groups on cell walls, producing extracellular polymeric substances, and inducing the formation of insoluble metal precipitates [15–18]. Consequently, *Bacillus spp.* is regarded as particularly suitable for applications requiring long-term stabilization of heavy metals in soils [16, 18].

Despite the growing body of research on microbial immobilization of heavy metals, immobilization efficiency remains strongly dependent on microbial strain selection, inoculum concentration, metal type, and incubation time [12, 19]. In addition, temporal fluctuations in metal mobility, including early-stage remobilization phenomena, are not yet fully understood [11, 19]. Addressing these aspects is essential for optimizing biological remediation strategies and ensuring their effectiveness under controlled and practical conditions.

In response to the need for sustainable soil bioremediation technologies, the present study investigates the ability of *Priestia sp. J4* to immobilize lead and copper in artificially contaminated soil. The study focuses on evaluating the effects of microbial inoculum concentration and incubation time on the mobility of Pb2+ and Cu2+, thereby providing experimental evidence to support the application of bacterial-based immobilization approaches for long-term risk reduction in contaminated soils.

Unlike many previous studies that focus on single-metal systems or short-term experiments, this study provides a systematic evaluation of Pb and Cu immobilization in both single and multi-metal contaminated soils using Priestia sp. J4 over an extended incubation period. Additionally, the combined application of ICP-OES and FTIR analysis allows not only quantification of metal immobilization but also identification of the underlying chemical interactions and functional groups involved in the process. This integrated approach provides new insights into the mechanisms of long-term microbial immobilization relevant to contaminated soil management in civil engineering applications.

The results of this study are particularly relevant to civil engineering, where stabilization of contaminated soils is essential for safe construction and land use. The proposed microbial immobilization approach offers a sustainable alternative to improve soil quality and reduce environmental risks associated with heavy metal contamination in engineering projects.

The aim of this study was to evaluate the efficiency of lead and copper immobilization in artificially contaminated soil using the bacterial strain *Priestia sp. J4* under controlled laboratory conditions.

2 Methodology

The experimental methodology was designed as a sequence of clearly defined phases, as illustrated in Fig. 1, including soil sampling and preparation, artificial contamination with heavy metals, and biological treatment using *Priestia sp. J4*, and analytical evaluation of heavy metal immobilization efficiency.

2.1 Soil Sampling and Preparation

Soil used for the experiment was collected from the Klebonai forest area (Siauliai district, Lithuania), selected to be far from major roads and potential sources of anthropogenic pollution to obtain uncontaminated background material. Prior to sampling, the surface organic layer was removed. Soil samples were collected from the topsoil layer (0–20 cm depth). Ten subsamples were taken from different locations within the site and combined to form a composite sample.

After collection, soil samples were air-dried at room temperature to constant mass, sieved through a 2 mm mesh, and homogenized to ensure uniform physicochemical properties. The initial concentrations of lead and copper were verified using X-ray fluorescence (XRF) analysis to confirm that the soil was suitable for controlled contamination experiments.

2.2 Artificial Contamination of Soil with Heavy Metals

Artificial contamination was performed under laboratory conditions using aqueous solutions of lead nitrate ($Pb(NO_3)_2$) and copper nitrate ($Cu(NO_3)_2 \cdot 3H_2O$). Target contamination levels were selected to exceed the limit values defined in the Lithuanian hygiene norm HN 60:2015 by a factor of 0.5, ensuring measurable contamination while maintaining conditions favorable for microbial activity.

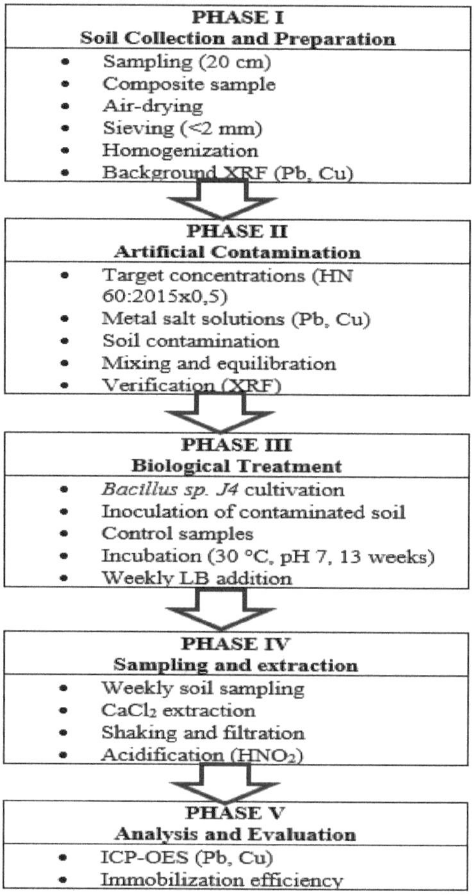

Fig. 1. Schematic representation of the experimental methodology divided into successive phases, from soil sampling and characterization to biological treatment with *Priestia sp. J4* and evaluation of heavy metal immobilization efficiency.

Portions of 300 g of homogenized soil were placed in plastic containers and contaminated with either individual metals (Pb or Cu) or a mixture of both. Metal salt solutions were prepared in distilled water and added to the soil, followed by thorough mixing to ensure uniform distribution. After contamination, soil samples were air-dried and mixed periodically to stabilize metal distribution prior to biological treatment.

2.3 Microorganism Treatment Using *Priestia sp. J4*

Biological immobilization of heavy metals was carried out using the bacterial strain *Priestia sp. J4*, obtained from the Life Sciences Center, Vilnius University. The strain was originally isolated from cereal rhizosphere soil and taxonomically identified based on 16S rRNA gene sequencing.

Bacterial cultures were grown in liquid LB (Lennox) medium at 30 °C with shaking at 180 rpm for 24 h until reaching the late exponential growth phase ($OD_{600} \approx 4.0$). Following cultivation, microbial suspensions were introduced into contaminated soil samples at three different inoculum volumes (100, 60, and 40 mL), corresponding to different microbial concentrations. Soil samples contaminated with heavy metals but without microbial inoculation were maintained as controls.

All samples were incubated at 30 °C for 13 weeks. To sustain microbial activity, 20 mL of sterile LB medium were added weekly to each inoculated sample. Soil samples were collected at weekly intervals to monitor the temporal dynamics of heavy metal immobilization.

The 13-week experimental period was selected to evaluate both short-term and longer-term dynamics of microbial immobilization. Previous studies indicate that microbial adaptation, biofilm formation, and stabilization processes require extended incubation periods. Therefore, this duration was chosen to capture both initial fluctuations and long-term stabilization trends.

2.4 Evaluation of Heavy Metal Immobilization

To assess the bioavailable fraction of heavy metals, the collected soil samples were dried at 30 °C. From each sample, 1 g of soil was extracted with 10 mL of 0.01 M $CaCl_2$ solution (soil-to-solution ratio 1:10). Suspensions were shaken for 2 h at 80 rpm, followed by filtration. Filtrates were acidified with concentrated HNO_3 to stabilize dissolved metals. The 2-h extraction time was selected based on commonly used $CaCl_2$ extraction protocols that target the mobile (bioavailable) fraction of heavy metals rather than total leaching potential. While longer extraction times (e.g., 24 h) are used in standard leaching tests, shorter durations are considered more representative for assessing readily available metal fractions influenced by microbial activity. Therefore, the selected extraction time allows evaluation of biologically relevant immobilization processes.

Concentrations of Pb^{2+} and Cu^{2+} in the extracts were determined using inductively coupled plasma optical emission spectrometry (ICP-OES). All measurements were performed in triplicate, and results are presented as mean values. The comparison of metal concentrations in inoculated and non-inoculated samples was used to evaluate the efficiency of microbial immobilization over time.

3 Research Results

3.1 Copper Leaching from Single-Metal-Contaminated Soil

After completion of the thirteen-week experimental study, heavy metal concentrations in the leaching solutions were determined using inductively coupled plasma optical emission spectrometry (ICP-OES). The obtained results are presented graphically in the corresponding figures.

Copper leaching concentrations in soil contaminated with Cu only and treated with *Priestia sp. J4* are presented in Figs. 2, 3 and 4. The results demonstrate that copper mobility varied over the experimental period and depended on both the microbial inoculum concentration and the incubation time.

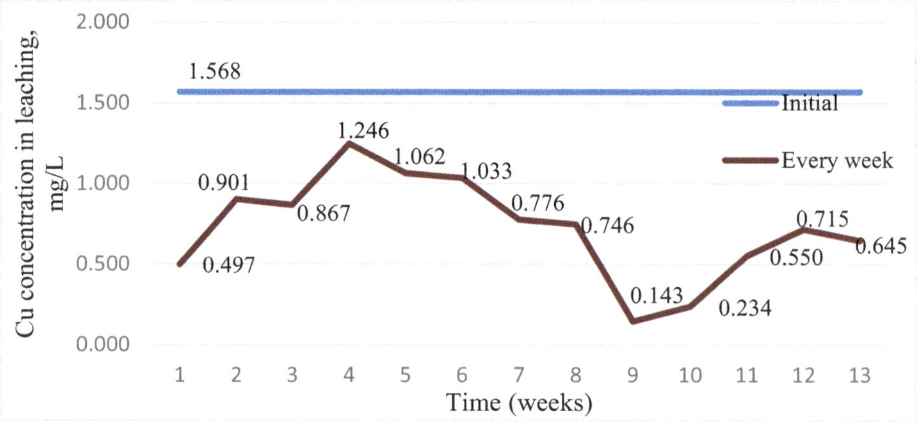

Fig. 2. Copper leaching concentration in single-metal-contaminated soil inoculated with 3.20×10^{11} microorganisms, mg/L.

At the highest microbial concentration (3.20×10^{11} cells), copper leaching initially increased during the early experimental phase, peaking around the fourth week (see Fig. 2). A gradual decrease in copper concentration was observed thereafter, with the lowest leaching values in the ninth week. Following this period, copper concentrations stabilized or slightly increased toward the end of the experiment.

When a medium microbial concentration (1.92×10^{11} cells) was applied, copper leaching followed a similar temporal pattern (see Fig. 3). The lowest copper concentrations were observed in the ninth week, while higher concentrations were detected during the early experimental stage. After reaching its minimum, copper leaching showed a slight increase in the later weeks.

At the lowest microbial concentration (1.28×10^{11} cells), copper leaching showed greater variability throughout the experimental period (see Fig. 4). The highest concentrations were observed in the fourth week, whereas reduced copper mobility was recorded after prolonged incubation, particularly in the ninth week.

Overall, copper immobilization in single-metal-contaminated soil became more pronounced with extended incubation time, regardless of microbial inoculum concentration.

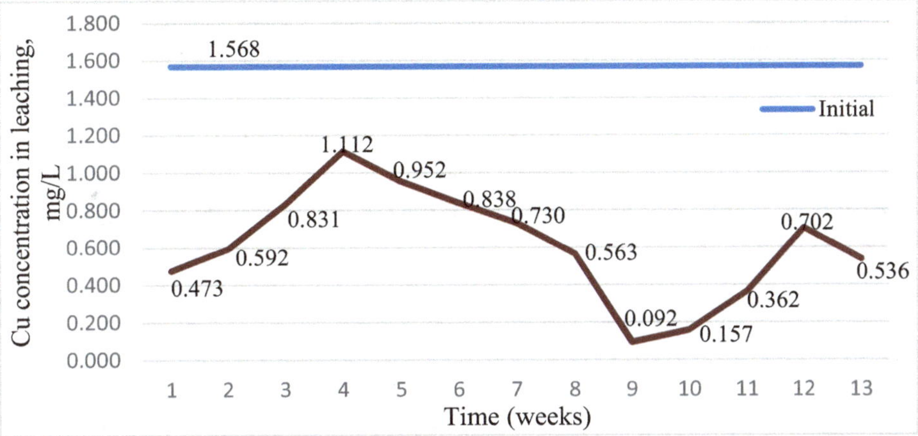

Fig. 3. Copper leaching concentration in single-metal-contaminated soil inoculated with 1.92×10^{11} microorganisms, mg/L.

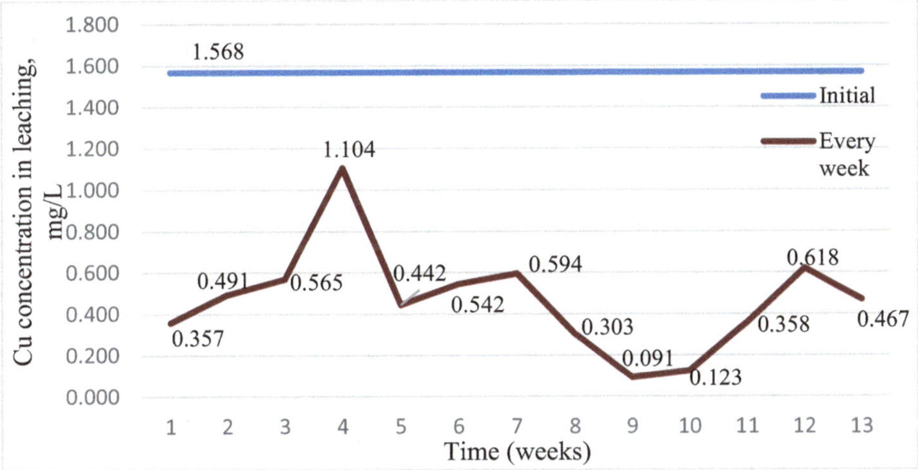

Fig. 4. Copper leaching concentration in single-metal-contaminated soil inoculated with 1.28×10^{11} microorganisms, mg/L.

3.2 Lead Leaching from Single-Metal-Contaminated Soil

Lead leaching concentrations in soil contaminated with Pb only and treated with *Priestia sp. J4* are shown in Figs. 5, 6 and 7. In contrast to copper, lead mobility exhibited a more stable temporal pattern.

At the highest microbial concentration (3.20×10^{11} cells), lead leaching gradually decreased during the experimental period, indicating progressive immobilization (see Fig. 5). Lower lead concentrations were consistently observed after the initial incubation phase.

For the medium microbial concentration (1.92 × 10^{11} cells), lead leaching initially increased during the second week, then declined substantially in subsequent weeks (see Fig. 6). The lowest lead concentrations were recorded in the later stages of the experiment.

At the lowest microbial concentration (1.28 × 10^{11} cells), lead leaching exhibited short-term fluctuations; however, a general downward trend was observed after the third week (see Fig. 7). Reduced lead mobility persisted during prolonged incubation.

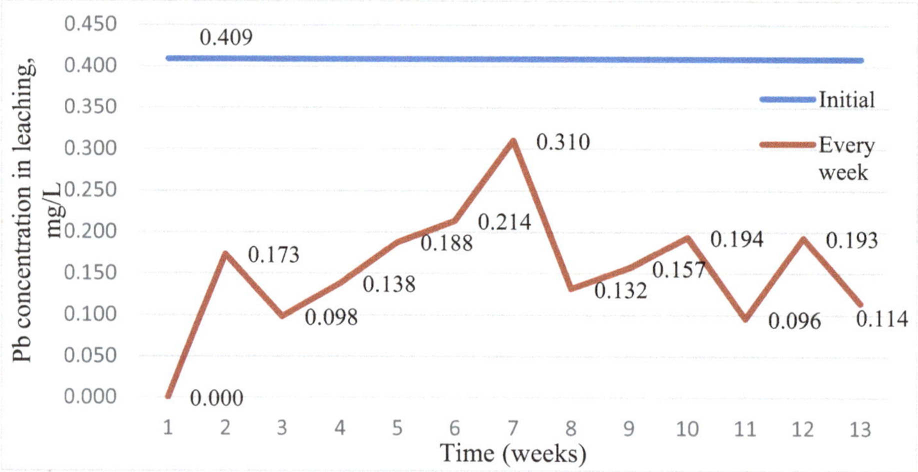

Fig. 5. Lead leaching concentration in single-metal-contaminated soil inoculated with 3.20 × 10^{11} microorganisms, mg/L.

These results indicate that lead immobilization occurred more consistently over time compared with copper, particularly at higher microbial concentrations.

3.3 Lead and Copper Leaching from Multi-metal-Contaminated Soil

The leaching behavior of lead and copper in soil contaminated with both metals and treated with *Priestia sp. J4* is presented in Figs. 8, 9 and 10. In multi-metal systems, metal mobility exhibited more complex temporal dynamics.

At the highest microbial concentration (3.20 × 10^{11} cells), lead leaching increased temporarily around the sixth week, followed by a gradual decrease toward the eleventh week, when the lowest lead concentrations were recorded (see Fig. 8). Copper leaching decreased rapidly during the first week, followed by a period of relative stabilization. The lowest copper concentrations were observed around the tenth week.

When a medium microbial concentration (1.92 × 10^{11} cells) was applied, lead leaching remained relatively stable throughout the experimental period, with minor fluctuations (see Fig. 9). The lowest lead concentrations were observed during the fifth and eleventh weeks. Copper leaching reached higher concentrations during the early phase, followed by a decrease after prolonged incubation.

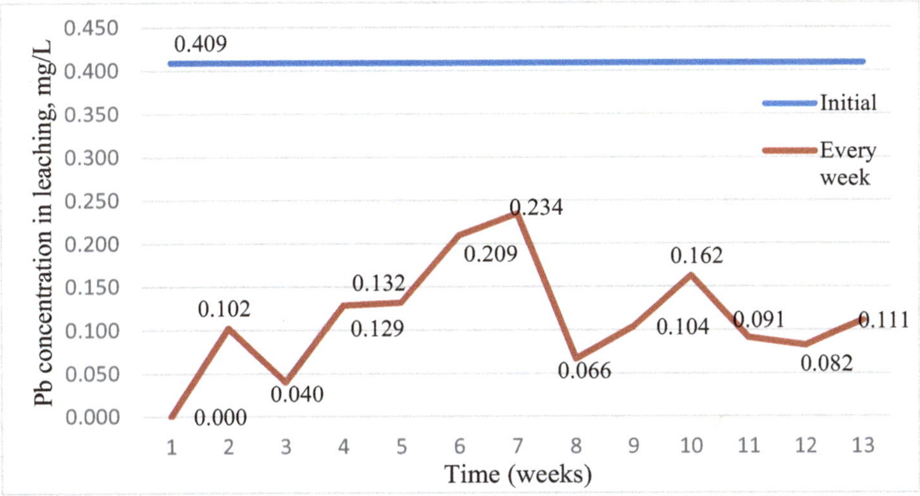

Fig. 6. Lead leaching concentration in single-metal-contaminated soil inoculated with 1.92×10^{11} microorganisms, mg/L.

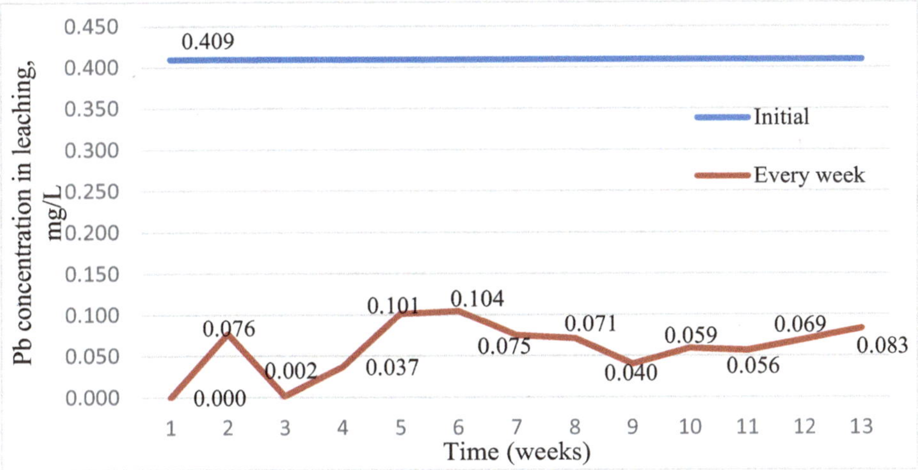

Fig. 7. Lead leaching concentration in single-metal-contaminated soil inoculated with 1.28×10^{11} microorganisms, mg/L.

At the lowest microbial concentration (1.28×10^{11} cells), both lead and copper leaching showed pronounced variability (see Fig. 10). Elevated metal concentrations were detected during the mid-experimental period, while reduced mobility was observed after extended incubation, particularly after the tenth week.

The experimental data were evaluated and presented in two stages. First, the mean leached concentrations of copper (Cu^{2+}) and lead (Pb^{2+}) were calculated using data from all experimental weeks. These mean values provide an overall assessment of microbial

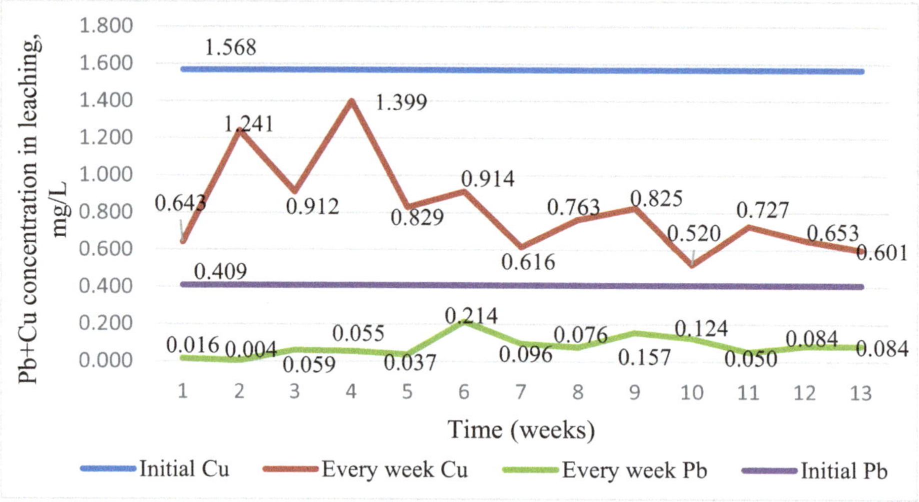

Fig. 8. Lead and copper leaching concentration in soil contaminated with two metals and inoculated with 3.20×10^{11} microorganisms, mg/L.

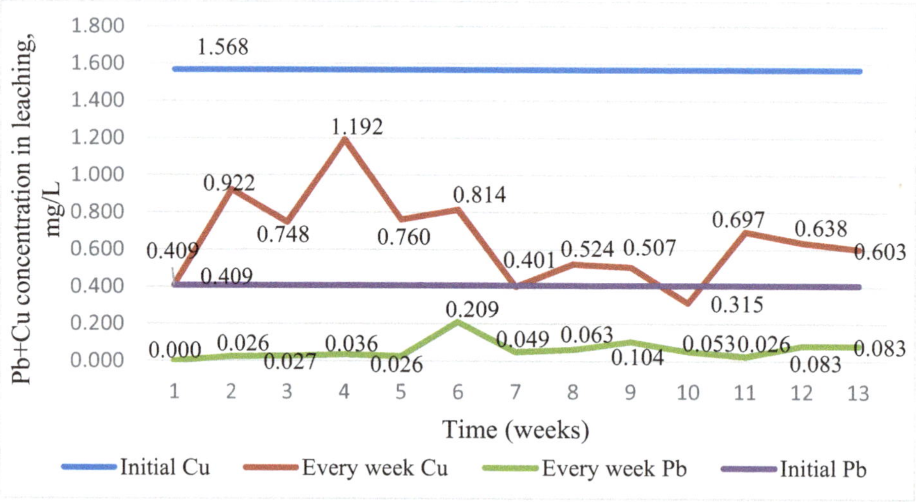

Fig. 9. Lead and copper leaching concentration in soil contaminated with two metals and inoculated with 1.92×10^{11} microorganisms, mg/L.

effects on heavy metal mobility in soil throughout the entire experimental period (see Table 1).

In the second stage, data from the optimal experimental weeks were analyzed separately. The first two weeks were excluded from this analysis, as this period corresponded to microbial adaptation to the soil environment and stabilization of metal-microorganism interactions. The optimal weeks (week 10 for Cu^{2+} and week 11 for Pb^{2+}) were selected

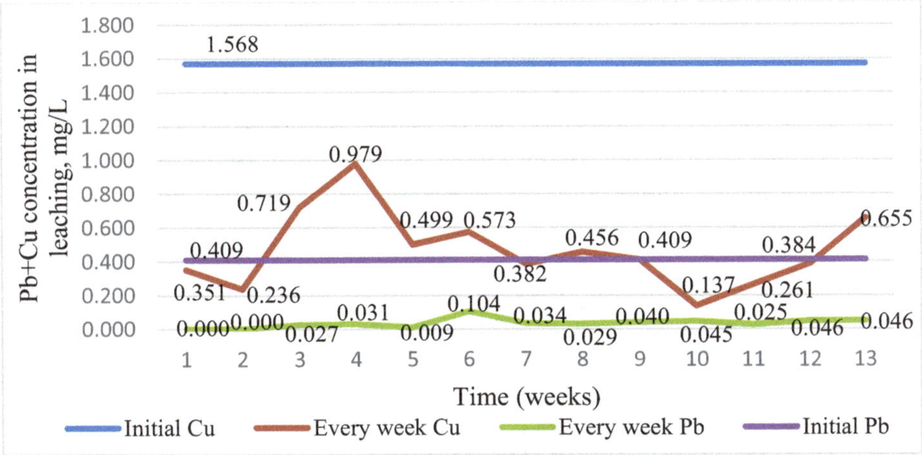

Fig. 10. Lead and copper leaching concentration in soil contaminated with two metals and inoculated with 1.28×10^{11} microorganisms, mg/L.

based on the lowest observed metal leaching concentrations, indicating the highest immobilization efficiency (see Table 1).

Immobilization efficiency was calculated by comparing leached metal concentrations with the initial metal concentrations in the soil.

The results demonstrate that increasing microbial concentration led to higher immobilization efficiency for both copper and lead. Lead was immobilized more effectively than copper under all experimental conditions, with Pb^{2+} immobilization efficiency exceeding 93% during the optimal experimental weeks. The selected optimal-week results indicate that, following the microbial adaptation period, metal mobility in soil was significantly reduced, suggesting stable and effective long-term immobilization.

Across all experimental variants, immobilization efficiency increased with incubation time. Higher microbial inoculum concentrations resulted in more stable reductions in metal mobility, particularly for lead. Copper leaching exhibited greater temporal variability, especially during the early stages of microbial treatment. In both single-metal and multi-metal systems, the most pronounced reductions in Pb and Cu^{2+} mobility were observed after prolonged incubation.

Table 1. Immobilization efficiency of copper (Cu^{2+}) and lead (Pb^{2+}) in soil using different microbial concentrations.

Microorganism concentration (mg/L)	Mean Cu^{2+} leached concentration (mg/L)*	Cu^{2+} immobilization efficiency (%)*	Mean Pb^{2+} leached concentration (mg/L)*	Pb^{2+} immobilization efficiency (%)*	Cu^{2+} leached concentration at week 10 (mg/L)**	Cu^{2+} immobilization efficiency (%)**	Pb^{2+} leached concentration at week 11 (mg/L)**	Pb^{2+} immobilization efficiency (%)**
3.20×10^{11}	0.819	0.819	0.819	0.819	0.819	0.819	0.819	0.819
1.92×10^{11}	0.656	0.656	0.656	0.656	0.656	0.656	0.656	0.656
1.28×10^{11}	0.465	0.465	0.465	0.465	0.465	0.465	0.465	0.465

*Mean values calculated using leached metal concentrations from all experimental weeks

**Selected optimal-week data after excluding the initial adaptation period

Initial concentrations: Cu – 1.568 mg/L; Pb – 0.409 mg/L

4 Discussion

The results of the present study demonstrate that *Priestia sp. J4* is capable of effectively reducing the mobility of lead and copper in artificially contaminated soil; however, the immobilization process was strongly dependent on incubation time, metal type, and microbial inoculum concentration. The observed temporal trends indicate that microbial immobilization is not an instantaneous process but rather a gradual phenomenon governed by cumulative biological and physicochemical interactions within the soil–microorganism system.

One of the most significant findings of this study is the pronounced influence of incubation time on immobilization efficiency. For both Pb^{2+} and Cu^{2+}, the most substantial reductions in metal mobility were observed after prolonged incubation, particularly between the ninth and eleventh weeks. This delayed immobilization suggests that initial microbial adsorption alone is insufficient to ensure long-term stabilization and that secondary mechanisms, such as biofilm development, extracellular polymeric substance (EPS) production, and biomineralization, play a critical role over time.

Similar time-dependent immobilization patterns have been reported in previous studies, where extended incubation allowed microorganisms to modify the soil microenvironment, alter metal speciation, and promote the formation of less soluble metal compounds [1 , 8, 9]. The results obtained here support the concept that effective microbial remediation requires sufficient time for biological processes to reach equilibrium, emphasizing the importance of long-term monitoring in bioremediation studies.

Distinct differences were observed between the immobilization behavior of lead and copper. Lead exhibited more stable and consistent immobilization across all experimental variants, particularly at higher microbial inoculum concentrations. This behavior can be attributed to the strong affinity of Pb^{2+} ions for negatively charged functional groups present on bacterial cell walls and soil organic matter, as well as the tendency of lead to form insoluble precipitates and stable complexes in soil environments [10, 11].

In contrast, copper immobilization exhibited greater temporal variability, particularly during the early stages of the experiment. Transient increases in copper leaching observed around the fourth week may reflect temporary solubilization of copper-organic complexes or competition with other cations present in the soil solution. Copper is known to exhibit complex redox behavior and strong interactions with organic ligands, which can result in fluctuating bioavailability under changing microbial and chemical conditions [9, 12]. Nevertheless, sustained microbial activity ultimately promoted copper stabilization, particularly at higher inoculum levels.

The comparison of different microbial inoculum concentrations highlights the importance of microbial density in achieving effective and stable immobilization. Higher inoculum concentrations resulted in more consistent reductions in the mobility of Pb2+ and Cu2+, likely due to increased availability of binding sites and enhanced EPS production. Lower microbial concentrations, although capable of initiating immobilization, exhibited greater variability and less stable long-term performance.

These findings are consistent with previous reports indicating that insufficient microbial biomass may limit biosorption capacity and reduce the efficiency of long-term stabilization mechanisms [8, 10]. Therefore, optimization of inoculum concentration represents a key parameter in the design of microbial-based soil remediation strategies.

The combined results of ICP-OES analysis and previously obtained FTIR data provide insight into the immobilization mechanisms induced by *Priestia sp. J4*. FTIR spectra indicated increased intensities of functional groups associated with metal binding, including carboxyl, hydroxyl, amino, and phosphate groups, following microbial treatment. These functional groups are known to participate in biosorption and complexation processes, facilitating the immobilization of metal ions on bacterial surfaces and within soil matrices. In addition to surface adsorption, *Bacillus spp.* are known to promote bioprecipitation and biomineralization, leading to the formation of insoluble metal compounds and further reducing metal mobility [10–12]. The gradual decrease in metal leaching observed over time supports the involvement of these mechanisms, which typically require prolonged microbial activity to become effective. Among bacteria, *Bacillus species* are distinguished by their capacity to immobilize heavy metals [13]. *Bacillus spp.* has applications in the food and pharmaceutical industries [14], agriculture [15], and the remediation of contaminated soils and water bodies [16]. These microorganisms exhibit high resistance and survival capacity in environments contaminated with heavy metals [17]. *Bacillus spp.* are Gram-positive, rod-shaped, spore-forming bacteria that are either aerobic or facultatively anaerobic [18]. Their ability to form spores under unfavorable environmental conditions confers resistance to high temperatures, drought, moisture fluctuations, radiation, and heavy metal stress [19].

From a practical perspective, the findings of this study demonstrate that microbial immobilization using *Priestia sp. J4* represents a viable and environmentally sustainable approach for reducing lead and copper mobility in contaminated soils. Unlike conventional remediation techniques that often involve physical removal or chemical stabilization, microbial immobilization preserves soil structure and biological integrity while achieving long-term risk reduction.

Studies indicate that bacteria belonging to the genus *Bacillus* effectively immobilize copper via biosorption, ion-exchange reactions, and complexation with functional groups on the cell wall, such as carboxyl, phosphate, and hydroxyl groups, as well as by promoting copper precipitation into less soluble forms [20].

The observed time-dependent nature of immobilization underscores the need to account for remediation timeframes when implementing biological approaches. Rather than expecting immediate contaminant removal, microbial-based strategies should be evaluated in terms of their capacity to achieve stable immobilization over extended periods. This approach aligns with current trends in sustainable soil management, prioritizing long-term ecological stability over rapid but disruptive remediation methods.

The results suggest that immobilization efficiency tends to stabilize after prolonged incubation; however, potential remobilization cannot be ruled out due to environmental changes such as pH fluctuations, organic matter degradation, or a decline in microbial activity. Further long-term studies are required to assess stability beyond 13 weeks under field conditions.

The main advantages of the proposed method include its low cost, environmental friendliness, and ability to reduce heavy metal mobility without disturbing soil structure. However, limitations include dependence on environmental conditions, variability in microbial activity, and the time required to achieve stable immobilization.

For practical application, scaling up would require optimization of microbial dosage, environmental conditions, and delivery methods. Field-scale implementation may involve the use of bioaugmentation or biostimulation techniques. Despite these challenges, the method shows strong potential for sustainable remediation of contaminated soils in engineering practice.

5 Conclusions

This study demonstrated that *Priestia sp. J4* is capable of effectively reducing the mobility of lead and copper in artificially contaminated soil, confirming its potential application as a biological immobilization agent for soil remediation. The results indicate that microbial treatment can significantly reduce the bioavailable fractions of heavy metals without requiring invasive physicochemical interventions.

Immobilization efficiency was strongly dependent on incubation time. For both Pb and Cu, the most pronounced reductions in metal mobility were observed after prolonged microbial activity, particularly between the ninth and eleventh weeks, highlighting the importance of long-term processes in microbial-based remediation strategies.

Distinct differences were identified in the behavior of lead and copper in soil. Lead exhibited more stable and consistent immobilization across all experimental variants, especially at higher microbial inoculum concentrations. In contrast, copper showed greater temporal variability, with transient remobilization during the early stages of treatment, emphasizing the need for extended monitoring when assessing copper-contaminated soils.

Microbial inoculum concentration was found to be a critical factor influencing immobilization performance. Higher concentrations of *Priestia sp. J4* resulted in more stable and sustained reductions in metal mobility, while lower inoculum levels led to greater variability in immobilization efficiency.

The combined experimental evidence supports the conclusion that microbial immobilization using *Priestia sp. J4* represents a sustainable, environmentally friendly approach to mitigating lead and copper contamination in soils. This strategy is particularly suitable for applications where long-term risk reduction and preservation of soil structure and functionality are prioritized over rapid contaminant removal.

References

1. Vasarevičius S, Paliulienė V (2025) Immobilization of cadmium, lead, and copper in soil using bacteria: a literature review. Land 14(8):1547. https://doi.org/10.3390/land14081547
2. Liu Z, Mo L, Liang J, Shi H, Yao J, Lun X (2025) Heavy metal pollution and health-ecological risk assessment in agricultural soils: a case study from the yellow river bend industrial parks. Toxics 13(10):834. https://doi.org/10.3390/toxics13100834
3. Campillo-Cora C, Rodríguez-Seijo A, Pérez-Rodríguez P, Fernández-Calviño D, Santás-Miguel V (2025) Effect of heavy metal pollution on soil microorganisms: influence of soil physicochemical properties. a systematic review. Eur J Soil Biol 124:103706. https://doi.org/10.1016/j.ejsobi.2024.103706
4. Rashid A et al (2023) Heavy metal contamination in agricultural soil: environmental pollutants affecting crop health. Agronomy 13(6):1521. https://doi.org/10.3390/agronomy13061521

5. Hou D et al (2020) Metal contamination and bioremediation of agricultural soils for food safety and sustainability. Nat Rev Earth Environ 1:366–381. https://doi.org/10.1038/s43017-020-0061-y
6. Song P, Xu D, Yue J, Ma Y, Dong S, Feng J (2022) Recent advances in soil remediation technology for heavy metal contaminated sites: a critical review. Sci Total Environ 838(3):156417. https://doi.org/10.1016/j.scitotenv.2022.156417
7. Wang L, Rinklebe J, Tack FMG, Hou D (2021) A review of green remediation strategies for heavy metal contaminated soil. Soil Use Manag 37(4):936–963. https://doi.org/10.1111/sum.12717
8. Atuchin VV, Asyakina LK, Serazetdinova YR, Frolova AS, Velichkovich NS, Prosekov AY (2023) Microorganisms for bioremediation of soils contaminated with heavy metals. Microorganisms 11(4):864. https://doi.org/10.3390/microorganisms11040864
9. Usmonkulova A et al (2026) Mapping research on microbial remediation of metals in soil (2020–2025). Microbiol Res 17(1):10. https://doi.org/10.3390/microbiolres17010010
10. Wang T et al (2014) The immobilization of heavy metals in soil by bioaugmentation of a UV-mutant *Bacillus subtilis* 38 assisted by NovoGro biostimulation and changes of soil microbial community. J Hazard Mater 278:483–490. https://doi.org/10.1016/j.jhazmat.2014.06.028
11. Qiao W et al (2019) Bioimmobilization of lead by *Bacillus subtilis* X3 biomass isolated from lead mine soil under promotion of multiple adsorption mechanisms. R Soc Open Sci 6(2):181701. https://doi.org/10.1098/rsos.181701
12. Radziemska M, Gusiatin MZ, Cydzik-Kwiatkowska A, Majewski G, Błażejczyk A, Brtnický M (2022) New approach strategy for heavy metals immobilization and microbiome structure in long-term industrially contaminated soils. Chemosphere 308(2):136332. https://doi.org/10.1016/j.chemosphere.2022.136332
13. Alotaibi BS, Khan M, Shamim S (2021) Unraveling the underlying heavy metal detoxification mechanisms of *Bacillus* Species. Microorganisms 9(8):1628. https://doi.org/10.3390/microorganisms9081628
14. Kharwar A, Bazaz MR, Dandekar MP (2022) Quantitative and qualitative characterization of commercially available oral suspension of probiotic products containing *Bacillus Clausii* spores. BMC Microbiol 22:217. https://doi.org/10.1186/s12866-022-02631-w
15. Mattei D, Pamphile JA, dos Santos de Oliveira JA, Battistus AG (2022) Bacillus: a eco-friendly source for plant development and sanity. Res Soc Dev 11(15). https://doi.org/10.33448/rsd-v11i15.37199
16. Wu C, Zhi D, Yao B, Zhou Y, Yang Y, Zhou Y (2022) Immobilization of microbes on biochar for water and soil remediation: a review. Environ Res 212(B):113226. https://doi.org/10.1016/j.envres.2022.113226
17. Schommer VA et al (2023) Biochar-immobilized *Bacillus* spp. for heavy metals bioremediation: a review on immobilization techniques, bioremediation mechanisms and effects on soil. Sci Total Environ 881:163385. https://doi.org/10.1016/j.scitotenv.2023.163385
18. Zocca VFB et al (2022) The CRISPR toolbox for the gram-positive model bacterium *Bacillus subtilis*. Crit Rev Biotechnol 42(6):813–826. https://doi.org/10.1080/07388551.2021.1983516
19. Qin Y, Angelini LL, Chai Y (2022) *Bacillus subtilis* cell differentiation, biofilm formation and environmental prevalence. Microorganisms 10(6):1108. https://doi.org/10.3390/microorganisms10061108
20. Wang H et al (2024) Improving prediction of soil heavy metal(loid) concentration by developing a combined co-kriging and geographically and temporally weighted regression (GTWR) model. J Hazard Mater 468:133745. https://doi.org/10.1016/j.jhazmat.2024.133745

Sustainable Additive Manufacturing in Concrete Technology

Hybrid Additive Manufacturing for Small-Scale Modular Housing: Enhancing Sustainability Through Concrete Waste Valorization

Marvin Abstoß(✉), Henrik Funke, Ferdinand Senf, and Sandra Gelbrich

Steinbeis Innovation Center FiberCrete, Chemnitz 09117, Germany
marvin.abstoss@stw.de

Abstract. Low-carbon concrete is increasingly important for reducing the environmental impact of construction. This study develops a low-carbon concrete based on ground granulated blast furnace slag (GGBFS) cement, optimized for a hybrid additive manufacturing process combining 3D concrete printing (3DCP), shotcrete, and mechanical post-processing. To achieve precise dimensions and functional joints for tiny house precast modules, the printed components undergo milling and brushing in the green state. A central aspect of this study is the ecological optimization of the material loop by characterizing the resulting milling residues and reusing them as a substitute for aggregates (sand and gravel) in the original mixture.

Comparative analyses of raw material density and frost resistance confirmed the suitability of the recycled material. Mechanical testing, including compressive strength, showed stable order and persistent performance. The modules utilize a force-flow-optimized, formwork-free design with minimized wall thicknesses and integrated reinforcement structures.

To enhance thermal performance, internal cavities are filled with a mineral-bound insulation material. The integration of calcium sulfoaluminate (CSA) cement proved essential, as its rapid-setting kinetics enable immediate overprintability and reduce production cycles. Furthermore, CSA cement reduces the carbon footprint by requiring lower calcination temperatures than Portland cement. Finally, the successful implementation of the recycled mixture in a full-scale module demonstrates the feasibility of a closed material cycle under real-world additive manufacturing conditions.

Keywords: Hybrid Additive Manufacturing · Post-Processing · Circular Reuse

1 Introduction

The global failure to meet Paris Climate Agreement targets is underscored by record-breaking greenhouse gas levels, with the construction sector alone accounting for 37% of worldwide CO_2 emissions [1]. To align with upcoming sectoral benchmarks, the industry must transition from Ordinary Portland Cement (OPC) to ecological alternatives and minimize cement content. Beyond material substitution, reducing carbon emissions

Z. Perkowski et al. (Eds.): ECCE 2026, LNCE 929, pp. 101–116, 2027.
https://doi.org/10.1007/978-3-032-39515-3_8

requires a shift toward force-flow-optimized designs that prioritize resource efficiency and weight reduction. In this context, additive manufacturing, specifically 3D Concrete Printing (3DCP) and shotcrete, emerges as a transformative solution, which, in addition to enabling expanded design freedom, simultaneously addresses the urgent shortage of skilled labor through high-level automation.

The design flexibility of additive manufacturing enables the production of topologically optimized components, where material is placed only where structurally necessary. This digital process facilitates the functionalization of concrete elements—such as integrating thermal regulators or sensors—through high-precision, fresh-state robotic milling [2]. A central objective of this research is to establish a closed-loop cycle where milling waste is not discarded but systematically recovered and reintegrated. Achieving this circularity requires a rigorous characterization of reclaimed grain sizes and their influence on mechanical performance.

This study focuses on fabricating modular "Tiny House" precast components using hybrid additive manufacturing. By combining 3DCP and shotcrete with parametric programming, the workflow enables a fully digital chain from design to localized post-processing [3, 4]. Furthermore, integrating carbon-positive biochar into the mixture enables the production of carbon-neutral modules [5]. Ultimately, this research demonstrates how a force-flow-oriented, circular production model can address both the environmental impacts of construction and the challenges posed by a shrinking labor market.

2 Material and Methods

2.1 Material Development and Investigations

This research focuses on the development of a low-carbon concrete mix using ground granulated blast-furnace slag cement (GGBFS - CEM III/B 42.5 N-LH/SR/LA). To further reduce cement content, a maximum aggregate size of 5.6 mm was implemented, specifically tailored for the hybrid printing process. Additionally, a mineral-bonded insulation composite was developed using calcium sulfoaluminate (CSA) cement, selected for its rapid setting to accelerate production [6]. To quantify the environmental impact, the CO_2 intensity of the mixture was evaluated relative to a reference OPC mixture.

Concrete properties were characterized following European standards: flowability via the Haegermann Flow Table (DIN EN 1015–3), volumetric stability through shrinkage channel monitoring (DIN EN 12390–16), and fresh properties using an air entrainment meter (DIN EN 12350–6/7) [7–10]. Particle densities of primary aggregates and milling byproducts were determined by pycnometer [9]. Particle size distributions were established using a Retsch sieve shaker (DIN EN 12620), with fine fractions (< 1 mm) analyzed via laser diffraction and dynamic image analysis (ISO 13320 [11, 12]).

A direct substitution strategy was implemented by comparing the grading curves for ground material with those for primary aggregates (sand and gravel). An average grading curve—representative of continuous robotic processing—was calculated from material processed between three and five hours. Unlike previous studies that required drying or re-fractionation [13], this approach allows for immediate reuse. Analysis targeted three critical ranges: 0–0.25 mm (rheology), 0.25–2 mm (packing density), and 2–6 mm (coarse skeleton). Statistical validation via Mean Absolute Error confirmed that a 60/40 sand-to-gravel ratio best replicates the original design, providing superior consistency in the critical fine and coarse fractions compared to 70/30 or 50/50 alternatives.

To evaluate mechanical performance, specimens were stored in accordance with DIN EN 12390–2 [14]. Compressive strength was assessed using a ToniNORM 3000 kN load frame on 150 mm cubic specimens (DIN EN 12390–3) [15]. For the recycling trials, the milled material was reintroduced as a 10%, 20%, and 30% volumetric substitution of the primary aggregate mix at intervals of 4 h, 6 h, 8 h, 24 h, 3 d, 7 d, and 28 d. Additionally, the 4-point bending tensile strength (4PBTS) was determined on $160 \times 40 \times 40$ mm^3 specimens (DIN EN 196–1 / DIN EN 12390–5) [16, 17]. To verify the underlying chemical mechanisms, Scanning Electron Microscopy (SEM) is employed to monitor the formation of hydration phases (ettringite and calcium silicate hydrate (C-S-H)) within the recyclate, providing evidence of the material's inherent strength development with milled material during the initial 24-h hydration phase.

2.2 Hybrid Additive Manufacturing Process

The 3DCP and shotcrete processes were executed using a KUKA KR120 6-axis industrial robot, integrated with an external linear axis (7th axis) to extend the operational workspace, as illustrated in Fig. 1.

Fig. 1. Robotic complex for hybrid additive manufacturing process.

The concrete was conveyed using a screw conveyor pump system. The robotic setup featured specialized nozzle assemblies engineered for diverse applications, including a forced-mixing unit for additive integration and a multi-tool nozzle designed for hybrid printing and shotcrete technologies [13, 18, 19].

The additive manufacturing processes operations were conducted using the KUKA KR120 industrial robot. Notably, the subsequent milling trials were executed using the same robotic system; the printing periphery was substituted with a high-speed milling spindle attached to the robot. This configuration enabled a seamless transition between additive and subtractive phases within a single, synchronized coordinate system.

2.3 Digital Workflow and Robotic Path Control

The digital workflow utilizes Rhinoceros 3D and Grasshopper to link complex geometric design directly with robotic fabrication. Rhinoceros 3D provides the CAD environment for freeform Non-Uniform Rational B-Splines (NURBS) modeling, while Grasshopper handles the computational logic through parametric scripts. This setup allows for rapid design iterations and automated toolpath generation based on mathematical relationships rather than manual input.

The programming for 3DCP, shotcrete, and robotic processing follows a standardized three-step pipeline. First, the design intent is defined, with parametric constraints applied to material behavior under hydrostatic pressure. Second, the geometry is decomposed into movement curves embedded with fabrication parameters such as layer height, feed rate, and nozzle orientation. Finally, these paths are translated into KRL (KUKA Robot Language) via specialized post-processors. This stage includes kinematic simulations to detect singularities and collisions, ensuring safe execution before transferring the code to the robot controller.

2.4 Material Properties of Basic Mixture

Table 1 specifies the concrete material composition; the mixture was prepared with a water-to-cement ratio of 0.42 to ensure suitability for both 3DCP and shotcrete applications.

Table 1. Composition of concrete

Component	Proportion (kg/m^3)
CEM III/B 42.5 N	590
Quartz sand 0–2 mm	570
Quartz gravel 0–5.6 mm	550
Mineral powder	200
Stabiliser	2.1
Shrinkage reducer	3.5
Superplasticiser	4.5
w/c	0.42

The integration of a superplasticizer reduced water content while, in combination with a stabilizer, ensured a homogeneous mixture with increased segregation resistance. With a slump flow diameter of 13–14 cm, the material proved easy to print and spray while maintaining high workability.

Significant environmental optimizations were achieved through the substitution of OPC with GGBFS cement. Specifically, CO_2 emissions were reduced by approximately 66%, decreasing from 529 kg CO_2/t for OPC to 180 kg CO_2/t for the GGBFS variant [20, 21]. By utilizing an aggregate with a maximum grain size of 5.6 mm, the cement content was reduced to 590 kg/m^3 at a w/c ratio of 0.42, further contributing to the reduction of the carbon footprint.

The investigation of mechanical performance highlights the high efficiency of the developed GGBFS-based mixture. Despite the slower hydration kinetics inherent to GGBFS compared to OPC, the mixture achieved a compressive strength of 59.3 MPa and a 4PBTS of 8.5 MPa after 28 days.

The durability of the concrete under frost-thaw conditions was evaluated using the Capillary Suction, De-icing Chemicals, and Freeze-thaw (CDF) test [22]. After 28 standardized cycles, the cumulative surface scaling was measured at 1149 g/m^2 (see Fig. 2), which is well below the BAW acceptance criterion of 1500 g/m^2, providing a quantitative assessment of the material's weathering resistance [23].

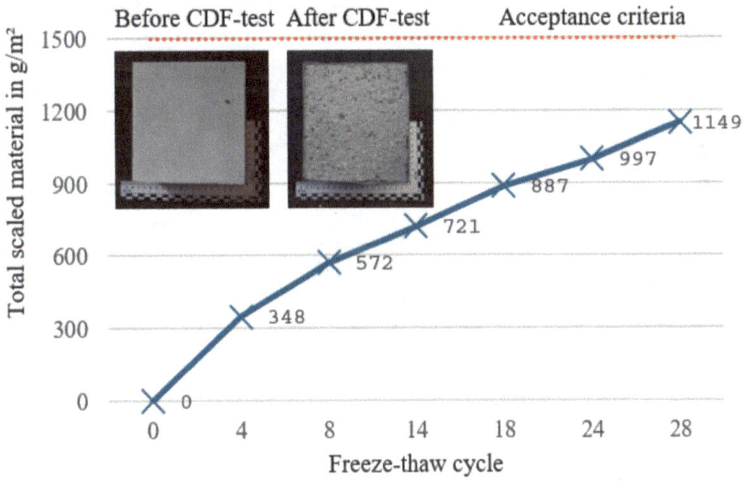

Fig. 2. Progression of weathering in the CDF test.

The material successfully met the freeze-thaw resistance thresholds established by the BAW criteria [23]. This validation ensures that components manufactured via the combined robotic spraying and printing processes possess the required durability for application in frost-exposed environments.

The architectural design employs force-flow-oriented reinforcement, resulting in the formation of internal cavities within the concrete modules. These gaps are subsequently filled with a mineral-bound insulating material. Experimental evaluations were conducted using various fillers; however, optimal automated conveyability was achieved with a specialized mixture of water, air-entraining agents, GGBFS, and CSA cement.

The synergistic combination of GGBFS and CSA cement facilitated rapid setting and stiffening characteristics while maintaining a low carbon profile. This accelerated hydration was critical for the prompt formation and closure of layers in both the 3DCP and shotcrete processes, as the use of CSA cement mitigated production delays that would otherwise occur with standard binders. Furthermore, CSA cement improves the environmental balance sheet due to its lower calcination temperatures than OPC.

The resulting mortar is characterized by its cost-effectiveness and structural stability. The fresh concrete density was precisely controlled through the dosage of the air-entraining agent, yielding values between 0.4 and 0.6 kg/l. While integrating biochar enables the modules to reach CO2 neutrality, it increases the insulation density to approximately 1 kg/l. This results in a relative reduction in thermal insulation performance—a trade-off deemed acceptable depending on the module's specific functional requirements.

2.5 Post-processing Parameters

To establish the optimal operational threshold for fresh-state post-processing, a comprehensive series of experimental tests was conducted (see Fig. 3). These investigations focused on characterizing the influence of specific milling and refinement parameters on the resulting surface topography and material stability of the green concrete.

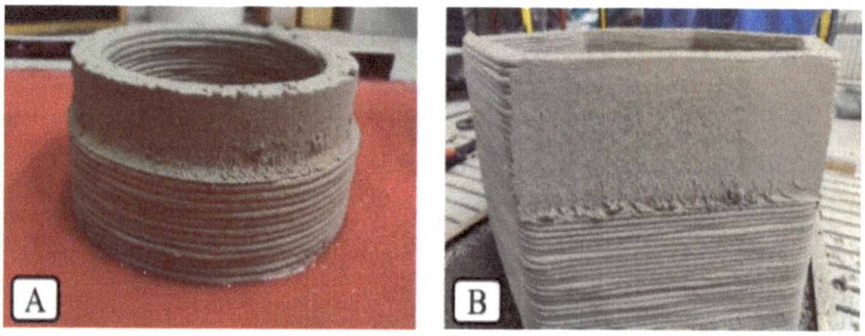

Fig. 3. Milling (A) and brushing (B) results [13].

Experimental results indicated that the optimal processing window for milling of the printed GGBFS cement-based mixture begins approximately four hours post-printing. Operations performed prior to this threshold resulted in material smearing rather than defined cutting edges. However, for large-scale components, the viable processing window can be extended to a range of three to five hours to accommodate the longer fabrication durations required for fresh-state machining.

Beyond six hours, the concrete reaches a level of strength and hardness that not only accelerates tool wear but also risks dislodging the specimen. At this stage, the lateral forces exerted by the milling cutter can exceed the frictional bond between the component and the substrate, causing the part to shift. While conventional machining would address this through mechanical clamping, such techniques are not yet feasible in the green state without causing structural damage.

To mitigate this risk and increase resistance to slipping, two primary strategies were identified. Utilizing a structured rather than a smooth formwork base significantly enhances the mechanical interlock between the first printed layer and the formwork. Printing an integrated base plate (raft) directly beneath the actual component increases the total contact surface area, thereby maximizing frictional resistance to lateral milling forces.

The depth of cut (infeed) was determined based on tool geometry. For milling, a depth of half the cutter diameter or less is recommended, with a maximum advisable infeed of 10 mm. For brushing, an infeed of up to 5 mm proved suitable; exceeding this value caused excessive tool penetration, potentially leading to spindle stalls. Process speeds were optimized accordingly: milling was performed at 120–240 mm/s (9,000–12,000 rpm), while brushing required lower rates of up to 60 mm/s (6,000 rpm).

2.6 Analysis and Reuse of Recyclate

Analysis of Sieving Curve. To support the circular material concept, the milling material generated during the surface refinement process was evaluated for reintegration into the primary cementitious matrix. A critical factor for successful reuse is the particle size distribution of the recovered material. Consequently, sieve analyses were conducted to establish grading curves for material removed at intervals of 3, 4, and 5 h post-printing (see Fig. 4). This comparative analysis allowed assessment of how the degree of hydration at the time of milling influences the grain size and quality of the resulting recyclate.

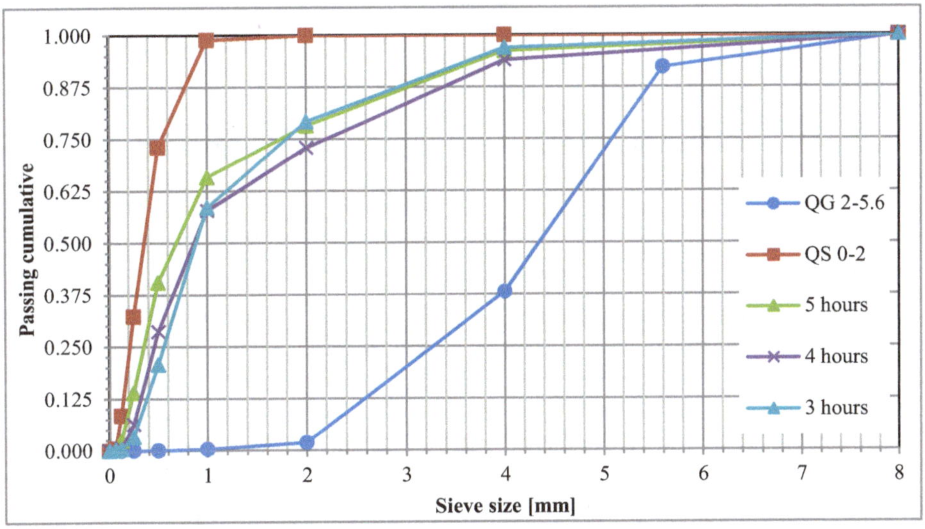

Fig. 4. Sieve lines for quartz sand (QS), quartz gravel (QG), and milling material.

Analysis of grading curves showed that longer delays between 3DCP and milling lead to a finer recyclate consistency. Due to significant deviations from primary aggregates (quartz sand 0–2 mm and gravel 2–5.6 mm), a systematic reprocessing strategy was required. While previous methods replicated grading curves by re-fractionating dry material, the high moisture content of fresh recyclate makes sieving unfeasible. To enable immediate reuse, a simplified substitution method was developed based on an average grading curve from material recovered between 3 and 5 h. This window represents the continuous generation of material during the process. The evaluation focused on three key ranges: 0–0.25 mm, 0.25–2 mm, and 2–6 mm (see Fig. 5).

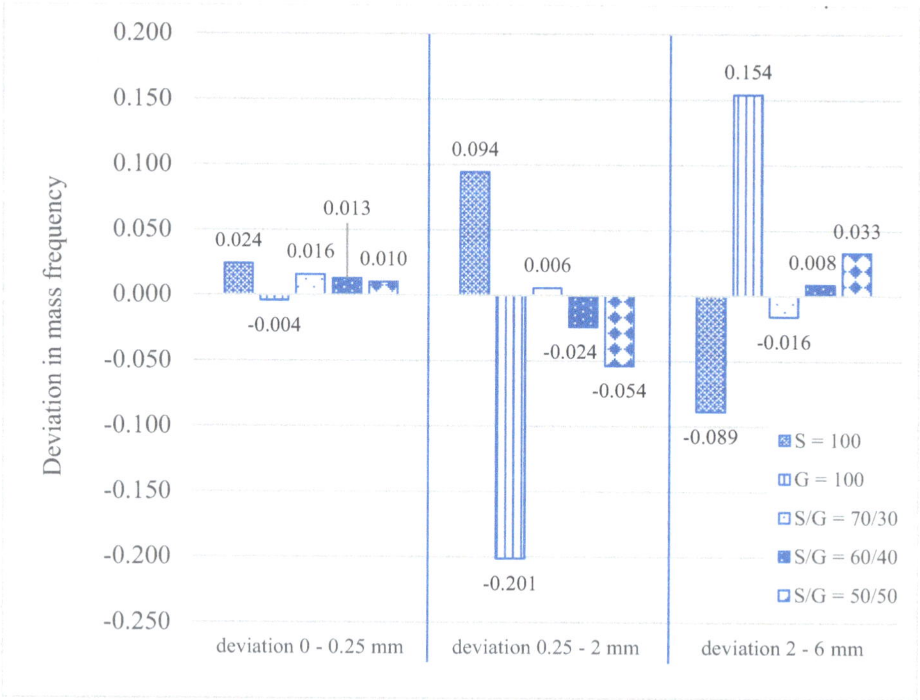

Fig. 5. Deviation in mass frequency of sand and gravel (mixture) compared to milling material

The particle intervals were defined by their functional roles within the concrete matrix. The 0–0.25 mm fraction (fines/silt) is critical for rheology and "internal lubrication"; while an excess increases water demand, a deficiency compromises coherence and pumpability. The 0.25–2 mm sand fraction contributes to the structural stability and packing density of the fresh paste, which are vital for 3DCP buildability. Lastly, the 2–6 mm coarse aggregate provides the structural "skeleton" of the hardened material, requiring careful monitoring to prevent excessive grain crushing during milling. By analyzing deviations within these boundaries, a 60/40 sand-gravel mixing ratio was validated to balance workability and mechanical strength, supported by the mean absolute deviations calculated for various ratios (see Fig. 6).

This statistical validation confirmed that while the 70/30 ratio exhibits a marginally lower mean deviation, the 60/40 ratio was ultimately selected. This decision is based on the fact that when evaluating the individual fractions, the 60/40 ratio shows significantly lower deviations in the critical fine and coarse ranges, thereby ensuring the highest consistency with the original aggregate design across all relevant scales.

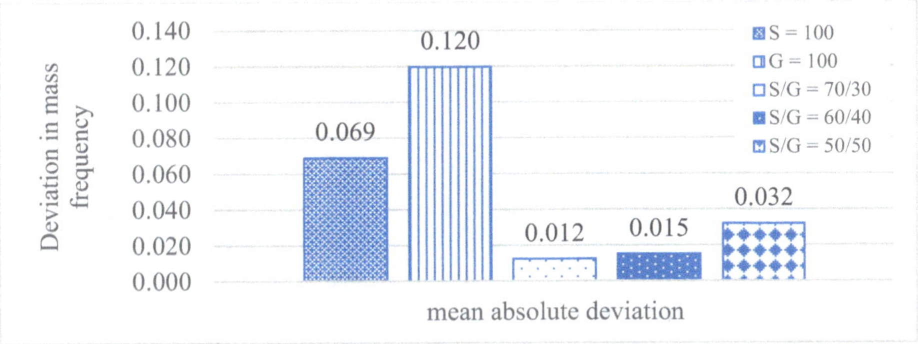

Fig. 6. Mean deviation in mass frequency of sand and gravel (mixture) compared to milling material.

Examination of Compressive Strength. Following the determination of the optimal mixing ratio, the milled recyclate was used to replace the primary aggregate mix (60/40 sand and gravel) at 10%, 20%, and 30% by volume, calculated based on the respective bulk densities. To evaluate the influence of hydration progress on the recycling potential, the material was recovered and reintroduced after 4 h, 6 h, 8 h, 24 h, 3 d, 7 d, and 28 d (see Figs. 7 and 8).

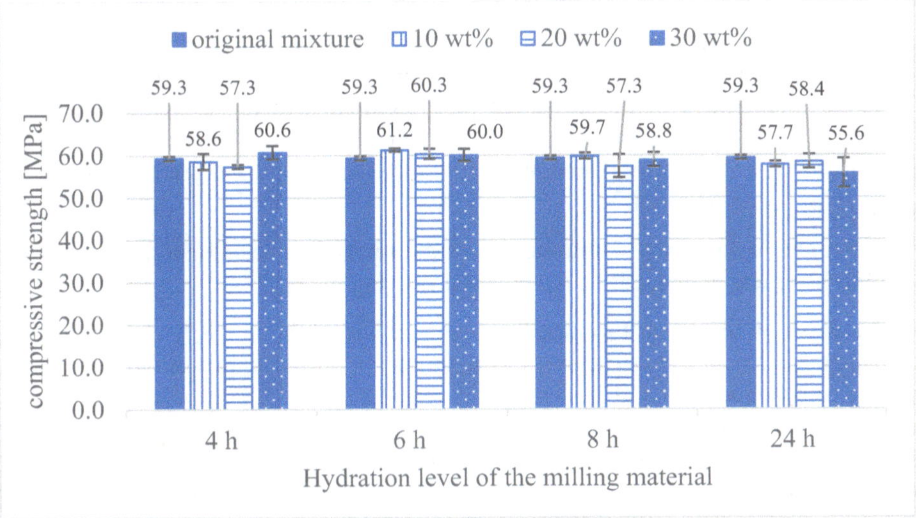

Fig. 7. Compressive strength of the substitution mix compared to the original mix, classified according to the age of the milled material (4 to 24 h)

The primary objective of this series was to assess the impact of the substituted milled material on the mechanical performance of the hardened concrete. Consequently, all specimens were tested for compressive strength after 28 days to identify potential

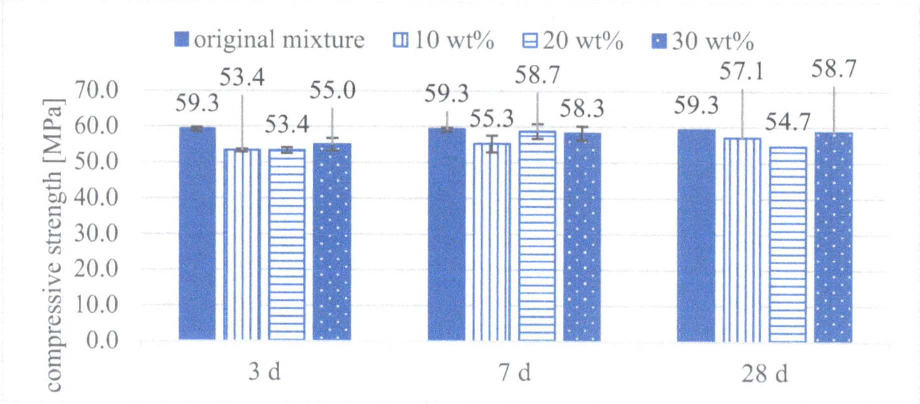

Fig. 8. Compressive strength of the substitution mix compared to the original mix, classified according to the age of the milled material (3 d, 7 d, and 28 d)

correlations between the age of the recycled material and the final strength of the new mix.

The experimental results indicated that the modified mixes achieved compressive strengths comparable to the reference mixture. While the overall performance remained stable, minor strength losses were observed in specific series, particularly at the 72-h, 7-day, and 28-day marks.

It is hypothesized that during the first 24 h, the milled material actively contributes to the strength development of the new mix due to the presence of unhydrated cement within the recyclate. Conversely, the observed fluctuations in later stages may be attributed to a different grain shape of the milled particles compared to the primary aggregates, which can negatively influence bond strength and packing density. However, detailed investigations regarding the influence of particle morphology are still pending and remain a subject for future research.

Microstructural Analysis. To further verify the strength development of the recycled material, SEM was conducted on the material (see Fig. 9).

The SEM confirms that for material recovered within the first 8 h, a clear strength development is evident, characterized by the formation of typical hydration phases from both the cement and the GGBFS.

The SEM image clearly shows the formation of needle-like ettringite crystals and fibrous C-S-H phases originating from both the cement and the GGBFS. These phases interlock to bridge the pore space, providing the structural integrity observed in the early-stage samples. Consequently, these microscopic observations provide clear evidence of the material's inherent strength development after milling in the initial hydration phase. In contrast, milling material recovered after 72 h did not achieve sufficient structural integrity to be prepared for SEM analysis. While it is likely that minor chemical bonds are still formed at this stage—as the cement within the milled material has not yet reached its final set—the remaining reactive potential is insufficient to produce a solid, cohesive sample using the milled material alone. This supports the observation that the

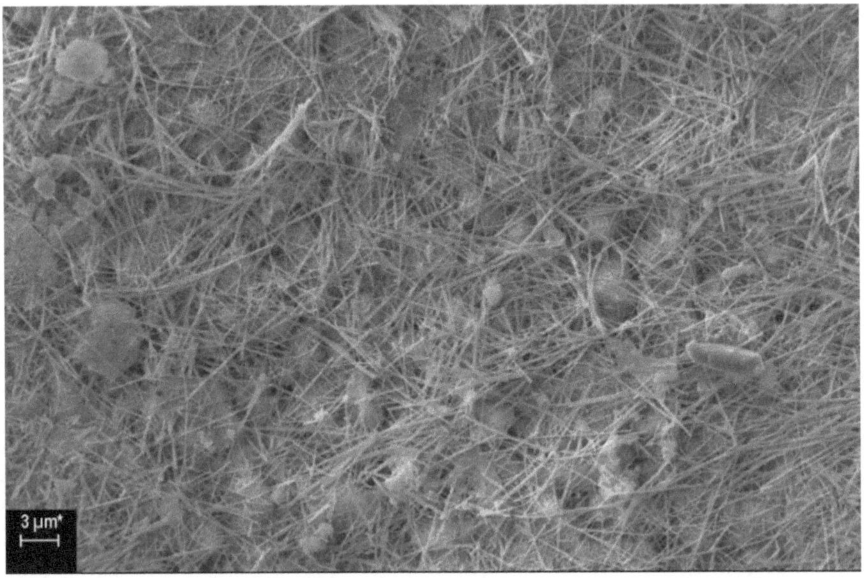

Fig. 9. SEM image of hydration phases (ettringite and C-S-H) of the milling material mixed with water after eight hours; examined after 28 days.

primary contribution of the recyclate to the final strength occurs within the early stages of hydration.

2.7 Integrated Hybrid Manufacturing

Parameter Characterization. This study identified the parameters of 3DCP by correlating printing speeds (60–240 mm/s) with material feed rates (1–4 l/min) to produce layer widths of 20–40 mm and heights of 5–20 mm. For the shotcrete application, process limits were established under a constant air pressure of 5.5 bar. To optimize surface quality and eliminate voids, the lateral path distance was capped at 60 mm, maintaining a nozzle distance of 150–160 mm. Spraying speeds ranged from 120 to 240 mm/s, with feed rates of 2 to 4 l/min, resulting in applied layer thicknesses of 5 to 10 mm. These optimized parameters effectively mitigated typical additive manufacturing defects, such as segregation, discoloration ("blue coloration"), and topographical irregularities, including "mountain and valley" formations (see Fig. 10). The final element quality was verified through quantitative 3D scanning analysis (see Fig. 10B).

Hybrid Additive Manufacturing in Practice: Production of Prototype Modules. The practical viability of the hybrid additive manufacturing process was demonstrated through the fabrication of several components, including three wall modules for a bicycle garage and two modular roof elements. These units validated the strategic integration of 3DCP and shotcrete techniques within a unified workflow. Notably, one of the wall modules was produced using the mixture incorporating 10 wt% of recycled material recovered from the milling process of a previous element (see Fig. 11).

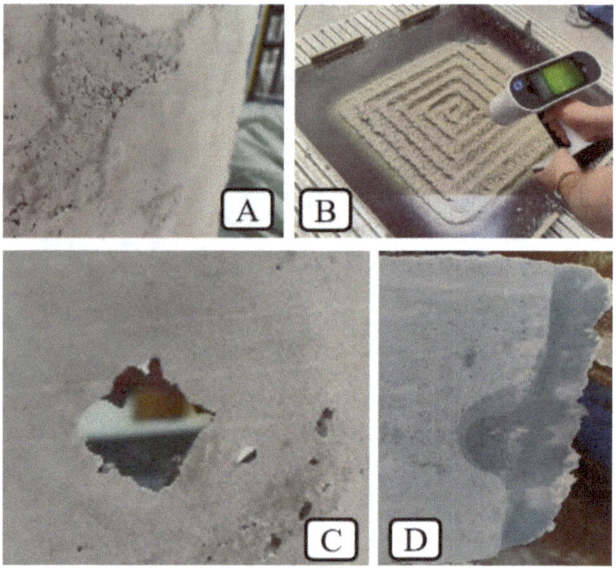

Fig. 10. Material defects in additive manufacturing: segregation phenomena (A), mountain-valley (B), defects (C) and blue coloring (D) [13].

To optimize production efficiency across these units, a multi-tool nozzle was utilized, eliminating the need for tool changes between individual manufacturing steps. This approach effectively bypassed time-consuming cleaning cycles for the delivery pipes and nozzle adapters, ensuring a continuous and streamlined fabrication process.

The production sequence commenced with the fabrication of formwork edges utilizing the 3DCP process, followed by the application of a shotcrete layer to establish the base surface. Subsequently, a lattice of structural reinforcement was printed onto the flat surface. The geometry of this reinforcement was pre-calculated to withstand dead, wind, and snow loads specific to Load Zone 2 in Central Germany (accounting for altitude-dependent variations).

Following the reinforcement phase, an insulating mixture was. Due to its high flowability, the mixture leveled itself to form a flush, integrated surface with the 3DCP reinforcement structure. Fabrication was finalized by applying a concluding 3DCP layer to seal the module. To ensure precise interlocking and structural integrity during final assembly, the edge geometries were post-processed using a specialized milling tool, creating high-tolerance contact surfaces for subsequent adhesive bonding (see Fig. 12). While this study utilized 3DCP for the final closure, the application of a secondary shotcrete layer remains a viable alternative and will be evaluated in subsequent stages of the research project.

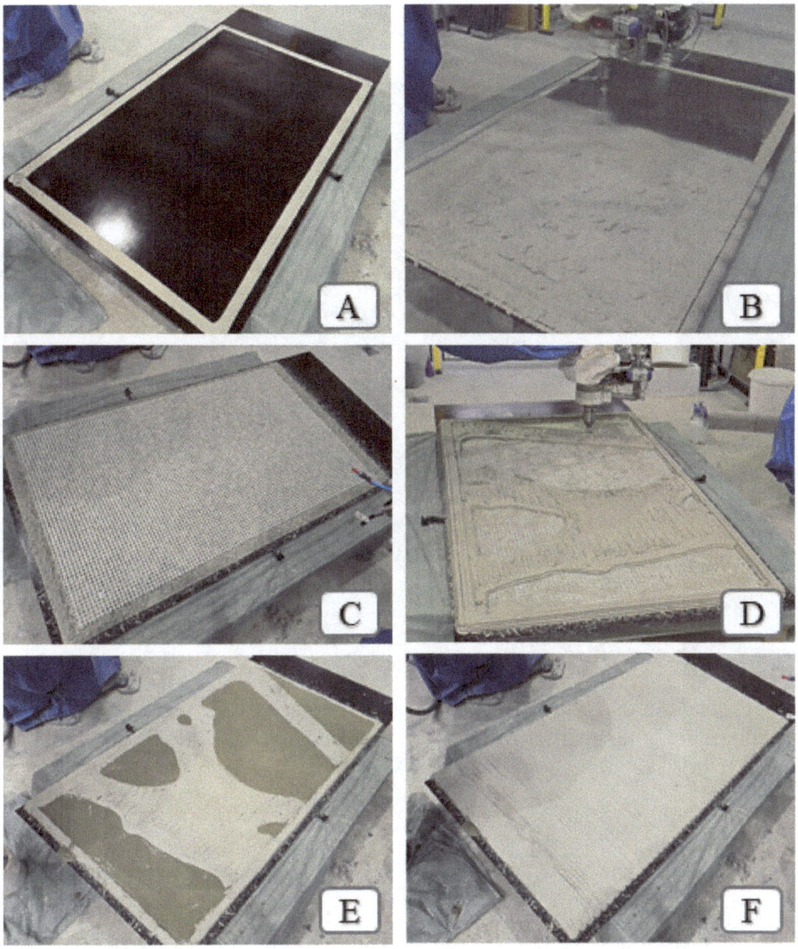

Fig. 11. Steps to produce a module with hybrid additive manufacturing: printing edge (A) and spraying surface (B), integrating glass-fiber reinforcement, printing reinforcement structure (D), filling gaps with insulation mixture (E), and closing surface via printing (F).

This prototype served to test the seamless integration of 3DCP, shotcrete, post-processing, and the direct reuse of the milling material under realistic production conditions.

Fig. 12. Steps for assembling the demonstrator (bicycle garage): Connecting the wall modules (A), connecting and gluing the roof module (B), finished garage – adhesive drying (C), and test for storing bicycles (D)

3 Conclusions

This paper presents a hybrid additive manufacturing process integrating sustainable 3DCP, shotcrete, and robotic post-processing. Using a low-carbon GGBFS-based cement with 5.6 mm maximum aggregate, the study established a direct recycling loop for milling bypass material. A systematic grading curve analysis identified an optimal 60/40 sand-gravel ratio, allowing for the direct substitution of primary aggregates with moist recyclate up to 30 wt% without energy-intensive drying.

Compressive strengths remained comparable to the reference, with SEM analysis confirming that unhydrated cement and GGBFS in early-stage recyclate (up to 24 h) actively contribute to strength through ettringite and C-S-H phase formation. The system was validated via a bicycle garage demonstrator, proving that complex, formwork-free modules can significantly reduce carbon footprints. This research marks a vital step toward a circular economy in digital construction. Future work will scale these findings to modular housing units to evaluate structural performance in real-world contexts.

References

1. United Nations Environment Program (2022) Global status report for buildings and construction: towards a zero-emission, efficient and resilient buildings and construction sector. https://www.unep.org/resources/publication/2022-global-status-report-buildings-and-construction. Last accessed 15 June 2026
2. Gliniorz R, Lindner M, Gelbrich S (2025) Multi-functional integration in clay-cardboard sandwich structure with semi-automated manufacturing. In: Perkowski Z, Beben D, Zembaty Z, Massimino MR, Oliveira MJ (eds) Environmental challenges in civil engineering III. ECCE 2024, Lecture notes in civil engineering, vol 615. Springer. https://doi.org/10.1007/978-3-031-73776-3_9
3. Gelbrich S, Abstoß M, Funke H (2023) Complex concrete elements for the production of recyclable formwork by extrusion and milling. ce/papers 6(6):729–734. https://doi.org/10.1002/cepa.2815
4. Abstoß M, Rudolph E, Funke H, Gelbrich S (2021) 3D printing and milling of complex concrete elements for the production of casting resin molds. CPI–Concr Plant Int 4:138–148. https://www.wasa-technologies.com/_Resources/Persistent/9/8/b/f/98bfcbc4d58a584d76e6430b7ef6c15f6c112d0a/3-D-Druck-f%C3%BCr-komplexe-Betonfertigteile_TU-Chemnitz_2104_abstoss_en.pdf
5. EcoLocked GmbH. https://www.ecolocked.com. Last accessed 15 June 2026
6. Senf F, Abstoss M, Funke H, Gelbrich S (in press) Advanced 3D concrete printing 7DCP. with low-carbon calcium sulfoaluminate (CSA) cement blends. In: Proceedings of the 2nd international conference on net-zero built environment: innovations in materials, structures, and management practices
7. DIN EN 1015-3:2007-05: methods of test for mortar for masonry—Part 3: determination of consistence of fresh mortar (by flow table). German version EN 1015-3:1999+A1:2004+A2:2006
8. DIN EN 12390-16:2019-12: testing hardened concrete—part 16: determination of the shrinkage of concrete. German version EN 12390-16:2019
9. DIN EN 12350-6:2019-09: testing fresh concrete—part 6: density. German version EN 12350-6:2019
10. DIN EN 12350-7:2022-05: testing fresh concrete—part 7: air content—pressure methods. German version EN 12350-7:2019 + AC:2022
11. DIN EN 12620:2008-07: aggregates for concrete. German version EN 12620:2002+A1:2008
12. ISO (2020) ISO 13320:2020: particle size analysis — laser diffraction methods. International Organization for Standardization (ISO), Geneva, Switzerland
13. Abstoss M, Senf F, Gliniorz R, Funke H, Gelbrich S (in press) Sustainable precast concrete construction using hybrid addivite manufacturing: 3D concrete printing and shotcrete with low-carbon concrete. In: Proceedings of the 2nd international conference on net-zero built environment: innovations in materials, structures, and management practices
14. DIN EN 12390-2:2019-10: testing hardened concrete—part 2: making and curing specimens for strength tests. German version EN 12390-2:2019
15. DIN EN 12390-3:2019-10: testing hardened concrete—part 3: compressive strength of test specimens. German version EN 12390-3:2019
16. DIN EN 196-1:2016-11: methods of testing cement—part 1: determination of strength. German version EN 196-1:2016
17. DIN EN 12390-5:2019-10: testing hardened concrete—part 5: flexural strength of test specimens. German version EN 12390-5:2019
18. Lindner M et al (2021) Kalibrierverfahren für konstante Materialstränge bei robotergestützter Betonextrusion. Beton- und Stahlbetonbau 116(S2):42–47. https://doi.org/10.1002/best.202100051

19. Lindner M, Mandel K, Gelbrich S, Funke H, Kroll L (2021) Kontinuierliche Fertigung von dünnwandigen faserverstärkten Präzisions-Betonelementen für freigeformte Modulbauweisen (KoBeMo). BetonWerk International Nr 1:28–29. ISSN 1437-9023
20. Verein Deutscher Zementwerke e.V (2022) Umwelt-Produktdeklaration–Portlandzement (CEM I), EPD-VDZ-20220153-IAG1-DE
21. Verein Deutscher Zementwerke e.V (2022) Umwelt-Produktdeklaration–Hochofenzement (CEM III/A), EPD-VDZ-20220154-IAG1-DE
22. DIN CEN/TS 12390-9:2017-05; DIN SPEC 91167:2017–05: testing hardened concrete—part 9: freeze-thaw resistance with de-icing salts—scaling. German version CEN/TS 12390-9:2016
23. BAW (2004) Merkblatt Frostprüfung von Beton (BAW Merkblatt Frostprüfung). Bundesanstalt für Wasserbau. Eigenverlag, Karlsruhe, Hamburg, Ilmenau

Application of Circular Economy Principles to Recycled Concrete Components from Additive Manufactured and Milled Structures

Ralf Gliniorz[(✉)], Marvin Abstoß, Henrik Funke, and Sandra Gelbrich

Chemnitz University of Technology, Reichenhainer Str. 31, Chemnitz 09126, Germany
ralf.gliniorz@mb.tu-chemnitz.de

Abstract. The integration of circular economy principles into the construction industry is increasingly discussed as a potential approach to address ecological and economic challenges in the sector. In particular, 3D concrete printing (3DCP) of individual components offers opportunities to improve resource efficiency and reduce material waste. However, near-net-shape 3DCP and subsequent mechanical post-processing steps, such as brushing or milling, inevitably generate residual materials and rejects. If these materials are not recycled and reintroduced into the production cycle, they remain unused despite their potential value as secondary raw materials.

Reintegrating these materials into the production cycle may help reduce material losses and improve resource efficiency. One possible framework for this is the application of circular economy strategies, such as the 10R's, which range from 'rethink' to 'recover'. In this context, recycling surplus concrete could support efforts to reduce production costs and decrease demand for primary raw materials.

This study examines approaches to returning residual materials from 3DCP processes back into the material cycle. By exploring recycling and reuse strategies for concrete residues, the work examines their potential to reduce waste generation and associated emissions. The implementation of circular economy concepts in the 3DCP of individual components requires technological adaptations and the development of suitable processing strategies that enable efficient recovery and reuse of the material. The 10R principles can be applied almost in their entirety, and a guide is being developed to align this approach with the goals of the circular economy. In the long term, such approaches may improve the resource efficiency of construction processes.

Keywords: Circular Economy · 3D Concrete Printing · 10R's · Efficient Recycling · Individual Components

1 Introduction

The construction industry faces the challenge of becoming more sustainable. One potentially beneficial approach would be to implement the 10R circular economy model [1], from 'rethink' to 'recover', which are discussed in more detail later in this work. Actually, the carbon footprint of the construction sector is projected to double globally by

Z. Perkowski et al. (Eds.): ECCE 2026, LNCE 929, pp. 117–128, 2027.
https://doi.org/10.1007/978-3-032-39515-3_9

2050. In 2022, cementitious materials, bricks, and metals accounted for over half of the construction industry's carbon emissions. Glass, plastics, chemicals, and bio-based materials contributed 6%, while the remaining 37% originated from transport, services, machinery, and on-site activities [2].

For this reason, the initial focus should be on minimizing general resource consumption, reducing waste, and minimizing environmental impact as much as possible. An important strategy for achieving these goals is implementing the circular economy, which promotes a transition from a linear model (take, make, dispose) to one in which valuable materials remain in the economic cycle for as long as possible.

A particular focus on 3D concrete printing (3DCP) in the construction industry shows how these principles can be implemented to reduce the ecological footprint.

In the 3DCP of buildings or parts of them, for example, materials are used only where needed, which drastically reduces waste [3]. Compared to conventional construction techniques, 3DCP in the construction industry wastes less material. This is due to the precise layer-by-layer construction method, which leaves no unnecessary residues. Traditional construction methods generate substantial waste, especially in the formwork construction, due to scrap and inefficiency, whereas 3DCP significantly reduces this waste.

For functional elements or highly precise geometries, such as specific surfaces, corners, and edges, 3DCP alone is not sufficient, and subsequent milling is required. The resulting milling waste plays a significant role in this study. It is collected by type, stored to cure, processed, and then returned to the material cycle. This approach gives 3DCP the overall advantage of working with recycled materials, such as construction industry waste. This development is particularly relevant in the context of waste management: Every year, 2.6 million tons of construction waste are produced in Europe, of which less than 5% is recycled [4].

3DCP in the construction industry can also help reduce CO_2 emissions by using recycled or locally available materials, as traditional construction is responsible for significant CO_2 emissions, mainly due to the use of cement, whose production accounts for around 8% of global CO_2 [5].

An advantage of the circular economy with 3DCP is the ease of repairing, redesigning, or replacing components. Around the world, construction projects suffer damage due to incorrect material selection and wear and tear after just a few decades, making them liable for repair or even demolition. The modular design of 3DCP, which allows building components to be reused or adapted, extends the service life of buildings and reduces resource consumption.

The shortage and high demand for valuable resources are major problems worldwide. However, many raw materials are currently in use in their respective application scenarios. In recent years, research has focused intensively on recycling, but this is only one component of the 10R strategy [6].

The implementation of the circular economy, particularly the use of the 10R's with recycled materials from 3DCP, represents a critical step toward more sustainable, resource-efficient construction methodologies. This study shows the enormous potential to reduce waste and significantly lower CO_2 emissions. The use of recycled materials, precise material selection, and the ability to easily repair and adapt building elements are key to this forward-looking strategy. Using waste material from 3DCP can thus play a central role in transforming the construction industry toward a circular economy while advancing global climate goals.

2 Materials and Methods

2.1 Concrete from Milled 3DCP Material

The special ingredient of the investigated concrete material is waste from milled 3DCP elements, generated when the milling cutter shapes the printed element to the desired final contour. 3D concrete printers require a distinct mixture that must satisfy the high requirements of the process, including pumpability and non-Newtonian fluid properties [3]. The mixture consists of White Portland cement (CEM I 42,5 R), quartz sand 0/2 mm, limestone powder, PCE-based superplasticizer, and methylcellulose-based stabilizer. As delineated in Table 1, the precise proportions are hereby demonstrated. The coloration of the subject can be produced in a range of tones, from white to black and all colors in between.

In the following explanations, the quartz sand content of the mixture mentioned above is replaced with recycled material at approximate wt. % of 0, 10, 20, 50, and 100. It was investigated as casting concrete, made flowable by vibration. However, it is important to note that the same mixture can be reapplied to 3DCP.

Table 1. Concrete Components.

Component	kg/m^3
CEM I 42,5 R	600
Quartz sand 0/2 mm	880
Limestone powder	450
PCE-based Superplasticizer	9
Methylcellulose-based Stabilizer	3,7
Color powder	6
Water	280

2.2 Preparation of Samples

Robot Arena Printing and Milling. A significant number of ongoing projects at Chemnitz University of Technology focus on the 3DCP technology and process. A state-of-the-art facility featuring two collaborative KUKA KR90 R2700 robots serves as the basis for numerous studies (see Fig. 1). The first robot is responsible for extrusion within 3DCP, while the second performs specialized tasks such as fiber laying, insert integration, or milling. In this study, a novel process has been developed that involves milling concrete in a direct sequence after the 3DCP, even before the concrete has reached its final strength. There is a narrow time window for this. If milling is carried out too early, the concrete smears uncontrollably. If milling is carried out too late, the concrete is too hard for the milling cutter, and it wears out very quickly. In addition, the larger forces could push the element to the side, requiring additional securing to the base.

Fig. 1. 3DCP Arena of Chemnitz University of Technology: Robot A extrusion (on the left), Robot B milling (on the right), some references (middle).

Recycled Material Extraction. Effective material recovery requires the systematic collection and segregation of milling waste by type. Consequently, the work area is carefully covered with a synthetic material, and the edges are bordered with supportive barriers. This ensures that the material removed by the milling machine is collected and can be gathered up once the milling process is complete. The material is then stored in containers and can harden separately (see Fig. 2).

Fig. 2. Collected material from milling in a receiving container.

Specimens: Flexural and Compressive Strength. As described in 2.1, 5 basic mixtures were prepared, varying the proportion of recycled material (0, 10, 20, 50, and 100 wt.%) used as a sand substitute. These were mixed in an "Eirich R05T." The flow properties were then examined in the slump test in accordance with DIN EN 13395–1:2002–09 [7].

The recycled material used as a sand substitute appears brown in color. To demonstrate the effect of the new mixture, a blue pigment was selected. This approach enables the visualization of both the distribution and the proportion of the material.

The five mixes are placed in special prismatic forms, each measuring 40 x 40 x 160 mm^3 for mechanical characterization of 4-point flexural strength (DIN EN 196–1:2016 [8]). The mechanical characterization of compressive strength is achieved using broken prismatic elements, which are employed for flexural strength characterization.

All specimens were stored in a climatic chamber in a dry state for 28 days.

2.3 Test Set-up

Jaw Crusher and Sieve Shaker. The recycled material obtained is crushed to the desired grain size of 3 mm in the jaw crusher "Retsch BB 250". The maximum grain size can be set on the machine (see Fig. 3). To check the process, the material passes through the sieve shaker "Retsch AS 400 control". To illustrate the crushing process, the material was tested both before and after passing through the jaw crusher.

Fig. 3. Jaw crusher (on the left), detail of the adjustment of the grain size 3 mm (middle), and sieve shaker (on the right) [9].

4-Point Flexural Strength. Flexural tensile tests are particularly important in concrete analysis, as concrete has a limited tensile capacity despite its high compressive strength. This is especially critical in the precast sector, particularly for applications such as facade elements. These models provide critical insights into the load-bearing behavior of components subjected to bending forces. To determine tensile strength, four-point flexural strength tests were conducted while maintaining the geometric proportions of the static scheme, in accordance with DIN EN 12390–5:2019 [10]. It was done this way because this method more realistically determines the material parameters, especially for plate-shaped samples. These parameters are less affected by possible sample inhomogeneity [11–14].

The test is executed using the Toni Technik TONINORM SERIE 2000 with a 20 kN load frame. The support with (L) is 135 mm, and the support distance (L/3) is 45 mm (see Fig. 4).

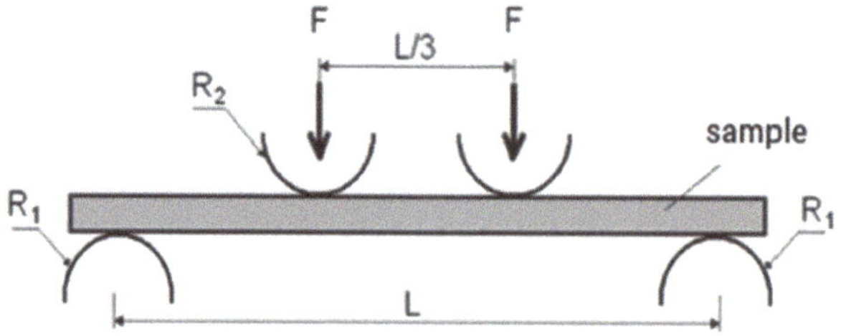

Fig. 4. Schematic overview of the 4-point flexural test (force F, support width L, support radius R_1, impactor radius R_2) [10].

Compressive Strength. As compressive strength represents the primary performance indicator for concrete, standardized compression tests serve as the basis for the material analysis in this study. They ensure that the concrete used can withstand the static requirements of a structure, such as its own weight and other loads. The samples were tested for each recycled material content, with 6 specimens per case. In accordance with DIN EN 196–1:2016 [8], the Toni Technik ToniPACT II was used to determine the compressive strength (see Fig. 5).

Fig. 5. Broken prismatic test specimens prior to the compressive strength test.

2.4 General Overview of the 10R Circular Economy

In contrast to the linear economy, the circular economy describes a model in which products and their resources are retained in the life cycle for as long as possible. This 10R strategy approach is anchored in the "United Nations Environment Program" and provides a groundbreaking roadmap for products and materials. This includes "Refuse, Rethink, Reduce, Reuse, Repair, Refurbish, Remanufacture, Repurpose, Recycle, and Recover" (see Fig. 9). The order listed also reflects the ranking of the strategies in terms of their contribution to resource conservation [1]. A detailed description of the individual stages and their connection to 3DCP elements is provided in Sect. 3.2.

3 Results and Discussion

3.1 Mechanical Characterization

Recycled Material. Following the 3DCP, a brief period exists during which milling can be performed optimally (3–4 h after printing). All process parameters, such as cutter size (end mill: 25 mm, ball-end mill: 12 mm), spindle speed (6000 rpm), and feed rate (10–60 mm/s), affect the finish of the machined material. To ensure optimal outcomes, the material should be systematically collected and stored according to its specific characteristics, facilitating subsequent analysis and evaluation of its effects on properties. Although the jaw crusher's gap was set to a maximum of 3 mm, oversized particles remained in the sieve. This is inherent to the crushing mechanism: material is primarily reduced in one dimension, allowing elongated or 'flaky' particles to pass through the crusher while potentially being retained during subsequent sieving due to their cross-sectional width. For further use, it is necessary to determine whether sieving should be implemented as a preliminary step, particularly when the concrete pump may be susceptible to blockage by larger particles. Additionally, it is possible to select a different shredding process that does not leave any oversized particles behind.

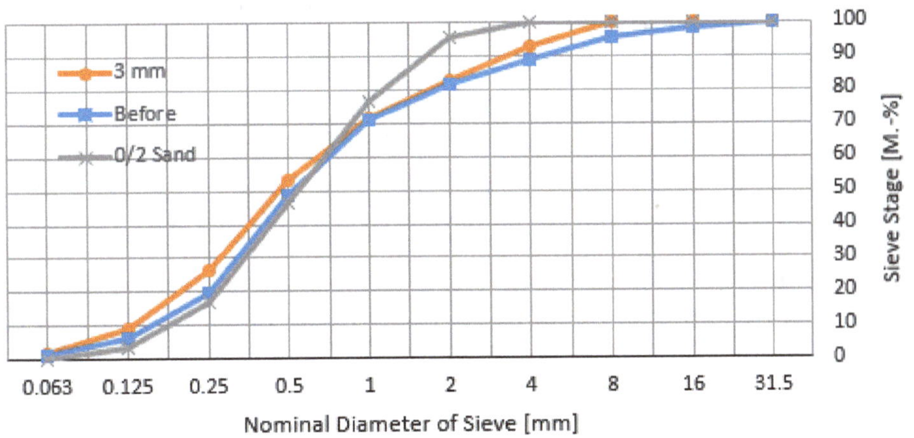

Fig. 6. Grain size distribution of quartz sand, the recycled material before and after jaw crushing.

As illustrated in Fig. 6, the sieve lines of the milling waste, collected prior to and following its processing through the jaw crusher, are compared to those of quartz sand. The proportion of large grain diameters after crushing has decreased. A marginal rise has been observed in the proportion of fine particles.

The curves for the recycled material and quartz sand do not differ significantly over the 0–1 mm range. Consequently, there is neither an alteration in the water-holding capacity of the aggregate, nor is there a necessity to modify the quantity of water and/or superplasticizer.

Flexural and Compressive Strength. Concrete samples are subjected to strength tests, a critical component of material analysis, as these tests yield essential data on the material's load-bearing capacity, quality, and durability. Concrete is a heterogeneous building material whose properties are contingent on its composition, production, and curing. Consequently, strength tests are used to verify that concrete meets the technical requirements of a structure. The charts clearly show that the use of 10 wt.% recycled material has no significant impact on strength (see Fig. 7). However, there are significant reductions at 20 wt.%, especially in compressive strength. The increase in compressive strength at 100 wt.% substitution compared to the 20 wt.% level can likely be attributed to a more favorable particle size distribution and a resulting higher packing density (filler effect) at full replacement. Pozzolanic reactivity of the recycled aggregate: Whilst at 20 wt.% the diluting effect (less primary aggregate) predominates, at 100 wt.% the chemical reactions between the recycled aggregate and the cement paste (secondary hydration) could consolidate the matrix to such an extent that the mechanical loss of quartz sand is partially compensated for. Notably, the development of strength at 100 wt.% replacement with recycled material is a remarkable achievement (see Fig. 7). These materials are nearly equivalent to the properties of the original mixture, provided that a slight loss in performance in the respective application is acceptable.

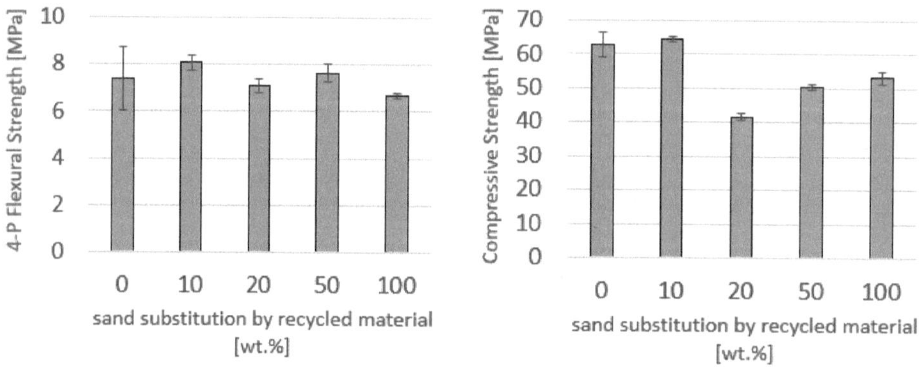

Fig. 7. Flexural and compressive strength diagrams depending on the proportion of recycled material used as sand substitution.

As described in Sect. 2.2, the coloring has a distinct visual effect that highlights the recycled material. Figure 8 shows the blue-colored concrete in two specimens: the one on the left contains no recycled material, while the one on the right contains 100 wt.% recycled material as a sand substitute. Here, the distribution, particle size, and proportion of the recycled material are clearly visible.

Fig. 8. Two examples of test specimens: One made from partially recycled material (on the right) and one made from non-recycled material (on the left), after the 4-point bending test.

3.2 Transferring 10R to Milled 3DCP Elements

As described in Sect. 2.4, the 10R strategy is an important tool for a resource-efficient future and can be applied to 3DCP elements and all following structures made of their recycled milling material (see Fig. 9). For these components, the first three R strategies—'Refuse, Rethink, and Reduce'—apply to all planning, design, and manufacturing processes. The 'Refuse' strategy focuses on avoiding production altogether, if possible, even before resource extraction. With 'Rethink' and 'Reduce', the focus in the design process should be on conserving resources, considering the materials used and the energy required for manufacturing. In detail, this means that for 3DCP elements, CO2-reduced cement, local sourcing of quartz sand and limestone powder, and near-net-shape 3D printing (material used only where needed) are used.

After production, the 3DCP elements are ready for consumption. The 'Reuse' strategy is important for 3DCP elements if they can be reused for their intended purpose, such as if they change location and/or owner. If possible, the elements should be inspected for damage using methods such as visual inspection, ultrasound, or CT/MRI.

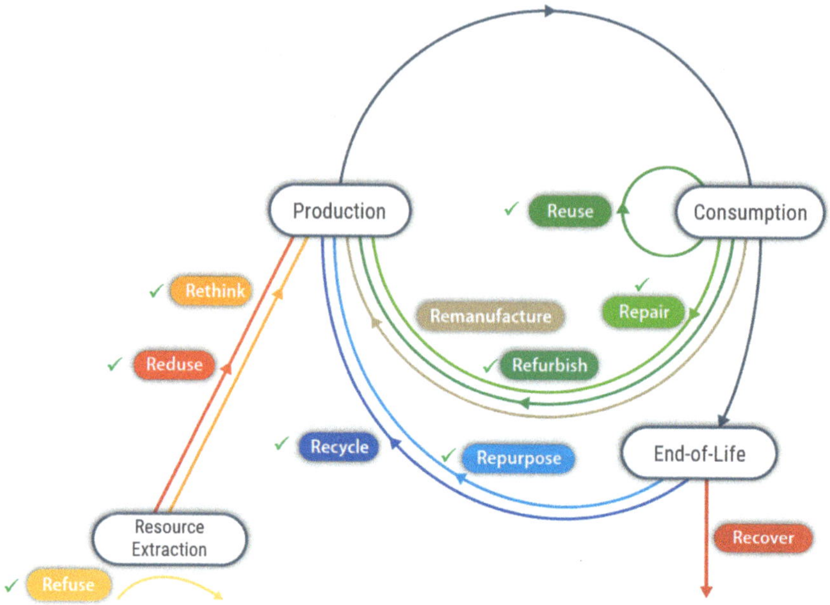

Fig. 9. Model of 10R strategies, according to [1], with marked application option.

In the event that the components are no longer functional, i.e., they are defective, the most suitable course of action is to perform 'Repair', thereby restoring the components to their original state of consumption. If the part is obsolete, 'Refurbish' allows it to be dismantled and examined. The part is then given a new location or purpose. 'Remanufacture' involves the complete renewal of all elements and functions, which is rarely possible for 3DCP elements.

Once the products have reached the end of their useful life, it should be checked whether a second life is possible with 'Repurpose'. The elements are then redesigned, similar to decommissioned wind turbine blades, which are reused to build bridges or bus stops. If all of this can be ruled out, 'Recycle' is applied. Materials are separated, isolated, and returned to production. For 3DCP elements, this involves using milling residues, shredding and grinding the components, and then adding them as additives or substitutes for sand and limestone powder to the new concrete mixture for production.

In this context, 'Recover' means the retrieval of energy or thermal utilization as the final stage of the 10R strategy. This makes sense for plastic elements and other combustible materials, but not as much for concrete elements unless fillers or thermoplastic force-transmission elements are used.

4 Conclusions

The circular economy is widely discussed as an approach to support more sustainable resource management in industry. It can help reduce resource consumption, lower energy demand, and minimize waste generation. Within this context, several objectives are commonly associated with circular economy strategies, including resource optimization, waste reduction, the development of durable products, the promotion of innovation, and the consideration of social and environmental responsibilities. While numerous projects currently focus on reducing CO_2 emissions, the circular economy offers a broader framework that addresses material and resource cycles more comprehensively.

This study examines the mechanisms underlying the production of 3D-printed and milled concrete components, as well as the resulting waste streams, to provide an overview of potential strategies for more efficient resource use. In particular, the reuse of recycled milling material in concrete mixtures for 3D printing is investigated. The mechanical properties of these mixtures were evaluated through mechanical testing and represent a significant contribution as a substitute material. Although the properties vary, they remain acceptable within certain limits, depending on the application scenario. In the near future, durability tests should be carried out to provide a more comprehensive overview. These tests should include frost resistance testing of mixtures with various recycled-material additions.

In the construction industry, the implementation of circular economy concepts is increasingly considered as a potential approach to addressing ecological and economic challenges. More efficient use of concrete and the recycling of residual materials may help reduce material demand and associated environmental impacts.

The application of such approaches requires technological developments and the adaptation of materials and production processes. These developments are often discussed as part of a longer-term transition toward more resource-efficient construction practices.

This work outlines an approach to integrating circular economy strategies into the production of 3D-printed and milled concrete components and other materials derived from them. The presented concept may serve as a basis for further investigations and projects addressing resource efficiency and circular economy in construction processes.

Acknowledgements. The authors would like to thank the following sponsors for their financial support: «BMWE – Bundesministerium für Wirtschaft und Energie» (Federal Ministry for Economic Affairs and Energy) within the initiative «ZIM – Zentrales Innovationsprogramm Mittelstand» (Central Innovation Program for Small and Medium-Sized Enterprises, SMEs) for the project "ScanAdd", funding indicator KK5038368KU. Many thanks also to our industry partner Betonstein Flöha GmbH.

References

1. Riedel L et al. Kunststoffe, https://www.kunststoffe.de/a/fachartikel/so-funktioniert-die-kreislaufwirtschaft%2D%2D6313618. Last accessed 09 Mar 2026
2. Li C, Pradhan P, Chen G et al (2025) Carbon footprint of the construction sector is projected to double by 2050 globally. Commun Earth Environ 6:831. https://doi.org/10.1038/s43247-025-02840-x
3. Abstoss M, Senf F, Gliniorz R, Funke H, Gelbrich S (in press) Sustainable precast concrete construction using hybrid additive manufacturing: 3D concrete printing and shotcrete with low-carbon concrete. In: Proceedings of the 2nd international conference on net-zero built environment: innovations in materials, structures, and management practices

4. European Commission Circular Economy. https://environment.ec.europa.eu/strategy/circular-economy_en. Last accessed 10 Mar 2026
5. WBCSD. Submission of Climate Change: Business Solutions to the United Nations Climate Summit, https://docs.wbcsd.org/2014/05/Submission_UN_Climate_Summit.pdf. Last accessed 11 Mar 2026
6. Lawcode Kreislaufwirtschaft: Prinzipien, EU-Aktionsplan & Umsetzung für Unternehmen https://www.lawcode.eu/blog/kreislaufwirtschaft/. Last accessed 14 May 2025
7. DIN EN 13395-1:2002-09: products and systems for the protection and repair of concrete structures—test methods—determination of workability—part 1: test for flow of thixotropic mortars. German version EN 13395-1:2002
8. DIN EN 196-1:2016-11: methods of testing cement—part 1: determination of strength. German version EN 196-1:2016
9. Retsch GmbH. https://www.retsch.com/products/sieving/sieve-shakers/as-400-control/. Last accessed 16 Jan 2026
10. DIN EN 12390-5:2019-10: testing hardened concrete—part 5: flexural strength of test specimens. German version EN 12390-5:2019
11. Lindner M, Gliniorz R, Funke H, Gelbrich S (2023) Anisotropic properties of additively manufactured concrete elements. Appl Res 3(3):e202300035. https://doi.org/10.1002/appl.202300035
12. Lindner M, Vanselow K, Gliniorz R, Gelbrich S (2020) Additiv gefertigte Leichtbaustrukturen mit Beton und Faserbewehrung. In: Fachtagung Werkstoffe und Additive Fertigung. https://additive-fertigung-2020.dgm.de/home
13. Lindner M, Rudolph E, Gliniorz R, Funke H, Gelbrich S (2022) Combined additive manufacturing technology for continuous production of precision concrete elements for modular construction. In: Proceedings of the 14th fib PhD symposium in civil engineering.; ISBN 978-2-940643-17-2, pp 787–794
14. Lindner M, Rudolph E, Gliniorz R, Funke H, Gelbrich S (2022) Tool development for robot assisted flow production processes for the fabrication of thin walled fiber reinforced concrete elements. In: Proceedings of the 14th fib PhD symposium in civil engineering, pp 771–778. ISBN 978-2-940643-17-2

Development of a Two-Component Mixing System for 3D Concrete Printing (3DCP) with Fast Setting Calcium Sulfoaluminate (CSA) Cement Blends

Ferdinand Senf(✉), Marvin Abstoß, Enrico Rudolph, Marco Lindner, Christian Müller, and Sandra Gelbrich

Chemnitz University of Technology, Chemnitz 09126, Germany
ferdinand.senf@mb.tu-chemnitz.de

Abstract. The substitution of Portland cement clinker by supplementary cementitious materials (SCMs) effectively reduces CO_2 emissions but is commonly associated with reduced early-age strength and prolonged setting and hardening times. In 3D concrete printing (3DCP) and precast concrete element production, this leads to extended holding periods and reduced process efficiency. Calcium sulfoaluminate (CSA) cement exhibits rapid hydration kinetics and pronounced early strength development. However, this very short setting time prevents its use in conventional 3DCP systems. To enable the application of fast-setting CSA cement blends in 3DCP while avoiding the use of setting-time retarders, which are known to negatively affect strength development at high SCMs contents, a novel two-component (2K) mixing system was developed and experimentally investigated. In this approach, the CSA cement is brought into contact with the base fresh concrete mixture immediately before extrusion. This process strategy decouples material storage from the initiation of hydration and allows precise temporal control of the setting reaction during printing. The optimized interaction between material formulation and processing technology results in accelerated setting and very early-age strength development while maintaining high long-term mechanical performance of the printed precast concrete elements. The rapid build-up of structural stability enables the printing of steep overhangs and vertically extended structures, thereby significantly increasing geometric freedom in 3DCP.

Keywords: Additive Manufacturing · Calcium Sulfoaluminate Cement · Two-Component Mixing System · 3D Concrete Printing

1 Introduction

Additive manufacturing processes, i.e., 3D Concrete Printing (3DCP), play a key role in the construction industry in the future, as they enable not only the desired automation of production workflows but also the efficient realization of complex structural components. For the concrete sector, particularly the precast concrete industry, 3DCP enables reduced material consumption through load-oriented design, a significant decrease in

Z. Perkowski et al. (Eds.): ECCE 2026, LNCE 929, pp. 129–144, 2027.
https://doi.org/10.1007/978-3-032-39515-3_10

formwork requirements, and a high degree of geometric and architectural design flexibility. Moreover, it enables reproducible component quality through precise, efficient process control and shorter production times. In addition, the emerging 3DCP technology offers an opportunity to address the growing shortage of skilled labor in the construction industry.

The 3DCP technology offers numerous ecological and economic advantages regarding global resource scarcity and raw material availability. In particular, the production of Portland cement clinker for the construction industry is associated with high CO_2 emissions. For this reason, clinker is increasingly substituted by supplementary cementitious materials (SCMs), although this often results in reduced early strength in concrete and precast concrete elements [1]. Moreover, extended setting and hardening times, such as those observed with ground granulated blast-furnace slag (GGBFS) as a cement substitute, increase the overall duration of the production process. Specifically, the low early strength development requires additional holding periods, which in turn lead to production delays. These limitations are addressed in this research using calcium sulfoaluminate (CSA) cement, whose rapid hydration kinetics enable accelerated setting and early strength development [2].

Since CSA cement initiates setting only a few minutes after mixing with water, its use in the existing 3DCP system is technologically infeasible. From a materials perspective, the use of setting-time retarders would therefore be required to ensure sufficient processability. However, such retardation is known to adversely affect strength development, particularly in binders with high SCMs content [2–4].

For this reason, the aim of this research is to further develop the 3DCP technology by enabling the use of CSA cement without the need for setting-time retarders. This allows CSA cement to be combined, for example, with blast-furnace cement (e.g., CEM III/B), so that the GGBFS can hydrate properly without being limited by retarders. At the same time, CSA cement can rapidly develop strength within the first few minutes after mixing with water. This very early strength development increases the geometric freedom of 3DCP, enabling the printing of more complex structures, such as overhangs and greater build heights. Ultimately, this approach aims to expand the design possibilities of 3DCP while reducing the overall production time for precast concrete elements, as rapid hardening allows elements to be produced sequentially.

In this context, a two-component (2K) mixing system is conceived, developed, and tested in the present study, in which the CSA cement is brought into contact with the material composition and the added water immediately before extrusion. The optimal interaction between the material formulation and the processing technology ensures both accelerated early-age strength development and high long-term strength for the printed concrete elements. At the same time, reducing clinker content increases CO2 savings and lays the foundation for a digitally controlled, geometrically flexible precast concrete production process that offers significant ecological and economic advantages [5–7].

2 Materials and Methods

2.1 Materials

To address the environmental impact of conventional binder systems, this study employs a ye'elimite-rich CSA cement, which emits approximately 20–30% less CO_2 than ordinary Portland cement due to its lower limestone requirement and reduced clinkerization temperature [7, 8]. This cement, containing approximately 45 wt.% ye'elimite and about 19 wt.% calcium sulfate, serves as the basis for various blended systems [9]. Additional cement used in this research is blast-furnace cement, CEM III/B 42.5 N-LH/SR/LA, and is mixed with the CSA cement to achieve increased early strength development. Moreover, in Sects. 4.3 and 4.4, the CSA cement-based mixtures are compared with reference mixtures in which Portland cement (CEM I 52.5 N) is used as the sole binder. Table 1 provides an overview of the material composition of the 3DCP mixture used for all subsequent experiments.

Table 1. Material composition for 3D concrete printing.

Component	Applied mixture
Cement (kg/m^3)	571
Quartz sand 0/2 mm (kg/m^3)	909
Limestone powder (kg/m^3)	317
Methylcellulose-based Stabilizer (kg/m^3)	2.8
PCE-based Superplasticizer (kg/m^3)	6.5
w/c (−)	0.50

2.2 Methods

To investigate the influence of the operating time of the 2K mixing system on early strength development and overall mechanical performance, the 4-point bending tensile strength of the developed concrete mixtures is determined in accordance with DIN EN 12390-2 [10] and DIN EN 196-1 [11].

The 3DCP process is carried out using a KUKA KR90 R2700 six-axis industrial robot equipped with a **PFT Bolero 230 V screw pump extruder for continuous material deposition**. All elements are printed at a pumping rate of 30%, corresponding to a printing speed of 0.06 m/s and a nozzle diameter of 20 mm. The specimens are produced in layers and subsequently tested for directional 4-point bending tensile strength. All mechanical tests are carried out in accordance with DIN EN 196-1 [11]. Geometric modeling, toolpath generation, and process control are conducted using Grasshopper, a parametric design environment integrated into the Rhinoceros 3D CAD platform [12–15].

3 Conceptual Design of the 2K Mixing System

The main advantage of CSA cement is its rapid strength development. However, when setting-time retarders are used to enable the application of CSA cement in conventional 3DCP, limitations in strength development may arise. In some cases, depending on the retarder dosage and the type of cement blended with CSA cement, not only is early-age strength reduced, but overall strength development is also adversely affected [2, 3].

Based on these findings, this research study deliberately refrains from using setting-time retarders. As a result, the processing window after CSA cement comes into contact with water is significantly shortened and, depending on the CSA cement content, may be limited to a few minutes. Consequently, using CSA cement in a conventional concrete mixing process, together with water and other components, is not feasible, as the time window between mixing and extrusion is insufficient.

For this reason, the objective is to add the CSA cement to the remaining mixture immediately prior to extrusion. To achieve this, a 2K mixing system is developed, in which the CSA cement is separated from the conventional mixing process and incorporated in a secondary mixing step.

Figure 1 illustrates the 2K mixing system within the overall concept of the robotic setup, in which the mixing unit and the screw conveyor are mounted on a six-axis industrial robot. The supply of the concrete mixture without CSA cement is carried out, as in conventional 3DCP, via a pipe connected to a screw pump, whereas the CSA cement is supplied through a second, independent screw conveyor.

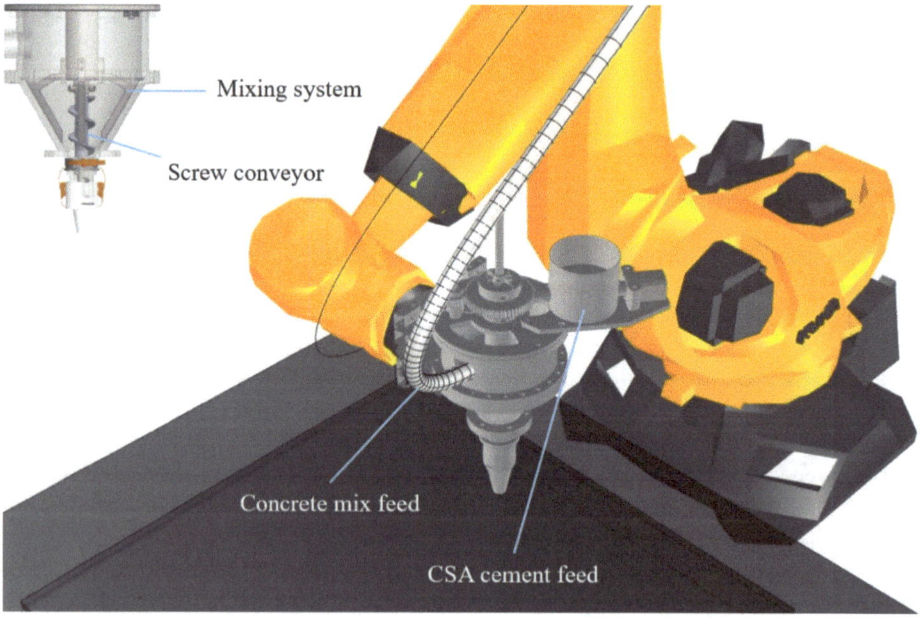

Fig. 1. Overview of the 2K mixing system in the 3DCP Process.

The mixing system, including the screw conveyor, is integrated into a container; the corresponding CAD design is shown in Fig. 2. A key feature of the system is that the drives of the conveyor screw and the mixer are mechanically independent. This is achieved by means of a coaxial double-shaft system consisting of two concentrically arranged drives: the inner shaft is mechanically connected to the screw conveyor and transmits the torsional moment from its motor. The outer hollow shaft is connected to the mixing elements and driven by a separate spur-gear stage. Both components are equipped with individual power units, allowing their rotational speeds to be controlled independently. This separation is essential for the successful experimental analysis, as it enables precise control of both the mixing time and mixing intensity via the drive of the mixing auger.

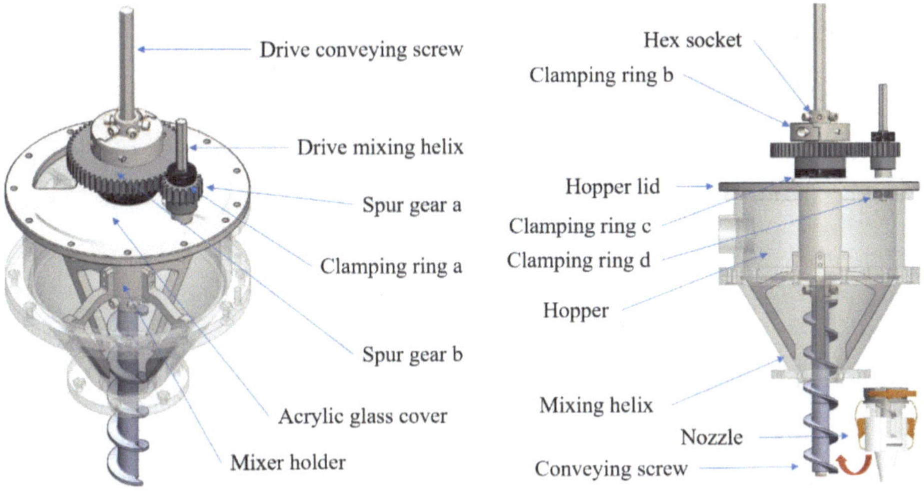

Fig. 2. Schematic overview of the 2K mixing system.

The conveying system is driven by a direct current (DC) motor (24 V, 500 W, 220 rpm). Motor power is regulated by a pulse-width modulation (PWM) controller (0–60 V, 0–30 A) in combination with a laboratory power supply (30 V, 30 A), allowing precise adjustment of the motor speed and the resulting volumetric flow rate. The screw conveyor used to deliver the fresh concrete mix has a pitch of 60 mm, a flight depth of 10 mm, and an overall length of 225 mm.

The mixing system is also actuated by a similar DC motor (24 V, 45 W, 220 rpm) and is supplied by a laboratory power supply (24 V, 10 A). Overall, the mixing system consists of four mixing augers, each individually designed. The basic geometry consists of an inclined paddle blade connected to a support shaft. This geometry generates a combined conveying and shearing motion during rotation: on the one hand, the material is recirculated both radially and axially; on the other hand, it is selectively divided by the sharp edge of the blade. As a result, adhesion to the walls of the mixing chamber is prevented, the mixing space is kept clear, and effective homogenization of the CSA cement and the fresh concrete mixture is ensured. In addition, the mixing auger is equipped with horizontal ribs arranged alternately in the upper and lower thirds of the mixing unit to deliberately influence the material flow. In these regions, there is an increased risk of bypass flows along the shaft or of the material bridging. The ribs generate additional transverse flows and turbulence. This improves mixing homogeneity and promotes the breakup of agglomerates. At the same time, the ribs prevent the formation of dead zones in the central region of the conveying unit.

Figure 3 illustrates the CSA cement conveying system, which is designed independently of the 2K mixing system shown in Fig. 2. Inside the CSA cement hopper, a delivery pipe is integrated in which the screw conveyor is mounted. The pipe is positioned in the central region of the hopper and is open at the top, allowing the cement to be fed directly into the screw conveyor. In the lower section, outside the hopper, an additional outlet opening is provided through which cement can be continuously dosed and conveyed into the mixing chamber of the 2K mixing system (see Figs. 1 and 2), where it comes into contact with the remaining fresh concrete mixture.

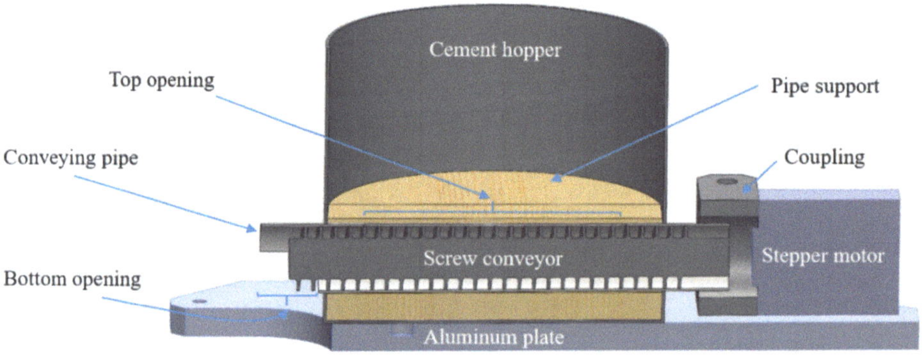

Fig. 3. Schematic sectional view of the CSA cement dosing unit.

The CSA cement dosage is carried out using a screw conveyor with a pitch of 15 mm, a flight depth of 5 mm, and an overall length of 150 mm. The screw conveyor is manufactured using 3D printing with polylactic acid (PLA). It is driven by a National Electrical Manufacturers Association (NEMA) 17 stepper motor (59 N·cm, 2.0 A). Control and regulation are implemented via an Arduino control unit, its software, and a digital stepper motor driver, DM556 (5.6 A, 0–48 V). This system enables precise adjustment of both the rotational speed and the direction of rotation of the screw conveyor, allowing the CSA cement dosage to be optimally adapted to the given processing conditions.

To ensure a continuous flow of cement into the delivery pipe and onto the screw conveyor, a DC vibration motor (24 V, 10 A), controllable up to 3800 rpm, is mounted on the outer wall of the hopper. Without such supportive measures, particularly fine-grained powders such as CSA cement are prone to bridging, caking, or void formation within the hopper, which can interrupt or make the conveying process discontinuous.

4 Results and Discussion

4.1 Influence of the Operating Time of the 2K Mixing System on Early Strength Development

To achieve early strength development using CSA cement, thereby enabling greater geometric flexibility and the subsequent manufacturing of precast concrete elements without waiting times for strength development (see Sects. 4.3, 4.4 and 4.5), the mixing behavior of CSA cement in fresh concrete within the 2K mixing system is analyzed in the following.

To investigate the temporal mixing behavior and to determine whether the ongoing 3DCP process with the 2K mixing system leads to differences in the mixing of fresh concrete and CSA cement, mixed concrete was discharged from the 2K mixing system and cast into prismatic test molds after 2, 4, 6, 8, 10, 12, and 18 min. Each specimen is tested individually, exactly 2 h after its respective casting time, to determine the 4-point bending tensile strength. In Fig. 4, the results are compared with the corresponding 2-h bending tensile strength of specimens produced using an EIRICH R05 T intensive mixer. In this early stage of hydration, strength development is exclusively governed by the CSA cement, allowing a direct correlation to be drawn with its incorporation and reaction with water [3].

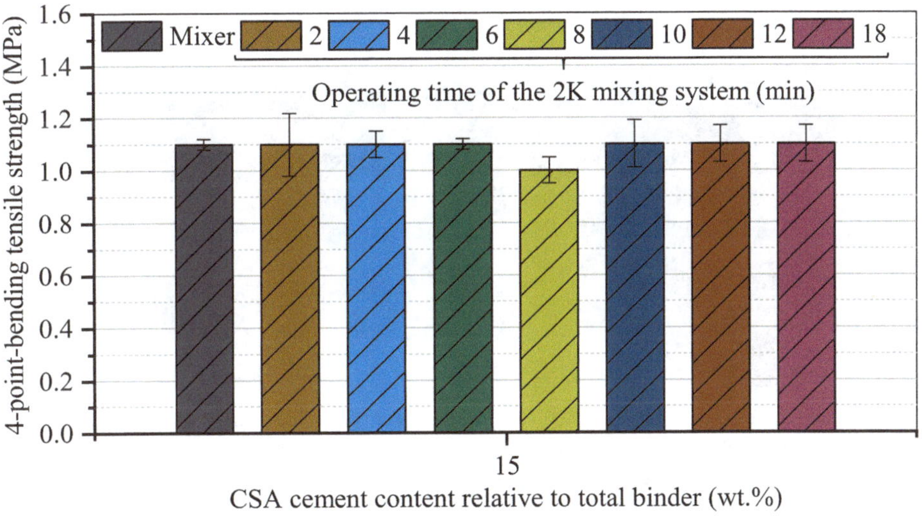

Fig. 4. Investigation of the 2-h 4-point-bending tensile strength of a CSA-CEM III/B blend (15 wt.% CSA cement and 85 wt.% CEM III/B) using the 2K mixing system as a function of operating time (2–18 min) to evaluate the homogenization behavior of the CSA cement.

The bending tensile strength results indicate that within the first two hours, consistent strength development is observed, with values ranging from 1.0 to 1.1 MPa over the entire operating time of the 2K mixing system, from 2 to 18 min. In comparison with specimens produced using a conventional concrete mixer, no reduction in strength is observed. Furthermore, the standard deviation remains within an acceptable range, with variations of approximately 10% during this very early phase of CSA cement hydration.

Overall, the 2K mixing system demonstrates consistent mixing performance for CSA cement throughout the entire processing time. The cement is brought into contact with the remaining mixture so that the CSA cement's strength development begins immediately after extrusion through the nozzle.

4.2 Anisotropic Strength Analysis

The primary objective is to achieve early strength development with CSA cement without compromising final strength. However, this objective is not achieved when CSA cement is combined with blast-furnace cement (e.g., CEM III/B) in the presence of setting retarders [2, 3]. By contrast, the use of the 2K mixing system avoids such retardation, ensuring not only increased early strength but also sustained overall strength development.

Figure 5 illustrates the printed plate geometry used for the investigation of bending tensile strength. The plate specimen has dimensions of $400 \times 400 \times 60$ mm^3 and serves as the basis for extracting $40 \times 40 \times 160$ mm^3 prismatic specimens for direction-dependent 4-point-bending tensile strength tests [3]. In addition to the plate specimen, a $600 \times 150 \times 120$ mm^3 beam geometry is shown in the figure and was used for compressive strength testing; however, compressive strength results are not within the scope of the present study.

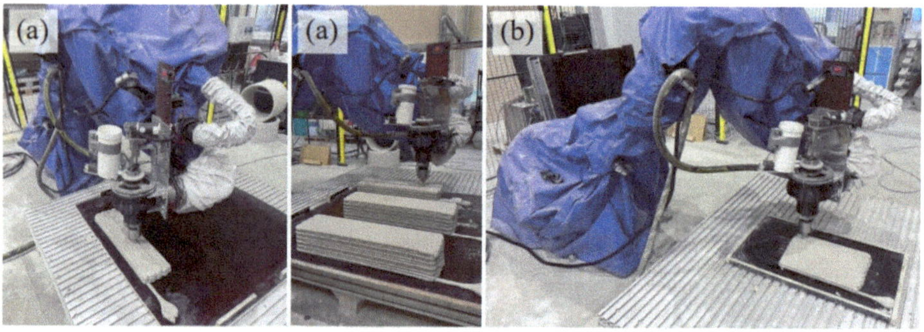

Fig. 5. Application of the 2K mixing system in 3DCP for the production of direction-dependent test specimens made from CSA cement: $600 \times 150 \times 120$ mm^3 beam specimen (a) and $400 \times 400 \times 60$ mm^3 plate specimen (b).

Figure 6 compares the directional 28-day 4-point-bending tensile strengths of two mixtures. The first mixture, based on Table 1, contains 30 wt.% CSA cement and 70 wt.% CEM III/B with the addition of setting-time retarders (0.60 wt.% tartaric acid and 0.60 wt.% phosphate relative to total binder mass) and is produced without the 2K mixing system (referred to as 'Setting time retarder' in Fig. 6) [2].

This is compared with an equivalent mixture containing 10 wt.% CSA cement and 90 wt.% CEM III/B, produced without setting retarders but using the 2K mixing system (referred to as '2K mixing system' in Fig. 6).

The principal material directions are defined as follows: the x-direction lies in the plane of the printing base and is parallel to the robot movement direction. The y-direction also lies in the plane of the printing base but is perpendicular to the robot movement direction. The z-direction is perpendicular to the printing base and thus also perpendicular to the robot movement direction.

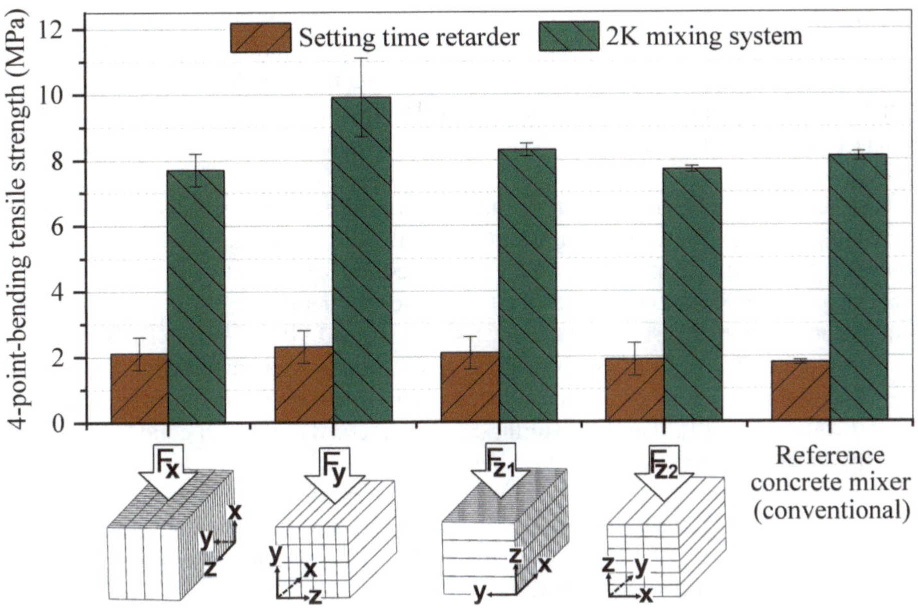

Fig. 6. Comparison of 28-day 4-point bending tensile strength in the principal material directions for the CSA-CEM III/B (30/70) mixture produced by 3DCP using setting time retardation and without the 2K mixing system and for the CSA-CEM III/B (10/90) mixture produced by 3DCP without using setting time retardation but with the 2K mixing system, with additional reference results from the same mixtures produced in a conventional concrete mixer.

The results clearly show that the use of setting retarders negatively affects the strength development of CSA cement and CEM III/B mixtures. In contrast, when the 2K mixing system is used, no reduction in strength is observed because no setting retarders are required. In particular, the comparison with the reference value of 8.1 MPa obtained for the same mixture (10 wt.% CSA cement and 90 wt.% CEM III/B) produced using a conventional concrete mixer further highlights the effectiveness of the 2K mixing system in maintaining long-term bending tensile strength.

In addition to these findings, Fig. 6 reveals a pronounced directional dependence of the 4-point bending tensile strength results. Differences are observed between loading applied parallel to the printed layers (x and z1) and loading applied perpendicular to the layer interfaces (y and z2). The 4-point bending tensile strengths in the y- and z1-directions (9.9 MPa and 8.3 MPa) are higher than those measured in the x- and z2-directions, which both reach 7.7 MPa.

This directional dependence reflects the anisotropic nature of the printed material and its layer interfaces, with particularly favorable performance under loading perpendicular to them.

4.3 Printing Tests for the Vertical Build-Up Height

The application of CSA cement, combined with the 2K mixing system designed in Sect. 3, enables new degrees of geometric freedom. Structurally, geometries such as hollow sections and overhangs can be realized without the need for molds or additional support structures. Moreover, the rapid strength development of the printed layers enables greater maximum build-up heights, as the layers attain sufficient load-bearing capacity early.

The following investigations are conducted to evaluate increased build-up heights in 3DCP. Two mixtures, as specified in Table 1, are compared. Mixture 1 employs 100 wt.% Portland cement (CEM I 52.5 N) as the sole binder and was processed using conventional 3DCP without the use of setting-time retarder or the 2K mixing system. In contrast, mixture 2 consists of 10 wt.% CSA cement and 90 wt.% CEMIII/B, contains no setting retarder, and was processed using 3DCP incorporating the 2K mixing system. The consistency of both mixtures is adjusted to a uniform Haegermann slump diameter of 14.0×14.0 cm^2 [2], thereby ensuring comparable and consistent conditions for the investigation of the maximum build-up height.

To investigate the maximum build-up height, a circular, hollow concrete element with a diameter of 150 mm is printed. The element consists of a single layer and is printed layer by layer, with a width of 20 mm and a height of 5 mm, until collapse occurs due to insufficient structural stability.

Mixture 1 reaches a maximum build-up height of 670 mm, corresponding to 134 printed layers, before collapse occurred (see Fig. 7).

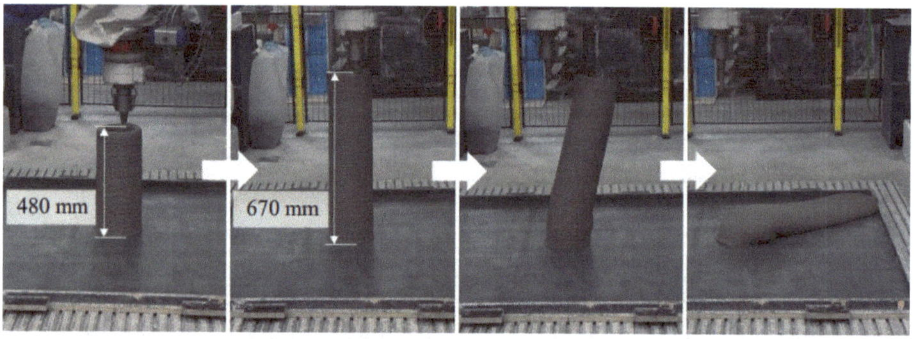

Fig. 7. Additive manufacturing of the CEM I mixture to determine the maximum build-up height (670 mm) in the 3DCP process using a KUKA KR90 R2700 industrial robot without the 2K mixing system.

By comparison, the mixture containing CSA cement exhibited markedly higher structural stability, enabling 250 printed layers without collapse (see Fig. 8). This corresponds to a build-up height of 1250 mm, which represents the maximum reach of the robotic arm. These results clearly demonstrate the positive impact on geometric design flexibility enabled by the early strength development of CSA cement, made accessible through the 2K mixing system.

Fig. 8. Additive manufacturing of the CSA-CEM III/B (10/90) mixture to determine the maximum build-up height in the 3DCP process using a KUKA KR90 R2700 industrial robot with the 2K mixing system.

4.4 Experimental Investigation of Overhang Printing

In addition to the maximum build-up height, the realization of overhangs plays an important role in geometric design flexibility. The schematic illustration of the geometries used to investigate printable overhangs is shown in Fig. 9. For each mixture from Sect. 4.3, four circular elements are printed, differing in the layer offset between successive layers and in the resulting overall external overhang.

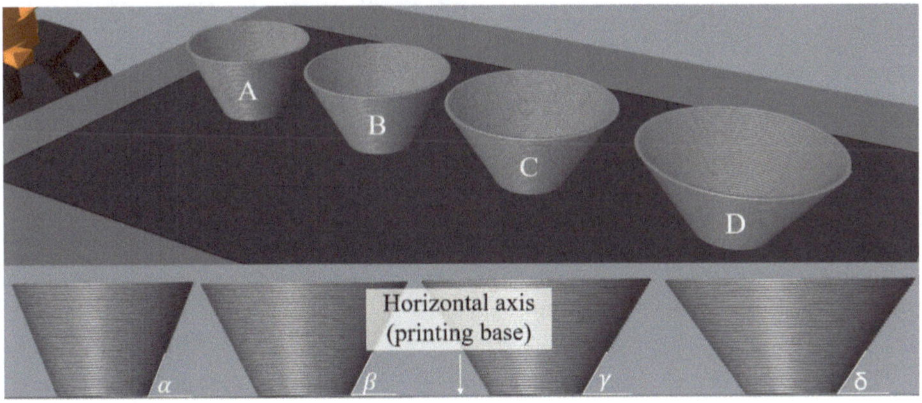

Fig. 9. Schematic representation of the test geometry A, B, C, and D with overhang angles α, β, γ, and δ for investigating the printing behavior of overhangs.

All elements (Fig. 9: A, B, C, and D) have a base diameter of 150 mm and a maximum printable height of 200 mm. Each layer has a height of 5 mm and a width of 20 mm. The layer geometry expands continuously outward with increasing height, resulting in an upper diameter of 310 mm for element 4.4A ($\alpha = 68.2°$), 350 mm for Element 4.4B ($\beta = 63.4°$), 390 mm for element 4.4C ($\gamma = 59.0°$), and 430 mm for element 4.4D ($\delta = 55.0°$). The overhang angle is defined with respect to the horizontal axis (printing base); consequently, smaller angles correspond to more pronounced overhangs (see Fig. 9). These progressively increasing overhangs are used to assess the performance limits of the respective concrete mixtures. The varying geometries allow specific differences in the printing behavior and structural stability to be identified.

The results shown in Fig. 10 indicate that element 4.4A can be printed without difficulty using the concrete mixture containing CEM I (mixture 1). For element 4.4B, with an overhang angle of 63.4°, the performance limit of mixture 1 is reached; although the element can still be printed, deformations and local loss of shape occur. Element 4.4C, with an overhang angle of 59.0°, collapses after 25 layers, corresponding to a height of 125 mm. Significant distortions are already observed during printing of the initial layers, indicating that this overhang cannot be printed stably with this mixture. Element 4.4D, with an overhang angle of 55.0°, cannot be printed using the CEM I mixture.

By comparison, mixture 2 (CSA cement and CEM III/B) in combination with the 2K mixing system exhibits significantly improved performance. However, mixture 2 fails during the 3DCP of element 4.4D, which collapses structurally at layer 39 out of 40, corresponding to a height of approximately 190 mm. To successfully print element 4.4D or geometries with even more pronounced overhangs, the current CSA cement content of 10 wt.% must be increased. In contrast, elements 4.4A to 4.4C can be printed without any difficulties using this mixture in combination with the 2K mixing system.

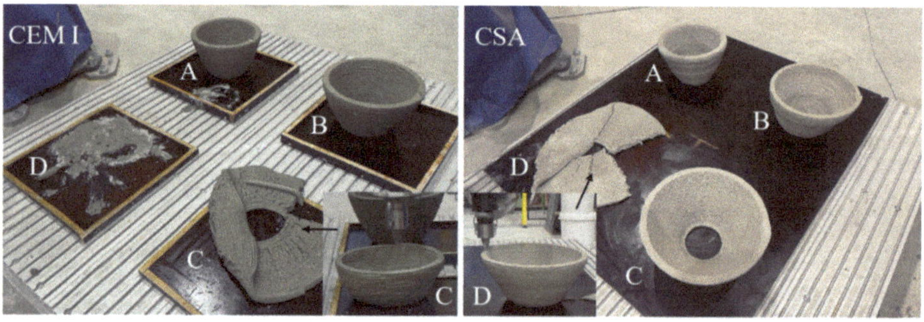

Fig. 10. Comparison of overhang printing in 3DCP using a CEM I reference mixture and a CSA cement mixture processed with the 2K mixing system.

The use of CSA cement in combination with the 2K mixing system not only increases the achievable maximum build-up height (see Sect. 4.3) but also improves the printability of overhangs. This enables enhanced geometric design flexibility and opens additional application possibilities for 3DCP. For small and geometrically demanding components with rapidly increasing overhangs, higher CSA cement contents are required to ensure sufficient early strength and structural stability. In contrast, for tall components printed at lower build-up rates, such as those investigated in Sect. 4.3, lower CSA cement contents are sufficient, as the slower geometric progression provides more time before critical stability limits are reached. This approach additionally supports sustainability by enabling a higher proportion of CEM III/B, which is associated with a reduced environmental footprint [4, 6, 7].

4.5 Fabrication of Reference Objects as Façade Elements

Based on the previous findings, three reference precast concrete elements in the form of lesenes (slender vertical wall elements) are modeled. The elements consist of different overhang geometries and are printed over the full working height of the robotic arm, reaching 1250 mm. An illustration of the three elements is shown in Fig. 11.

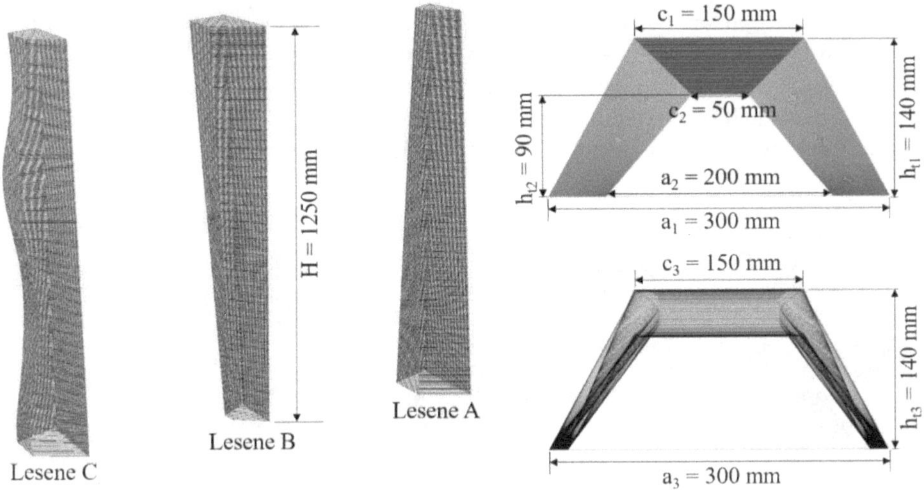

Fig. 11. Modeling of lesenes A, B, and C as reference objects in Grasshopper 3D.

Element 4.5A has a trapezoidal base geometry with side lengths of $a_1 = 300$ mm, $h_{t1} = 140$ mm, and $c_1 = 150$ mm (opposite side to a_1). Over the entire height of $H = 1250$ mm, the element exhibits an inward-directed overhang of 50 mm on three sides, corresponding to a wall inclination angle of approximately 87.7° with respect to the horizontal plane. The element is designed as a hollow component with a constant wall thickness of 20 mm.

Element 4.5B represents the inverse geometry of element 4.5A. It also features a trapezoidal base with side lengths of $a_2 = 200$ mm, $h_{t2} = 90$ mm, and $c_2 = 50$ mm. In contrast to element 4.5A, element 4.5B exhibits an outward-directed overhang of 50 mm on three sides over the full height of 1250 mm.

Element 4.5C represents a hybrid configuration that combines the geometric characteristics of Elements 4.5A and 4.5B. In the lower part of the element height, an inward-directed overhang is present, which gradually transitions into an outward-directed overhang toward the upper part. Both the inward and outward overhangs on three sides correspond to a wall inclination angle of approximately 87.7° with respect to the horizontal plane. The transition between the two overhang configurations follows a sinusoidal function, and the total height of Element 4.5C is likewise 1250 mm.

The geometries of the technology demonstrators are deliberately designed to realize both pronounced overhangs and large build-up heights. By manufacturing these precast concrete elements using the 2K mixing system in combination with CSA cement, the performance of the proposed technology is demonstrated, as components of this type cannot be successfully produced using conventional 3DCP processes with Portland cement as the sole binder. At the same time, the incorporation of internal hollow structures highlights a key advantage of 3DCP over conventional casting methods, in which realizing such voids would require complex, material- and cost-intensive formwork systems.

After the design and modeling of the façade elements for the technology demonstrators, and following the verification that the required early and final strengths could be achieved using CSA cement within the 2K mixing system, the façade elements are manufactured using 3DCP in combination with the 2K mixing system. The applied mixture is based on 10 wt.% CSA cement and 90 wt.% CEM III/B.

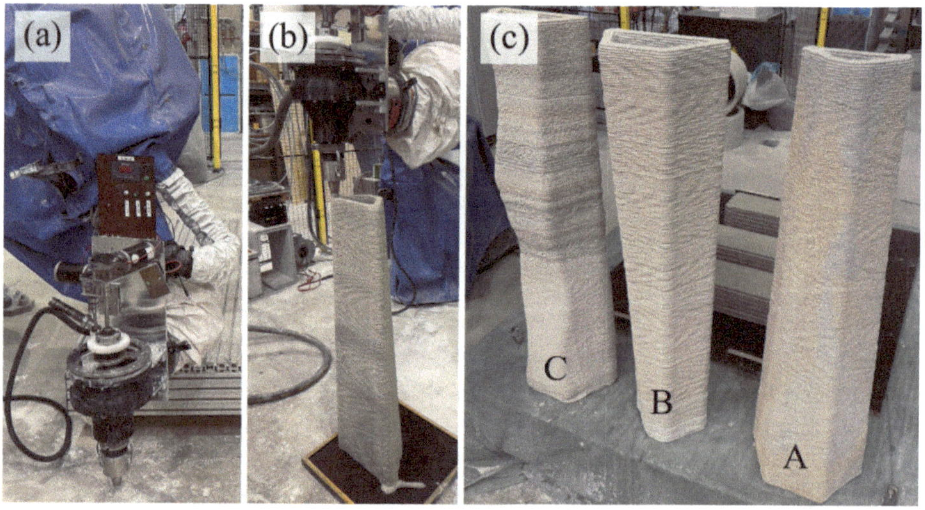

Fig. 12. 2K mixing system mounted on the 3DCP robot (a), 3DCP process of the lesene-shaped façade element A made from CSA cement using the 2K mixing system (b), lesene-shaped façade elements A to C after extrusion (c).

All printed reference elements remain structurally stable throughout the entire 3DCP and exhibit no signs of mechanical or structural failure. The CSA cement-based mixture demonstrated sufficient load-bearing capacity to reliably support overhangs and internal hollow structures in the investigated geometries. Figure 12 shows the 3DCP process of the lesene-shaped façade element A (see Fig. 11), produced using the 2K mixing system with CSA cement and CEM III/B as binders.

5 Conclusions

In this research, the 2K mixing system was conceived, developed, and tested, enabling the controlled and selective use of CSA cement in 3DCP, in combination with binders containing high proportions of SCMs, without the need for setting-time retarders. The central design principle of the system is the independent actuation of the CSA cement dosing unit, the mixing system that integrates CSA cement into the conventional fresh concrete mixture, and the conveying system that transports the resulting mixture to the extrusion nozzle. This functional decoupling enables precise synchronization of individual volumetric flow rates and fine-grained calibration of CSA cement incorporation as a function of the conveying screw and mixing auger geometries.

By initiating CSA cement hydration immediately before extrusion, early-age strength development and setting behavior can be precisely adjusted without the need for retarders, which are known to adversely affect strength development in GGBFS-rich systems. The 2K mixing system effectively homogenizes the CSA cement with the base mixture, resulting in consistent 2-h 4-point bending tensile strengths after extrusion. Furthermore, the measured 28-day 4-point bending tensile strengths indicate that the use of setting retarders adversely affects the strength development of CSA cement and CEM III/B mixtures. In contrast, when the 2K mixing system is used, no reduction in strength is observed because setting retarders are unnecessary.

The results further show that the very early strength development of CSA cement, enabled by the 2K mixing system, enables reliable printing of steep overhangs and increased build heights, thereby expanding geometric design flexibility in 3DCP.

Overall, the 2K mixing system developed in this research allows the use of CSA cement in 3DCP without the need for setting retarders. This approach also facilitates combining CSA cement with blast-furnace cement, enabling proper hydration of GGBFS without hindrance from retarders. The accelerated build-up of structural stability provided by CSA cement effectively compensates for the delayed hydration of blast-furnace cements, reducing process-related waiting times and improving the production efficiency of precast concrete elements, as rapid hardening allows elements to be manufactured in quick succession.

Acknowledgments. The authors would like to thank the sponsors for the financial support provided by the Department Bank of Saxony (SAB) via the European Regional Development Fond (ERDF).

Disclosure of Interests. The authors have no competing interests to declare that are relevant to the content of this article.

References

1. Kroboth K (1986) Zement/Herstellung-Eigenschaften–hydratation Gesundes Wohnen. Beton-Verlag, pp 336–348
2. Senf F, Javed R, Funke H, Gelbrich S (2025) Enhanced environmental sustainability of 3D-printed concrete mixtures based on calcium sulfoaluminate (CSA) and blast furnace cement (CEM III/B). In: MATEC web of conferences, vol 409, p 06004
3. Senf F, Abstoß M, Funke H, Gelbrich S (2025). Advanced 3D Concrete Printing (3DCP) with Low-Carbon Calcium Sulfoaluminate (CSA) Cement Blends. In The International Conference on Net-Zero Civil Infrastructures: Innovations in Materials, Structures, and Management Practices (NTZR) (pp. 522–534). Cham: Springer Nature Switzerland
4. Senf F, Rudolph E, Abstoß M, Funke H, Gelbric S, Kroll L (2026). Interaction between retarder-induced clinker hydration suppression and CSA hydration on GGBFS reactivity in ternary cementitious systems. Construction and Building Materials, 542, 148025
5. Cemex Deutschland AG CEM I 52,5 R (ft) [process dataset, ÖKOBAUDAT, DIN EN 15804+A2], ÖKOBAUDAT German database for life cycle assessment data of construction products, https://oekobaudat.de (dataset UUID: 37e62d96-7e02-466e-aaf4-99539b29d9a9). Last accessed 03 Jan 2026
6. Heidelberg Materials AG Engineering & Innovation: Blastfurnace cement CEM III/B 42,5 L-LH/SR (na) [Process dataset, ÖKOBAUDAT, DIN EN 15804+A2], ÖKOBAUDAT – German database for life cycle assessment data of construction products, https://oekobaudat.de (dataset UUID: f140d929-49c4-44f9-ac2a-7f8763b0c010). Last accessed 03 Jan 2026
7. Nie S et al (2022) Analysis of theoretical carbon dioxide emissions from cement production: methodology and application. J Clean Prod 334:130270
8. Verein Deutscher Zementwerke (VDZ) Dekarbonisierung von Zement und Beton – Minderungspfade und Handlungsstrategien, https://www.vdz-online.de/fileadmin/wissensportal/publikationen/zementindustrie/VDZ-Studie_Dekarbonisierung_von_Zement_und_Beton.pdf. Last accessed 15 Jan 2026
9. GmbH D (2019) Buzzi Unicem: next base – hydraulic binder based on calcium sulfoaluminate clinker (product data sheet), Wiesbaden, Germany
10. DIN EN 12390-2:2019-10 Testing hardened concrete—Part 2: Making and curing specimens for strength tests. German version EN 12390-2:2019
11. DIN EN 196-1:2016-11 Methods of testing cement—Part 1: Determination of strength. German version EN 196-1:2016
12. Senf F, Javed R, Funke H, Gelbrich S (2024) Concrete extrusion using sustainable cement alternatives. In: Perkowski Z, Beben D, Zembaty Z, Massimino MR, Oliveira MJ (eds) Environmental challenges in civil engineering III. ECCE 2024, Lecture notes in civil engineering, vol 615. Springer
13. Abstoss M, Senf F, Gliniorz R, Funke H, Gelbrich S (2025). Sustainable Precast Concrete Construction Using Hybrid Additive Manufacturing: 3D Concrete Printing and Shotcrete with Low-Carbon Concrete. In The International Conference on Net-Zero Civil Infrastructures: Innovations in Materials, Structures, and Management Practices (NTZR) (pp. 559–571). Cham: Springer Nature Switzerland
14. Lindner M, Gliniorz R, Funke H, Gelbrich S (2024) Anisotropic properties of additively manufactured concrete elements. Appl Res 3(3):e202300035
15. Abstoß M, Funke H, Gelbrich S (2025) Combined additive manufacturing process for the production of sustainable concrete modules for the construction of a tiny house. In: MATEC web of conferences, vol 409, p 06001

Infrastructure, Structural Engineering

Friction Coefficient of Cement Concrete Road Surface Measured Using the ViaFriction and SRT-3 Devices

Jan Czmiel[1(✉)], Zbigniew Perkowski[2], Daniel Frączek[2], and Robert Jurczak[3]

[1] Opole University of Technology, Doctoral School, Prószkowska 76, Opole 45-758, Poland
jan.czmiel@student.po.edu.pl
[2] Faculty of Civil Engineering and Architecture, Opole University of Technology, Katowicka 48, Opole 45-061, Poland
{z.perkowski,d.fraczek}@po.edu.pl
[3] Faculty of Civil and Environmental Engineering, West Pomeranian University of Technology, Al. Piastów 50a, Szczecin 70-311, Poland
robert.jurczak@zut.edu.pl

Abstract. This paper presents a comparative analysis of friction coefficient measurements on cement concrete pavements, conducted using two distinct methods: the ViaFriction (VF) device for continuous measurement at a partial slip ratio, and the SRT-3 device for spot measurement under full wheel lock. The motivation for undertaking this research is to facilitate measurements in this area for the national road administrator in Poland, as its permissible values for the surface friction coefficient have been precisely defined for acceptance and diagnostic tests in relation to the SRT-3. Recently, this has also been done for measurements harmonised with the Dynamic Friction Tester (DFT) point standard, but these requirements, including the recommended degree of test wheel slippage, are still not defined for the VF. The analyses presented here constitute a significant extension of the authors' earlier publication on friction on concrete road surfaces. The measurements were performed at various speeds and wheel slippages using the VF device. A new element of the research is the inclusion of test sections with more varied friction in different seasons (early spring, late summer/early autumn). This resulted in strong correlations between the friction coefficient values for the concrete road surface from both devices. Furthermore, it was demonstrated that the road surface temperature had no significant effect on the results of such measurements within the range of 11–36 °C.

Keywords: Road Surface · Skid Resistance · Friction Coefficient · ViaFriction · SRT-3 · Cement Concrete

1 Introduction

Ensuring the appropriate anti-skid properties of a road surface is crucial for the users' safety [1, 2]. For this reason, the assessment of the friction coefficient is necessary, among others, at the acceptance testing stage, when a decision is made whether a new road can

Z. Perkowski et al. (Eds.): ECCE 2026, LNCE 929, pp. 147–160, 2027.
https://doi.org/10.1007/978-3-032-39515-3_11

be approved for use [3], as well as during operation, when it is assessed whether the road surface can remain without repair or whether it should be repaired/renovated and to what extent [4]. The methodology for precisely, automatically, and accurately determining the friction coefficient is also important in the environmental context. The surface course repair or renovation process is always, to a greater or lesser extent, harmful to the natural environment [5], and the decision to carry it out at the appropriate time is often based on friction analysis. In this context, the paper focuses on the use of continuous measurement of the friction coefficient with variable slip of the test wheel using the ViaFriction (VF) device [6], which enables detailed monitoring of the technical condition of road surfaces.

Road surface friction is the net force that opposes the relative motion between a vehicle tyre and the road surface. This force varies depending on whether the tyre is rolling or sliding on the surface. The non-dimensional coefficient of friction is a measure of anti-skid performance, calculated using the formula [7]:

$$\mu = \frac{F}{F_w} = \frac{F_a + F_c + F_h + F_d}{F_w} \tag{1}$$

where: μ – friction coefficient; F – tangential friction force between the tyre and the road surface; F_w – vertical wheel load; F_a, F_c, F_h, F_d – friction force components, respectively, from: adhesion between the tyre rubber and the road surface, cohesion between the tyre rubber and rubber contaminants on the road surface, hysteresis in the stress-strain relationship during loading and unloading of the tyre rubber in contact with road surface irregularities, and aerodynamic and hydrodynamic drag. It should be noted, however, that F is primarily the result of two phenomena: adhesion and hysteresis. Adhesion results from the bonding/interlocking of tyre rubber with the road surface at the microscale and depends, first of all, on the surface's micro-irregularities, which in road engineering are described by the term microtexture. The roughness of the aggregate exposed on the road surface is important in this respect. In turn, hysteresis, in the stress-strain relationship, depends on the intensity of strain energy dissipation. Hence, this phenomenon occurs mainly in those parts of the tyre that are locally deformed in contact with macro-irregularities in the surface, which are described in road engineering as macrotexture. Macrotexture is shaped by appropriately selected production technology of the wearing course (material composition and surface finishing method). As a result, adhesion primarily accounts for friction on smooth, dry surfaces, while hysteresis accounts for friction on wet, rough surfaces.

At the national level, different methods for measuring μ of road surfaces are preferred depending on experience, with a large group of devices using a measuring wheel with a slip ratio (*SR*) less than 100% [8]. In Poland, until 2025, the minimum required friction coefficient values were explicitly defined only for the SRT-3 device with a full locked-wheel system (*SR* = 100%) [3]. Recently, this has also been done for measurements harmonised with the Dynamic Friction Tester (DFT) point standard [9], but these requirements, including the recommended degree of test wheel slippage, are still not defined for devices like the VF. Due to the significant wear and tear of the SRT-3, the Polish administrator, the General Directorate for National Roads and Motorways (GDNRM), intends to introduce new friction measurement standards. This motivated a comparison of the friction coefficient measurements from the SRT-3 device and the more modern VF one, with determination of their degree of correlation and confidence intervals. In particular, the subject of this paper is an extension of article [8] for cement concrete (CC) road surface. In [8], the relationship was initially determined based on measurements taken at the turn of summer and autumn, 2024, from sections of the national road No. 46 (Poland) with a total length of 8.5 km after 2 years of operation. However, a satisfactory correlation was not obtained due to insufficient variability of the measured friction. Therefore, six months later, during early spring, 2025, the additional measurements were conducted once again on the same sections of road No. 46, and during early autumn 2025 on a much more worn 10-km section of the A4 motorway CC surface (Poland). As a result, based on a much broader measurement database, strongly correlated relationships were obtained between the friction coefficients of the CC road surface as measured by the VF and SRT-3, including analysis of the seasonality impact on the results.

2 Cement Concrete Road Pavements

Roads with CC pavements currently constitute a smaller share of the road network, but it is steadily increasing as the network of motorways and expressways expands. The advantages of CC road pavements are evident in their long-term operational performance without the need for frequent repairs, even under heavy traffic on main communication routes. In Western Europe, the relative share of concrete pavements in the network of main national roads varies. The largest share, reaching approximately 35%, is held by Austria and Belgium, while Germany has a slightly smaller share, approximately 25%. In France, the Netherlands, Spain, Switzerland, and the United Kingdom, the share of concrete pavements does not exceed 10% [10]. In Poland, on the other hand, concrete pavements already account for over 20% of motorways and expressways, which indicates the importance of the issue discussed in this paper.

For road pavement concrete, the following binders are used: Portland cements CEM I 32.5 R or N and CEM I 42.5 R or N and Portland slag cement CEM II/A-S or CEM II/B-S [11]. The aggregate used shall have the following maximum sizes: $D \leq 31.5$ mm for single-layer and two-layer pavements constructed of the same mix; and $D \leq 8$ mm for the surface course of exposed aggregate concrete pavements. Additionally, the ratio of coarse to fine aggregate dimensions should not be less than 1.4. The aggregate should meet the requirements of EN 12620 [12]. Admixtures complying with EN 934-1 and 934-2 [13, 14] should be used to entrain air into the concrete mix. The preparation of concrete mixes with air-entraining admixtures and the determination of their air content should be in accordance with EN 12350-7 [15]. Mineral additives to concrete may be used according to the principles specified in EN 206-1 [16] for traffic categories KR1–KR2 (from 30,000 to 500,000 vehicles with a cumulative number of equivalent standard axles with a load of 100 kN in a 20-year design life [17]).

Concrete pavements shall exhibit the following functional properties [18]: adequate thickness, density, compressive and tensile strength, freeze-thaw resistance, bond between concrete layers, and resistance to fuel and oil penetration.

The surface course shall provide an adequate macrotexture depth, as specified in Poland by the National Requirements [19]. The surface course texturing may be executed using the following techniques:

- Longitudinal drag: using a burlap (hessian) drag in the direction of the carriageway axis.
- Transverse brushing: brushing the plastic concrete with a steel wire brush perpendicular to the carriageway axis.
- Transverse tining: forming grooves with a metal-tined tool perpendicular to the carriageway axis.
- Exposed Aggregate Concrete: applying a set retarder (e.g., glucose-based) followed by the removal of the unhydrated mortar skin using a mechanical brush or high-pressure water.

In the last case, an exposed aggregate surface with a macrotexture depth of up to 1.5 mm is obtained. This technology is typically applied in two-layer 'exposed aggregate' pavement systems [20]. Its use is recommended for pavements designed for traffic categories KR5–KR7 (exceeding 7.3 million vehicles, with a cumulative number of equivalent standard axles with a load of 100 kN in a 20-year design life [17]).

3 Friction Coefficient Measurements

Friction coefficient measurements were conducted on three sections of CC pavement located in the Opole and Lower Silesian voivodeships (Poland), designated as RS1–RS3. RS1 and RS2 (Figs. 1a and 2) represent the lanes of the single-carriageway national road No. 46 between Nysa and Kłodzko, which was opened to traffic in 2023. RS1 is located in the Kłodzko-bound lane, while RS2 is in the Nysa-bound lane. Section RS3 (Figs. 3 and 4) is situated on the A4 motorway in lane 1, starting at the Wrocław Południe junction and heading westbound. Notably, RS3 is located on one of Poland's busiest motorways, where the average daily traffic in 2024 was approximately 60,000 vehicles.

The fundamental data for the selected test sections are presented in Table 1. It should be noted that the texturing techniques and concrete properties of sections RS1–RS2 differ from those of RS3. However, these sections were selected for comparison due to logistical constraints, specifically the limited availability of pavements with concrete technologies yet exhibiting significantly different levels of wear and friction coefficients. Consequently, the scope of this study was limited to sections with concrete pavements of a similar strength class, although they have been in service for significantly different periods.

3.1 Experimental Programme

Measurements on each section were conducted using two devices: the SRT-3 and the VF (Fig. 1). The water-film thickness applied by the devices ahead of the test wheel was 0.5 mm (Fig. 2). Measurements for sections RS1 and RS2 were carried out in late summer and early autumn 2024, and in early spring 2025. For section RS3, the measurements were performed in early autumn 2025. Due to varying weather conditions across all measurement campaigns, the surface temperature ranged from 11 °C to 36 °C.

Table 2 presents the adopted vehicle speed (V) and the slip ratio (SR) of the test wheel. The SRT-3 was operated under standard conditions, with measurements every 100 m at 60 km/h with full wheel lock (100% slip). The VF device was used with a

Table 1. Characteristics of test sections.

Test section	Road/motorway No. Section description	Location in road: beginning/end (km + m)	Distance [m]	Technology	Operation time
RS1	46 single carriageway, Kłodzko-bound lane	21 + 750 26 + 000	4,250	CC class according to [16]: C35/45, aggregate: amphibolite, texturing method: exposed aggregate concrete	since 2023
RS2	46 single carriageway, Nysa-bound lane	26 + 000 21 + 760	4,250		
RS3	A4 lane 1 of westbound carriageway	153 + 800 143 + 500	10,300	equivalent CC class according to [16]: C30/37, aggregate: granite, texturing method: transverse tining	since 2006

Fig. 1. Left: test sections RS1 and RS2 on the national road No. 46, with a parked vehicle equipped with a VF device. Right: the SRT-3 device.

measurement reading frequency of every 10 m at slip ratios of 15%, 18%, and 21%. It should be noted that, while the full experimental programme using the VF device involved various speeds ranging from 40–60 km/h on the national road and 60–80 km/h on the motorway, for comparative purposes, this study considers only data obtained at 60 km/h. Consequently, the subsequent analyses are based on 188 SRT-3 measurements and 3 × 1,880 measurements from the VF.

Fig. 2. Surface of test sections RS1–RS2, showing the characteristic wetted track following a VF device measurement.

Fig. 3. Test section RS3 on the A4 motorway.

Table 2. Measurement parameters.

SRT-3	VF
Distance between measurements: 100 m	Distance between measurement readings: 10 m
$V = 60$ km/h $SR = 100\%$ (locked wheel)	$V = 60$ km/h $SR = 15\%$
	$V = 60$ km/h $SR = 18\%$
	$V = 60$ km/h $SR = 21\%$

Fig. 4. Pavement surface of section RS3.

3.2 Analysis of Results

Figures 5, 6 and 7 present the friction coefficient measurements as scatter plots. The coordinates represent measurement pairs from the SRT-3 (horizontal axis) and the VF (vertical axis), obtained over the same 100-m sections, yielding 10 points per section. The midpoint of these sections corresponds to the exact location of the SRT-3 measurement. The figures include a red solid line representing the linear regression approximation, which is defined by the following equation:

$$\mu_{SRT} = a\mu_{VF} + b \tag{2}$$

where: μ_{SRT}, μ_{VF} – friction coefficients from the SRT-3 and VF devices, respectively; a, b – the slope and intercept of the regression line, respectively. The standard deviation relative to the line defined by Eq. (2), i.e. $\mu_{SRT} \pm \sigma$, is indicated by red dashed lines, where:

$$\sigma = \sqrt{\frac{\sum_{i=1}^{n} \left(a\mu_{VF,i} + b - \mu_{SRT,i}\right)^2}{n-1}} \tag{3}$$

n – number of measurements; $\mu_{SRT,\,i}$, $\mu_{VF,\,i}$ – friction coefficients from the SRT-3 and VF devices, respectively, for the i-th measurement pair represented as points in Figs. 5, 6 and 7. Furthermore, the purple solid lines indicate the limit values $\mu_{SRT(0.05)}$ and $\mu_{SRT(0.95)}$, which satisfy the following conditions:

$$P\left(M_{SRT} < \mu_{SRT(0.05)}\right) = 0.05 \text{ and } P\left(M_{SRT} < \mu_{SRT(0.95)}\right) = 0.95 \tag{4}$$

where: $P(\ldots)$ – the probability of an event; M_{SRT} – a random variable following a normal distribution $\mathcal{N}(\mu_{SRT}, \sigma)$. The figure legends also provide numerical values for μ_{SRT}, σ, and the Pearson correlation coefficient R for each approximation. It was assumed that $|R| \leq 0.3$, $0.3 < |R| \leq 0.5$, $0.5 < |R| \leq 0.7$, and $|R| > 0.7$ indicate, respectively, no correlation, a weak, a moderately strong, and a strong correlation between the variables.

The results presented in Fig. 5 illustrate the range of variation in μ_{SRT} as a function of μ_{VF}. These data pertain to a pavement recently opened to traffic (in 2023). Across the combined 8.5 km length of sections RS1–RS2, the μ_{SRT} values ranged from 0.47 to 0.60, while the μ_{FV} ones ranged from 0.70 to 1.03. The range of friction coefficients remained independent of the *SR* applied in the VF device at a speed of $V = 60$ km/h. Notably, after the winter season, the friction coefficient decreased slightly due to road surface wear (with red points generally shifted further to the left and downward compared to black points). However, based on the *R* values (ranging from 0.107 to 0.256), it can be concluded that there is practically no correlation between μ_{FV} and μ_{SRT}. This results from the lack of variation in the degree of road surface wear, combined with a high σ relative to the data range, a phenomenon previously indicated in [8].

The results presented in Fig. 6 exhibit a wider range of variation in μ_{SRT} and μ_{VF} relative to σ. This is attributed to significant road surface wear on the A4 motorway over 20 years of service (since 2006). On the 10.3 km length of section RS3, the μ_{SRT} values ranged from 0.34 to 0.47, and the μ_{VF} ones from 0.35 to 0.76. Similar to the observations in Fig. 5, these values remained independent of the *SR* adopted in the VF device at a speed of $V = 60$ km/h. Consequently, the *R* values (ranging from 0.420 to 0.476) indicate a correlation between the μ_{VF} and μ_{SRT} measurements, albeit at a weak level. Concurrent with the increase in *R*, an approximately twofold increase in the slope of the linear approximation can be observed in Fig. 6 relative to the data in Fig. 5, which is more physically justifiable. However, the correlation remains too low to suggest any practical application of the $\mu_{SRT}(\mu_{VF})$ relationships presented in Fig. 6.

Given the low *R* values in Figs. 5 and 6, the correlation of measurements from all sections combined was investigated for late summer/early autumn, as illustrated in Fig. 7. Consequently, the results presented in Fig. 7 exhibit a broad range of variation, with μ_{SRT} ranging from 0.34 to 0.60 and μ_{VF} from 0.35 to 1.03, while maintaining the same σ levels as in Figs. 5 and 6. These data encompass friction coefficient changes for both recently commissioned pavements (black points for RS1–RS2) and highly worn surfaces (red points for RS3). This leads to an increase in the slope of the $\mu_{SRT}(\mu_{VF})$ function to approximately 0.3. Simultaneously, this results in high *R* values, ranging from 0.943 to 0.946, indicating a strong correlation between μ_{VF} and μ_{SRT}. These findings justify the practical application of the linear approximations for the measurement relationships between μ_{SRT} and μ_{VF} shown in Fig. 7. Furthermore, these relationships are not noticeably affected by the slip ratio (*SR*) levels adopted in the VF device at a speed of $V = 60$ km/h.

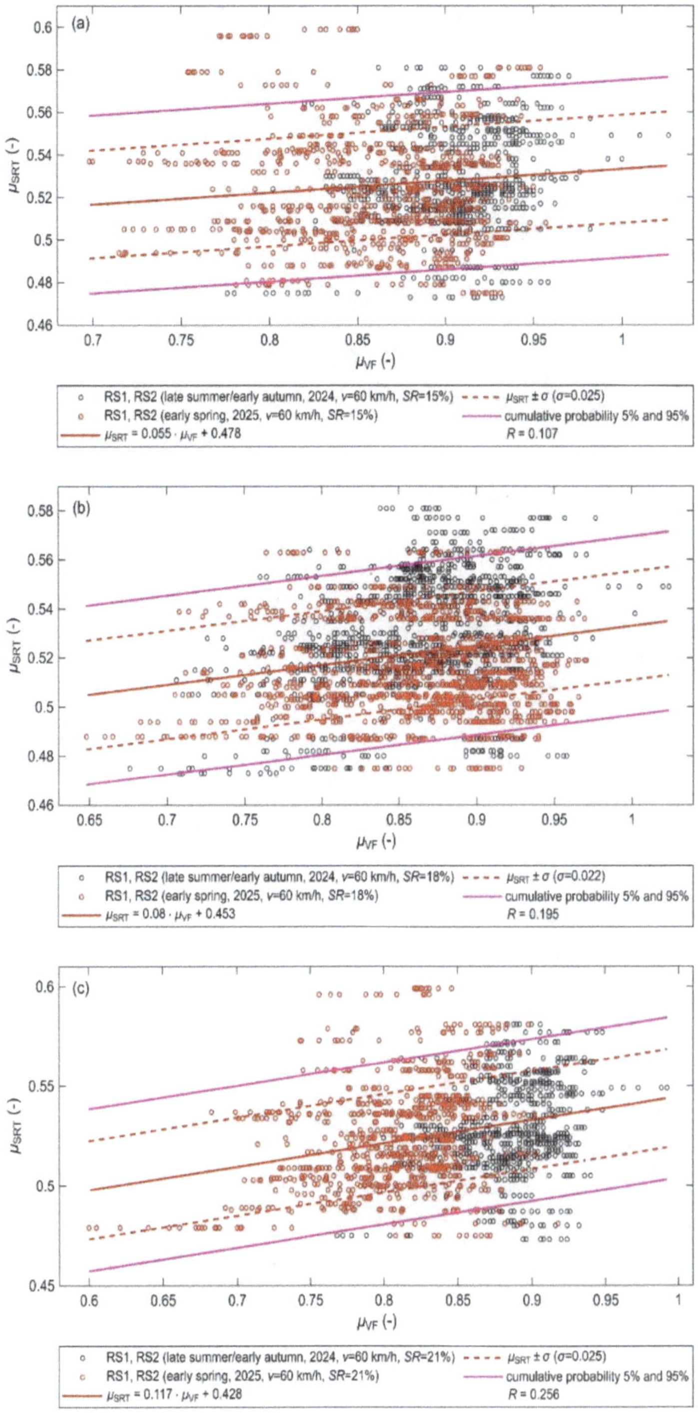

Fig. 5. Dependence between μ_{SRT} and μ_{VF} on RS1–RS2. a) $V = 60$ km/h, $SR = 15\%$; b) $V = 60$ km/h, $SR = 18\%$; c) $V = 60$ km/h, $SR = 21\%$.

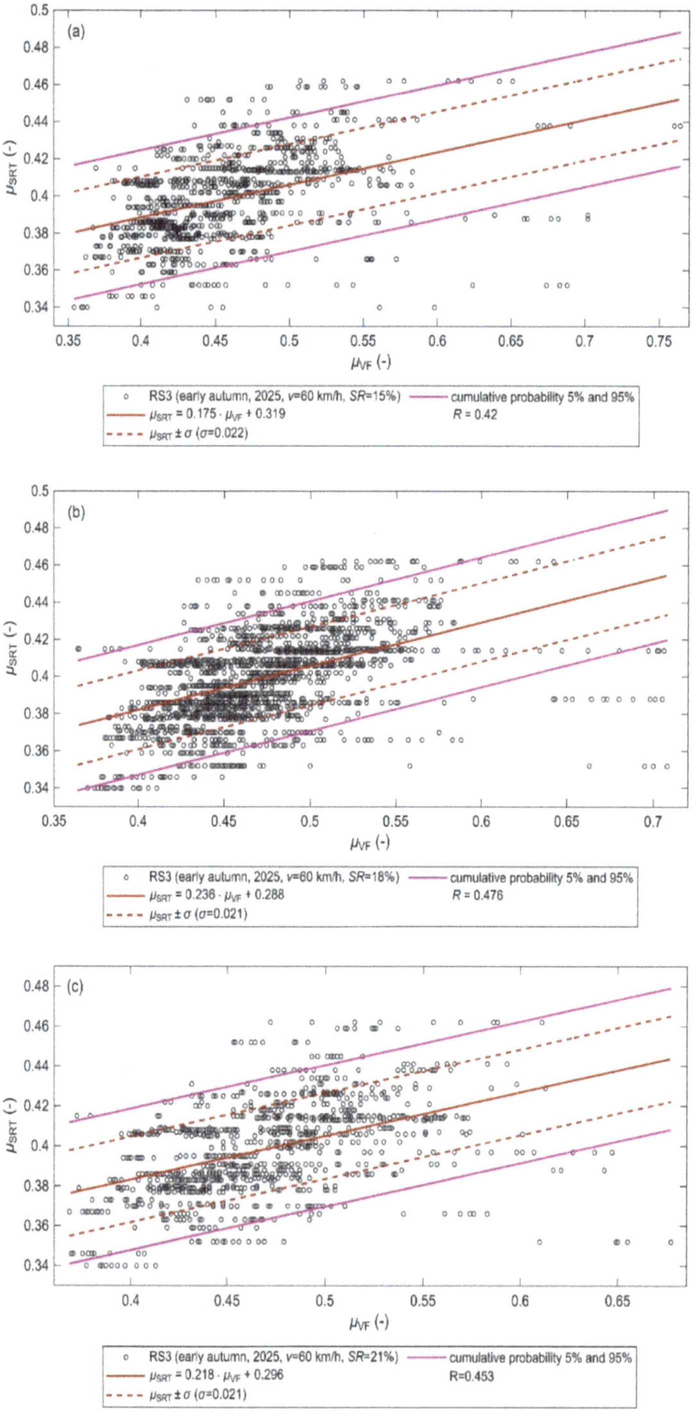

Fig. 6. Dependence between μ_{SRT} and μ_{VF} on RS3. a) $V = 60$ km/h, $SR = 15\%$; b) $V = 60$ km/h, $SR = 18\%$; c) $V = 60$ km/h, $SR = 21\%$.

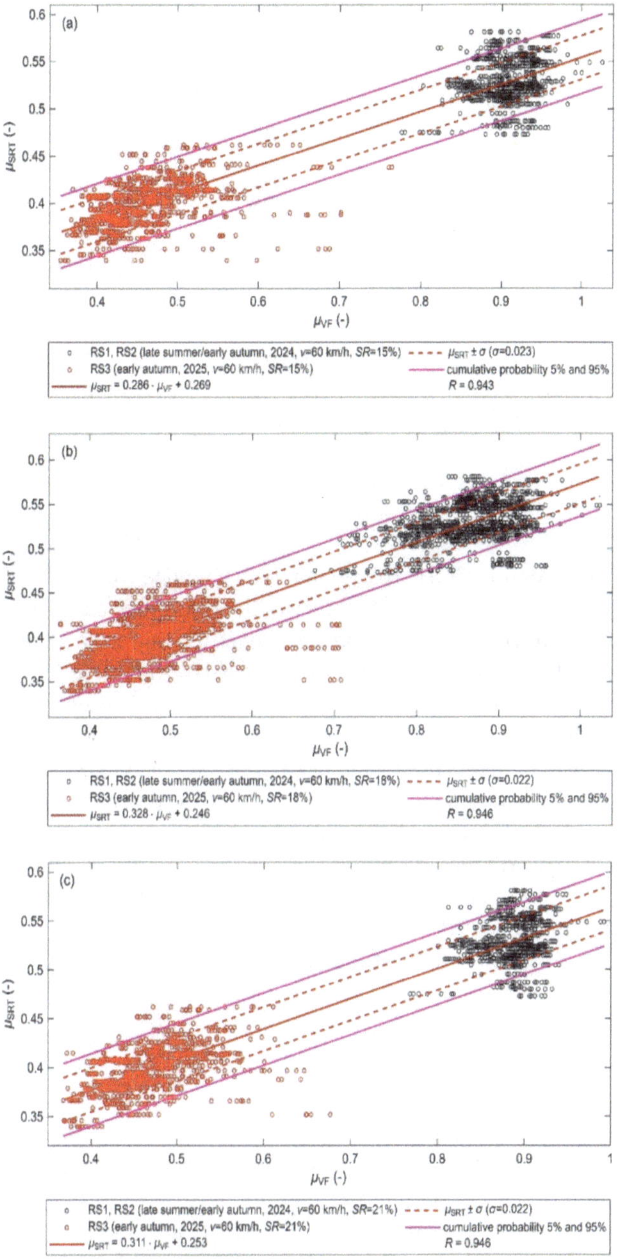

Fig. 7. Dependence between μ_{SRT} and μ_{VF} on RS1–RS3. a) $V = 60$ km/h, $SR = 15\%$; b) $V = 60$ km/h, $SR = 18\%$; c) $V = 60$ km/h, $SR = 21\%$.

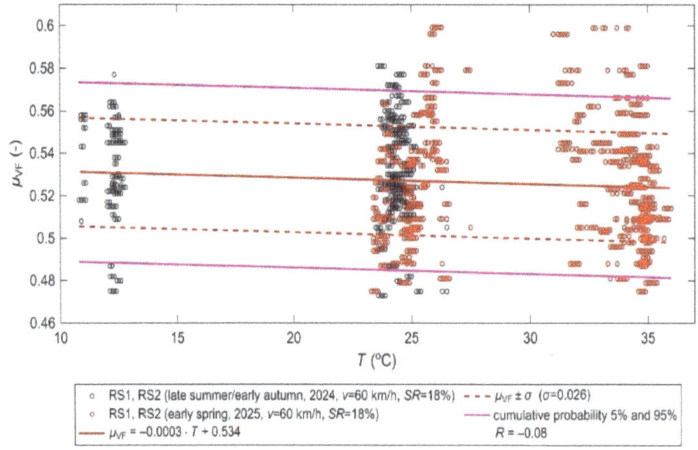

Fig. 8. Dependence between μ_{VF} and T on RS1–RS2; VF measurement: $V = 60$ km/h, $SR = 18\%$.

Measurements on sections RS1–RS2 using the VF device were conducted in different seasons at approximately a six-month interval (late summer/early autumn and early spring). This involved a wide range of road surface temperatures (T), ranging from 11 °C to 36 °C. Consequently, a correlation analysis was performed to determine whether the road surface temperature significantly influences the results presented in Figs. 5, 6 and 7. To develop the plot in Fig. 8, VF measurements at a speed of $V = 60$ km/h and a slip ratio of $SR = 18\%$ were used. The results in Fig. 8 were calculated analogously to Eqs. (2, 3 and 4), where the linear approximation $\mu_{\text{VF}}(T)$ is represented by a red solid line, $\mu_{\text{VF}}(T) \pm \sigma$ by red dashed lines, and $\mu_{\text{VF}(0.05)}(T)$ and $\mu_{\text{VF}(0.95)}(T)$ by purple solid lines. The figure legend also provides numerical values for $\mu_{\text{VF}}(T)$, σ, and R. The resulting near-zero slope for the $\mu_{\text{VF}}(T)$ approximation and a Pearson correlation coefficient of $R = -0.08$ indicate that, within the tested temperature range, there is no correlation between μ_{VF} and T. It means temperature does not affect the $\mu_{\text{SRT}}(\mu_{\text{VF}})$ relationship for CC road surfaces in the considered range.

4 Conclusions

A strong linear correlation was found between the friction coefficients of CC road surfaces measured using the SRT-3 and VF devices. This relationship was established within a temperature range of 11–36 °C, independent of the slip ratio (ranging from 15% to 21%) applied in the VF device. These findings provide a sound basis for the use of the VF device in Poland for both acceptance and diagnostic testing of skid resistance on CC pavements, in accordance with the strictly defined requirements for the SRT-3 device. Furthermore, it was demonstrated that road surface temperature within the 11–36 °C range has no significant effect on the friction coefficient measured by the VF device on CC surfaces. To further validate the presented conclusions, the authors plan to conduct additional testing on an alternative concrete pavement with friction coefficient values intermediate between those of sections RS1–RS2 and RS3.

References

1. Davies RD, Cenek PD, Henderson RJ (2005) The effect of skid resistance and texture on crash risk. In: International surface friction conference: roads and runways: improving safety through assessment and design, 1–4.05.2005, Christchurch, New Zealand
2. von Quintus H, Mergenmeier A (2015) Case Study: Kentucky Transportation cabinet's High Friction Surface Treatment and Field Installation Program, Report no. FHWA-SA-15-038, U.S. Department of Transportation, Federal Highway Administration. https://highways.dot.gov/sites/fhwa.dot.gov/files/2022-06/ky_hfst_15_038.pdf. Last accessed 20 Feb 2026
3. General Directorate for National Roads and Motorways (2024-2025) Model Conditions for the Execution and Acceptance of Construction Works, Warsaw, Poland. https://www.gov.pl/web/gddkia/wzorcowe-warunki-wykonania-i-odbioru-robot-budowlanych-wwiorb2. Last accessed 20 Feb 2026 (in Polish)
4. Working Group chaired by D. Sybilski (2013) Catalogue of Flexible and Semi-rigid Pavement Reconstruction and Repair, KPRNPP-2013, General Directorate for National Roads and Motorways & Road and Bridge Research Institute, Warsaw, Poland. https://www.archiwum.gddkia.gov.pl/userfiles/articles/p/prace-naukowo-badawcze-w-trakcie_3434/KPRNPP%20i%20Zalaczniki%202013.pdf. Last accessed 20 Feb 2026 (in Polish)
5. Elliot T, Carter A, Ghattuwar S, Levasseur A (2023) Environmental impacts of road pavement rehabilitation. Transp Res Part D: Transp Environ 118:103720. https://doi.org/10.1016/j.trd.2023.103720
6. CEN/TS 15901-14:2016 (2016) Road and airfield surface characteristics—Part 14: procedure for determining the skid resistance of a pavement surface using a device with longitudinal controlled slip (LFCN): ViaFriction (road analyser and recorder of ViaTech AS). European Committee for Standardization (CEN), Brussels, Belgium
7. Rado Z (2009) Evaluating Performance of Limestone Prone to Polishing, Pennsylvania Department of Transportation, Final Report. https://rosap.ntl.bts.gov/view/dot/17621. Last accessed 20 Feb 2026
8. Czmiel J, Perkowski Z, Frączek D, Jurczak R (2026) Road surface friction coefficient measured at different slip ratios—comparison of results from the ViaFriction and SRT-3 devices. Arch Civ Eng 3 (in press)
9. Wrocław University of Science and Technology (2025) General Directorate for National Roads and Motorways, The Road and Bridge Research Institute: Guidelines for determining the surface characteristics of carriageway pavements and other road components. Patterns and standards recommended by the Minister responsible for transport. WR-D-64. Ministry of Infrastructure, Warsaw, Poland. https://www.gov.pl/web/infrastruktura/wr-d. Last accessed 23 Feb 2026 (in Polish)
10. Glinicki MA (2019) Engineering of concrete road pavements. Wydawnictwo Naukowe PWN, Warsaw, Poland. (in Polish)
11. EN 197-1:2011 (2011) Cement—Part 1: composition, specifications and conformity criteria for common cements. European Committee for Standardization (CEN), Brussels, Belgium
12. EN 12620:2013 (2013) Aggregates for concrete. European Committee for Standardization (CEN), Brussels, Belgium
13. EN 934-1:2008 (2008) Admixtures for concrete, mortar and grout—Part 1: common requirements. European Committee for Standardization (CEN), Brussels, Belgium
14. EN 934-2:2009+A1:2012 (2012) Admixtures for concrete, mortal and grout—Part 2: concrete admixtures—Definitions, requirements, conformity, marking and labelling. European Committee for Standardization (CEN), Brussels, Belgium

15. EN 12350-7:2019 (2019) Testing fresh concrete—Part 7: air content—Pressure methods. European Committee for Standardization (CEN), Brussels, Belgium
16. EN 206-1:2000 (2000) Concrete—Part 1: specification, performance, production and conformity. European Committee for Standardization (CEN), Brussels, Belgium
17. Judycki J et al (2025) Catalogue of Typical Flexible and Semi-Rigid Pavement Structures. Patterns and Standards Recommended by the Minister Responsible for Transport. WR-D-61. Ministry of Infrastructure, Warsaw, Poland. https://www.gov.pl/web/infrastruktura/wr-d. Last accessed 02 June 2026 (in Polish)
18. EN 13036-1:2010 (2010) Road and airfield surface characteristics—Test methods—Part 1: measurements of pavement surface macrotexture depth using a volumetric patch technique. European Committee for Standardization (CEN), Brussels, Belgium
19. Szydło A, Mackiewicz P, Wardęga R, Krawczyk B (2014) Catalogue of typical rigid pavement structures. Wroclaw University of Science and Technology, Wrocław, Poland. (in Polish)
20. Michalski W (2016) Methods of texturing concrete pavements Drogownictwo, vol 4, pp 126–134. (in Polish)

Immediate and Final Reconstruction of Bridge Structures: Case Study of the Great Flood in Southern Poland in 2024

Mariusz Pustelnik(✉)

Pracownia Projektowa MOSTOPOL Sp. z o. o, Opole, Poland
mpustelnik@mostopol.pl

Abstract. The chapter presents both immediate and final actions for the reconstruction of bridges damaged during the September 2024 flood. The collaborative efforts of units at various administrative levels, the military, and private companies that led to the efficient reconstruction of bridge infrastructure are shown. Based on the example of the bridge over the Biała Głuchołaska River in Głuchołazy, Poland, along National Road No. 40, solutions to prevent future flood damage are presented. A commentary is provided on how to design and construct bridges built in flood zones to avoid their destruction, even in the event of high flood flows.

Keywords: Great Flood in Poland · bridge destruction · reconstruction of bridges · bridge damage · the Genoese low "Boris"

1 Introduction

The flood that occurred in Poland in September 2024 also affected many other countries: the Czech Republic, Serbia, Slovakia, Austria, Slovenia, Croatia, Romania, Moldova, Germany, and Italy. It was caused by heavy rainfall triggered by the Genoese low known as "Boris". Warm, humid air masses from the Mediterranean Sea, rising over the Alpine massif, cooled, leading to heavy rainfall [1]. Rainfall totals from September 12th to 15th reached extreme levels, with some weather stations recording values exceeding 500 mm per square meter [1]. The flood caused significant damage to transport infrastructure and property. A total of 26 people died, 9 of whom in Poland. This is the third catastrophic flood since the 1997 and 2010 floods. It should be noted that this is also the third time in 27 years that river flows have reached millennial levels. Bridge destruction due to flooding has occurred many times in history [3–8], and some regions experience floods particularly frequently [4, 7, 8]. The 2024 flood caused total or partial damage to almost 600 bridges, with 460 km of roads destroyed [6]. The chapter presents examples of the reconstruction of bridges after floods, both immediate and final, in the south of the Opole Voivodeship; in particular, the unusual solutions applied in the rebuilt bridge in Głuchołazy, intended to prevent future damage. Cooperation among the road administration, local administrations, private companies, and the army is also discussed.

Z. Perkowski et al. (Eds.): ECCE 2026, LNCE 929, pp. 161–168, 2027.
https://doi.org/10.1007/978-3-032-39515-3_12

2 Bridges Over the Biała Głuchołaska River

2.1 Bridge in Głuchołazy

At the time of the flood passage through Głuchołazy, the Civil Construction Company Himmel i Papesch Opole sp. z o. o. was carrying out a project in the city, implemented by the Opole branch of the General Directorate for National Roads and Motorways in Poland. The construction work involved the expansion of the city section of National Road No. 40, leading to the Czech Republic border. The road expansion was accompanied by the demolition of the old bridge and the construction of a new crossing in the same location. The new, three-span bridge was supported by two piers, founded directly within the sheet pile walls in the Biała Głuchołaska Riverbed. At the time of the flood, scaffolding and formwork were in place, and reinforcement work was underway on the structure. The final bridge was only a week away from completion before the concrete was poured and the scaffolding dismantled. A three-span, single-carriageway, two-lane temporary bridge was constructed for the duration of the construction work. The structure was set directly on the riverbed, using heavy rip-rap. Additionally, a structure was designed with spindle-shaped steel sections driven into the riverbed, clad with wooden beams to protect the bridge's steel supports (Fig. 1). These structures performed well during the flood (Fig. 2). However, the bridge collapsed on September 15th, failing to withstand the pressure of a 2,000-year flood, particularly the force of accumulated logs (originating, among other sources, from a sawmill located upstream) and floating debris (Figs. 3 and 4).

Fig. 1. Temporary bridge over the Biała Głuchołaska River. Condition before flooding.

Although both the permanent and temporary bridges were designed in accordance with applicable regulations, they failed to withstand the catastrophic 2024 flood. It should be noted, however, that a significant increase in floodwaters was largely due to the dam located near the bridges. Probably built for the former water mill (a branch of the Młynówka River lies upstream of the weir, but it is currently partially piped), it currently serves during periods of low water levels in the Biała Głuchołaska, maintaining biological

Fig. 2. Temporary bridge over the Biała Głuchołaska River. Condition during flooding.

life in the river. Unfortunately, during floods, it unnecessarily dams water, blocking the flow of floodwaters.

Fig. 3. Bridges after the disaster.

This situation has remained unchanged since the 1997 flood, when Głuchołazy suffered severe damage. It is hard to resist the impression that if the planned investments on the Biała Głuchołaska River had been implemented, including deepening the riverbed and converting the permanent weir into a flap weir, the flood would not have led to the collapse of 4 bridges and the destruction of the city. All the bridges mentioned above that were damaged were located above the weir. The single-span footbridge, designed by the chapter's author and located about 200 m below the weir and supported on the wells, survived (Figs. 5 and 6).

The flood destroyed the right-bank embankments at the watershed, simultaneously damaging the reinforcement at the abutments (Figs. 6 and 7). However, it did not damage the footbridge's superstructure. Designed in 2021–2022, in accordance with Eurocode, the superstructure accommodated emergency passage for an ambulance (12 tons), which was the only public crossing in the city in the initial days after the flood. 3 weeks after the disaster, the area around the damaged structures had been cleared, and new abutments in the form of "Cofferdamm" sheet pile boxes were designed and constructed for the single-span DMS-65 military bridge (Fig. 8). Parallel to the sheet pile installation and earthworks to reconstruct the access roads, work was underway on the superstructure of the two bridges. The two separate superstructures, each with a single lane, were designed to accommodate vehicles weighing 42 tons. After being joined on one bank, the bridges were lifted into their final location using lifting equipment. It should be noted, however, that the theoretical span of these structures, based on the given assumptions, is 35 m, and the resulting clear width is approximately 30 m, less than the previous temporary bridge. However, the structures have been appropriately elevated.

Fig. 4. Damaged pillar of the target bridge.

After the events of September 2024, the investor decided to rebuild the new bridge as a single-span structure that would be resistant to a future flood wave of similar strength. The single-span bridge, 52 m long and 19 m wide, was designed. The dynamic water pressure, as well as the possibility of impact from 10-ton tree limbs carried by water at a speed of 10 m/s, were considered. The newly designed reinforced concrete bridge is shown in Fig. 9.

2.2 Bridges in Skorogoszcz and Krapkowice

The reinforcement was constructed using openwork "Jombo" concrete elements. As the floodwaters flowed, backfill was washed out behind the abutment (Fig. 10). To restore

Fig. 5. The footbridge in Głuchołazy before the floodwaters cross.

Fig. 6. The footbridge in Głuchołazy after floodwaters passed through.

Fig. 7. The footbridge in Głuchołazy after floodwaters passed through.

Fig. 8. Reconstruction of the new temporary bridge.

Fig. 9. New single-span permanent bridge.

traffic for emergency services and the local military forces, the temporary bridge "Daglezja" (Fig. 11) was installed. Similar damage has been observed on many other structures, for example, on the bridge along National Road No. 45 in Krapkowice, where embankment cones and the area behind the abutment backfill were washed away. It seems that in the future, particular emphasis should be placed on the proper selection of reinforcements, adapted to the hydrological function of this bridge element. In addition to the bridges themselves, sections of roads have also been damaged. In these cases, the cause was often the improper selection or lack of construction materials that would resist the force of floodwaters flowing through the "at-risk" areas. Again, a more comprehensive analysis of the planned roads, including hydrological considerations, would help prevent similar damage in the future.

Fig. 10. Washed-away backfill behind the abutment.

Fig. 11. Temporary military bridge "Daglezja".

3 Summary

The disaster that occurred in September 2024, which caused the third tragic flood in Poland in the last 27 years, demonstrates that we should expect similar events in the future. Climate change has once again made its presence. We should hope that this time, as a society and the State, we will learn the lessons from this tragedy. Further investments in hydraulic structures are necessary. However, some conclusions can be implemented immediately.

When designing structures, especially on mountain rivers, supporting bridge foundations should be erected as intermediate foundations; direct foundations should be strictly avoided. If such options are unavailable, it seems necessary to base support on piles and take bed erosion into account. As with lowland rivers, where the pressure of ice floes is taken into account, in the case of mountain rivers, the impact of limbs carried by the river current on the superstructure should also be taken into account, as well as the additional pressure on the structure due to the accumulation of drifts in front of the superstructure, when water levels are exceeded. However, there are no appropriate guidelines or standard solutions. Single-span structures should be used on mountain rivers, although even the largest temporary bridges have limited spans. Bridge design should be supported by analysis not only of the structure itself but also of the entire hydrological system. In the case of the bridge along the National Road No. 40 in Głuchołazy, reconstruction should include not only the bridge itself but also the dam, which constantly dams the water. However, the problem lies in the lack of close cooperation and commitment between administrators to achieve a common goal. There is much room for improvement in this area. Water, road, and railway managers often do not cooperate closely, which is why many design tasks are not addressed comprehensively. A significant problem is the lack of joint financing for such investments.

One cause of bridge damage due to flooding is the erosion of embankment cone reinforcements. In the author's opinion, these reinforcements should be designed in collaboration with hydraulic engineers, who possess the appropriate expertise. To properly select the arrangements, a detailed analysis of water flow in the vicinity of abutments or bridge piers is often required.

References

1. https://pl.wikipedia.org/wiki/Powódź_w_Europie_Środkowej_ (2024). Last accessed 16 June 2026
2. https://opolska360.pl/nowy-most-w-glucholazach-bedzie-zmiana-projektu. Last accessed 16 June 2026
3. Scheer J (2011) Failed bridges: case studies, causes and consequences (trans: Wilharm L). Ernst & Sohn (Wiley)
4. Burghardt L, Klopries E-M, Schüttrumpf H (2025) Structural damage, clogging, collapsing: analysis of the bridge damage at the rivers Ahr, Inde and Vicht caused by the flood of 2021. J Flood Risk Manag 18:e13001
5. Pejović J, Serdar N, Pejović R (2022) Damage assessment of road bridges caused by extreme streamflow in Montenegro: reconstruction and structural upgrading. Buildings 12(6):810
6. Bohatkiewicz J, Świtała M, Królikowski R, Biskup M (2025) The aftermath of a flood: assessing its impact on roads and bridges with insights from the road and bridge research institute. Roads Bridges – Drogi i Mosty 24:127–144
7. Holemba GA, Matsumoto T (2019) Flood-induced bridge failures in Papua New Guinea. In: International conference on sustainable civil engineering structures and construction materials (SCESCM 2018), MATEC web of conferences, vol 258, p 03014
8. Shahbaz Shah SA (2021) Flooding impacts on the bridges – an overview. In: 3rd conference on sustainability engineering in civil engineering (CSCE'21). Department of Civil Engineering, Capital University of Science and Technology, Islamabad, Pakistan

Enhancing the Load-Bearing Capacity of a Suspended Roof System for an Exhibition Pavilion

Oleksandr Shymanovskyi[1] (✉) and Wiesław Baran[2]

[1] V. Shymanovskyi Ukrainian Institute of Steel Construction, V. Shymanovskyi St. 2/1, Kyiv 02125, Ukraine
{niipsk,ashim}@urdisc.com.ua

[2] Opole University of Technology, Faculty of Civil Engineering and Architecture, Katowicka 48, 45-061 Opole, Poland
w.baran@po.edu.pl

Abstract. The paper describes a study of the stress-strain state of a suspended long-span hybrid roof system for an exhibition pavilion, whose geometry is defined as a hyperbolic paraboloid. The primary load-bearing elements of the suspended roof structure are finite-stiffness cables supported by a closed, elastically compliant, concrete-filled steel tubular support frame, over which profiled decking is laid. The proposed roof structure fully meets the requirements of modern fabrication and installation technologies for suspended systems.

The development of the design model and topological layout of the roof system is presented. A verification of the load-bearing capacity of the roof structure was performed, and the results showed that the initial configuration does not meet regulatory requirements for strength and deformability and requires refinement. To this end, it is proposed, first, to introduce an additional vertical elastically compliant connection to reduce horizontal displacements along the longitudinal (cable-parallel) edges of the support frame contour, and second, to extend the configuration of the concrete-filled steel tubular support frame in the horizontal plane to enhance its stiffness.

Taking these recommendations into account, a revised structural analysis of the roof structure was carried out. The results demonstrate that their implementation ensures high load-bearing capacity and overall stiffness of the roof structure.

Keywords: Suspended Roof System · Design Model · Topological Layout · Ruled Constructions · Load-Bearing Capacity · Stiffness

1 Introduction

Shell structures in which the generating lines are straight lines have been described in the literature [1–3]. An example is the hyperbolic paraboloid, a shell often used as a roof covering, whose analytical solutions [4–6] can also be applied to rod structures of similar shape. Reichhart's work [7] discusses the principles of geometric and structural shaping of sheet metal shells.

Z. Perkowski et al. (Eds.): ECCE 2026, LNCE 929, pp. 169–178, 2027.
https://doi.org/10.1007/978-3-032-39515-3_13

This paper investigates the stress state of the suspended long-span hybrid roof for an exhibition pavilion. The roof geometry is defined as a hyperbolic paraboloid with a vertical elevation difference between the contour corners of 20 m, a spacing between column axes in plan of 94.34 m, and diagonal lengths of 100 m and 160 m (Fig. 1) [1]. The primary load-bearing elements of the suspended roof structure are finite-stiffness cables, spaced at 3.776 m intervals, supported by a closed, elastically compliant concrete-filled steel tubular support frame and fixed in their design position using channel-section struts, size 10. Profiled decking is laid over the cables, serving as the base for the flexible roofing membrane. The proposed roof structure ensures maximum reduction of material consumption and fully meets the requirements of modern fabrication and installation technologies for suspended systems [8].

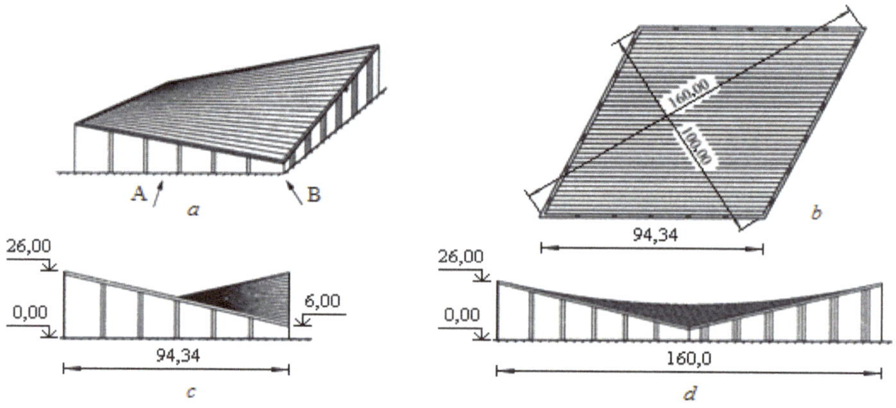

Fig. 1. Structural layout of the roof system: *a* – general view; *b* – plan; *c* – view A; *d* – view B.

Figure 2 presents the cross-sections of the primary load-bearing elements of the roof system – finite-stiffness cables made of channel sections, size 40, and the concrete-filled steel tubular support frame, consisting of a steel tube Ø1200 × 14 mm filled with concrete and containing a reinforcement cage of 12 bars Ø28, grade AIII. The finite-stiffness cables and the tube of the support frame are made of steel grade C46/33 with a yield strength of $\sigma_{y.s.} = 330$ MPa, while the concrete of the support frame is adopted as of class B45 with an ultimate compressive strength of $\sigma_{c.s.} = 60$ MPa.

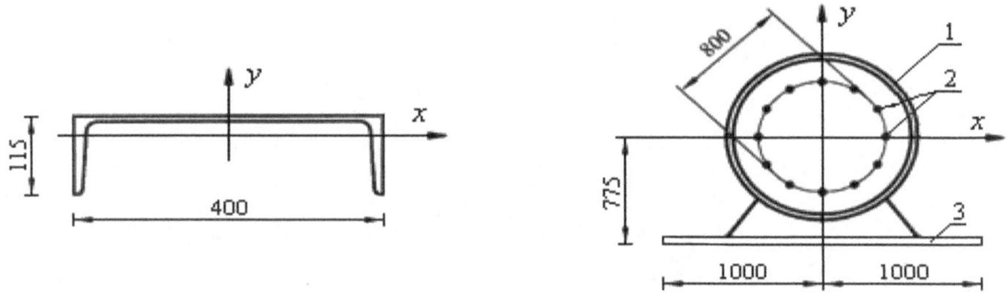

a *b*

Fig. 2. Cross-section: *a* – finite-stiffness cable; *b* – concrete-filled steel tubular support frame; 1 – steel tube Ø1200 × 14; 2 – reinforcement bars Ø28 AIII; 3 – plate 2000 × 25.

2 Design Model and Topological Layout of the Roof System

Figure 3 shows the design model of the roof system, in which the primary load-bearing elements are represented by bar-type finite elements with rectangular cross-sections, capable of resisting axial tension and compression forces as well as bending forces. The parameters of the finite elements (cross-section dimensions and equivalent modulus of elasticity) were determined taking into account the equivalent "smearing" of the bar-type elements of the roof structure, so as to ensure equivalence between the stiffness characteristics of the modeled and the actual structural components.

When describing the roof system's topological structure, its design model is mapped onto a rectangular mesh domain. Local numbers of nodes are introduced: from 1 to $M_1 = 2$, from 1 to $M_2 = 43$, and from 1 to $M_3 = 49$, as well as global numbers – from 1 to $M = M_1 \cdot M_2 \cdot M_3$. The node numbering order is shown in Fig. 3.

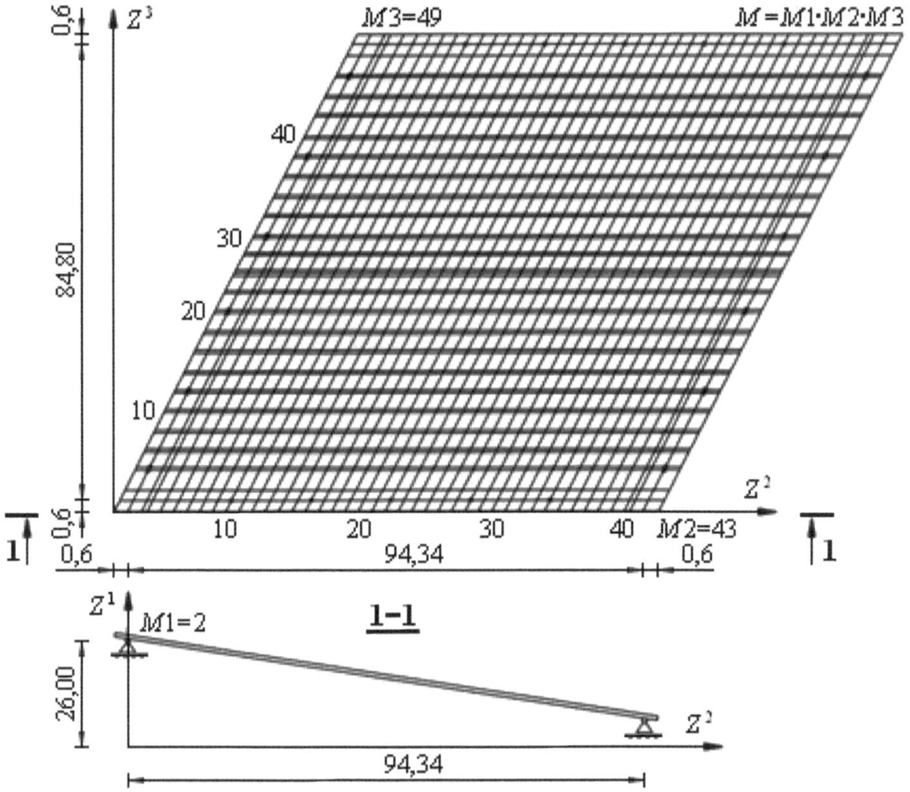

Fig. 3. Design model of the roof system with numbered coordinate lines in the mesh domain.

3 Loads and Actions on the Roof System

As an external action on the roof system, a uniformly distributed load over its surface is adopted, accounting for the self-weight of the structures together with the loads from the lightweight suspended ceiling and electrical equipment, $q_{ex.ac.} = 2.63$ kN/m^2, as well as the maximum snow load $q_s = 1.57$ kN/m^2.

4 Results of the Roof System Calculation

Based on the calculation results, diagrams of displacements, axial forces, and bending moments were plotted for the central finite-stiffness cable (Fig. 4) and for the lateral side of the tubular support frame, perpendicular to the cables (Fig. 5). The analysis of the obtained results revealed that the deformation curve of the cable has three inflection points, while the maximum deflections are quite significant, reaching the value $u^1 =$ 9.689 m. The diagrams of axial forces $N^{\ 2}$ and bending moments $M^{\ 3}$ exhibit a smooth distribution in the central part, whereas at the edges, sign changes and considerable absolute values of internal forces are observed.

A verification of the load-bearing capacity of the central finite-stiffness cable was carried out in the most stressed section 1–1 (Fig. 4), where the values of the tensile force and bending moment are $N^{2} = 4.75 \cdot 10^2$ kN and $M^{3} = -78.5$ kNm, respectively. The maximum stress in the cable corresponding to these force factors is $\sigma_{max} = 413.5$ MPa, which exceeds the allowable value.

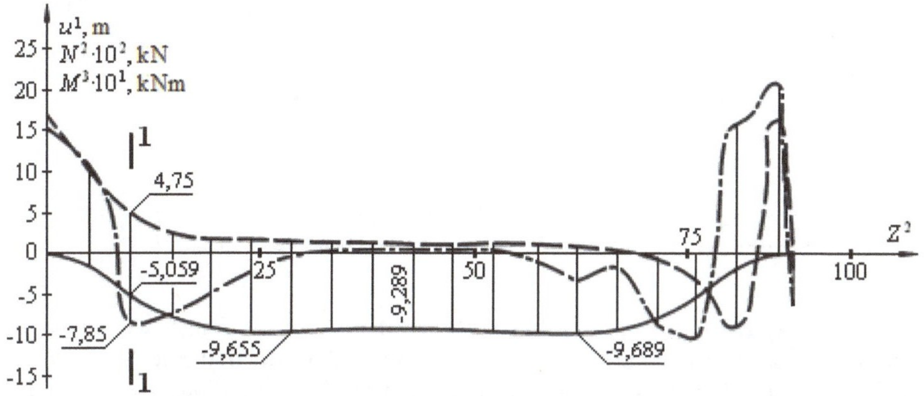

Fig. 4. Diagrams for the central finite-stiffness cable: ——— – displacements u^1; - - - - - - – axial forces N^{2}; ——— – ——— – bending moments M^{3}.

According to the diagrams obtained for internal forces in the support frame (Fig. 5), the most stressed section is 2–2, where the following forces act: $N^{3} = -3.604 \cdot 10^3$ kN, $M^{1} = -3.089 \cdot 10^4$ kNm, and $M^{2} = -0.385 \cdot 10^3$ kNm. Since the stresses in the support frame cross-section are redistributed between the steel tube and the reinforced concrete in proportion to their stiffness values, the maximum stresses in them are $\sigma_{max}^{tub} = -281$ MPa and $\sigma_{max}^{rfc} = -59$ MPa. The latter indicates that the normative requirements for the strength of the support frame are satisfied, whereas those for its deformability are not, since = 2.727/94.34 = 1/35 > 1/150. Thus, the analysis of the results obtained has shown that the adopted design of the main load-bearing elements of the roof system does not meet the normative requirements for strength and deformability [9] and therefore requires refinement.

Therefore, in order to reduce the horizontal displacements of the longitudinal (cable-parallel) sides of the support frame, an elastically compliant connection is applied in the direction Z^{3} with the unit stiffness $R^{3} = 1260$ kN (the location of which is shown by the red circle in Fig. 6). A new variant of the finite-stiffness cable cross section is also adopted, comprising channel section, size 40, a steel flat bar 540 × 20 and two angles 50 × 5 (Fig. 7a). In order to increase the stiffness of the concrete-filled steel tubular support frame, its configuration was developed in the horizontal plane and includes two concrete-filled tubes Ø1620 × 18 with a reinforcement cage of 24 bars Ø32, grade AIII in each. Their joint action is ensured by two tubes Ø600 × 10 and Ø660 × 10, a base plate 4400 × 40, a top plate 2000 × 20, and two longitudinal smooth bars Ø32 mm (Fig. 7b).

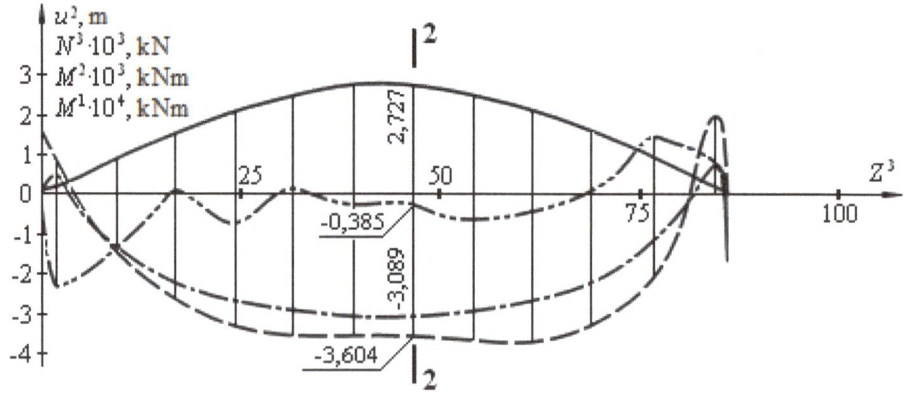

Fig. 5. Diagrams for the support frame lateral side: ——— – displacements u^2; - - - - axial forces N^3; — — — — – bending moments M^2; —— – —— – bending moments M^1.

Taking into account the new cross sections of the support frame and the finite-stiffness cables, a recalculation of the roof system was performed, the design model of which is shown in Fig. 6. Based on the calculation results, the ordinates for the initial and deformed states of the roof system were determined (Figs. 8 and 9), as well as the deflections, axial forces, and bending moments in the central finite-stiffness cable (Fig. 10) and in the lateral side of the support frame (Fig. 11).

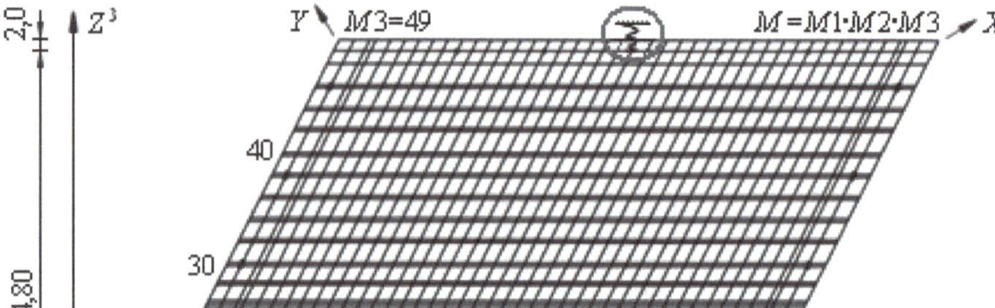

Fig. 6. Design model of the suspended roof system with numbering of the coordinate lines in the mesh domain and the additionally introduced elastically compliant connection.

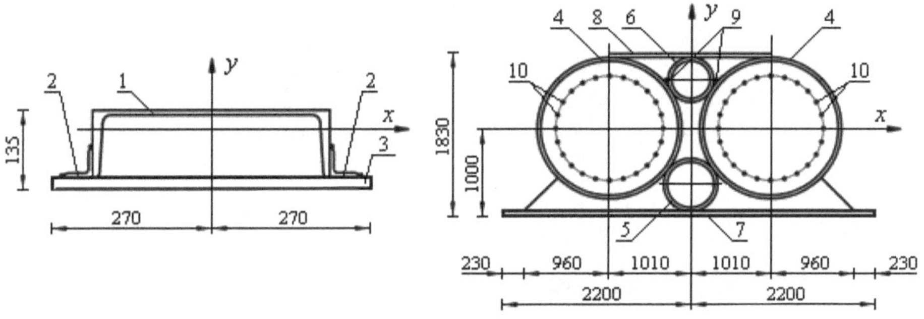

a *b*

Fig. 7. New cross section: *a* – finite-stiffness cable; *b* – concrete-filled steel tubular support frame; 1 – channel section, size 40; 2 – angle 50 × 5; 3 – plate 540 × 20; 4 – tube Ø1620 × 18; 5 – tube Ø660 × 10; 6 – tube Ø600 × 10; 7 – base plate 4400 × 40; 8 – plate 2000 × 20; 9 – bar Ø32; 10 – reinforcement Ø32 AIII.

The analysis of the roof system calculation results indicates that the displacements of the support frame in the horizontal plane are smaller than the permissible normative parameters $u^2_{max}/l_{frame} = 0.474/94.34 = 1/200 < 1/150$, and the deformed surface of the roof in the section along the secondary diagonal exhibits a depression with an amplitude of 0.21 m over a length of 50 m, causing a reverse slope of the roof surface with a value of $i = -\ 0.0084$. Therefore, to meet the structural requirements for a positive slope $i = +0.005$, a lightweight surface structure with a maximum amplitude of $h = 0.35$–0.40 m must be installed under the roof.

Based on the analysis of the data presented in Figs. 10 and 11, the most stressed cross sections were identified in the central finite-stiffness cable (section 3–3) and in the support frame lateral side (section 4–4).

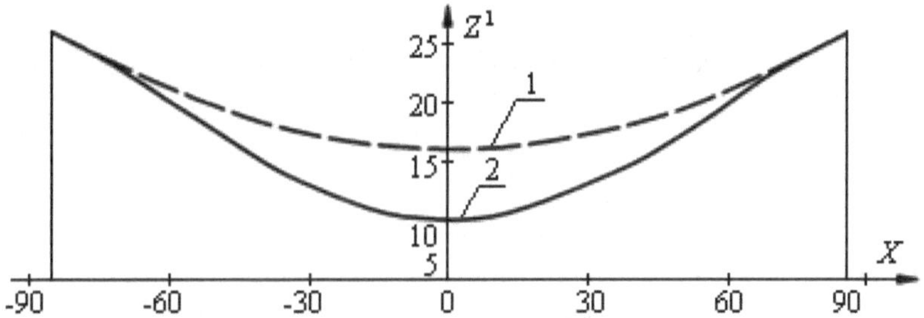

Fig. 8. Main diagonal states of the roof system: 1 – initial; 2 – deformed.

For the central finite-stiffness cable in section 3–3 (Fig. 10), the forces obtained are $N^2 = 2.888 \cdot 10^3$ kN and $M^3 = -\ 0.466 \cdot 10^2$ kNm, which produce a maximum stress of $\sigma_{max} = 256.7$ MPa. At the same time, in the edge cables at the middle (most stressed) cross-sections, the forces are $N^2 = 4.170 \cdot 10^3$ kN and $M^3 = -\ 0.419 \cdot 10^2$ kNm, resulting in the maximum stress of $\sigma_{max} = 318{,}6$ MPa, which does not exceed the limiting values

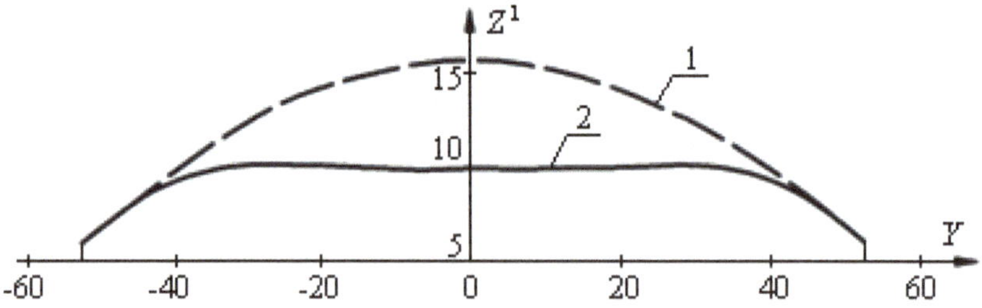

Fig. 9. Secondary diagonal states of the roof system: 1 – initial; 2 – deformed.

for steel grade C46/33. It should be noted that in the finite-stiffness cables, the stresses due to bending account for 25–30% of the total stress values.

The maximum internal forces in the support frame are observed in section 4–4 (Fig. 11) and amount to $N^3 = -1.8047 \cdot 10^4$ kN, $M^1 = -2.1996 \cdot 10^5$ kNm, and $M^2 = -0.74 \cdot 10^3$ kNm. Taking into account the redistribution of stresses between the steel tube and the concrete in the support frame in proportion to their stiffness values, the maximum stresses in them are $\sigma_{max}^{tub} = -195.7$ MPa and $\sigma_{max}^{rfc} = -41.5$ MPa, respectively, which do not exceed the limiting stress values for the adopted concrete of class B45 and steel grade C46/33.

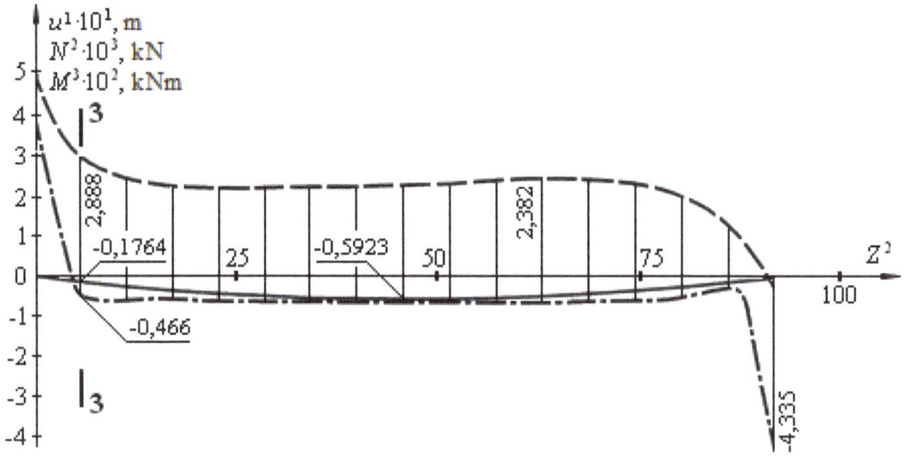

Fig. 10. Diagrams for the central finite-stiffness cable: ——— – displacements u^1; - - - - – axial forces N^2; —— - —— – bending moments M^3.

The calculation performed to determine the design of the elastically compliant connection has shown that its cross-section should consist of six 40-size channel sections and a 1400 × 32 mm steel flat bar (Fig. 12).

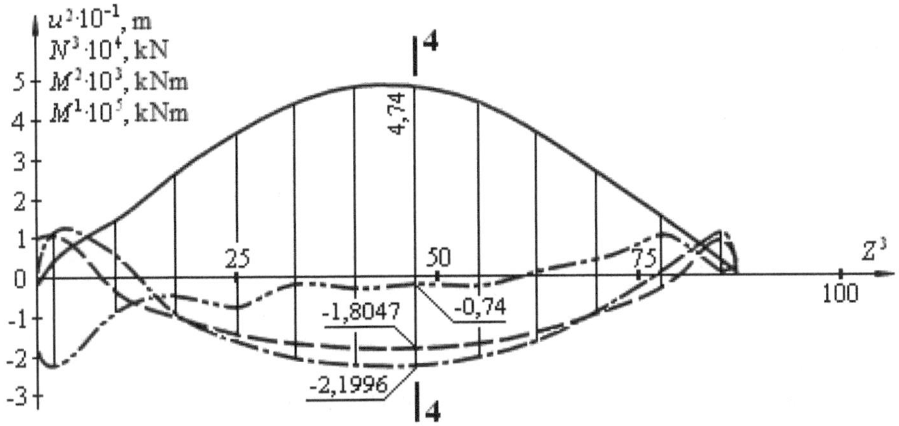

Fig. 11. Diagrams for the support frame lateral side: ——— – displacements u^2; – – – – axial forces N^3; ——– – – —— – bending moments M^2; ——– – —— – bending moments M^1.

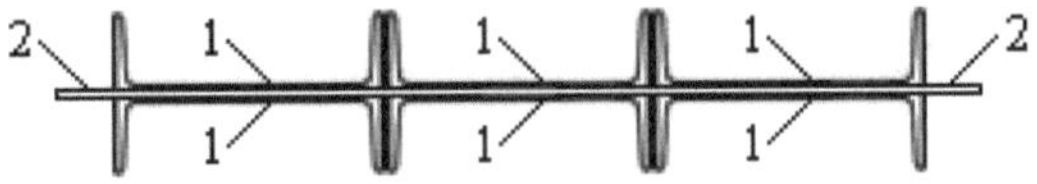

Fig. 12. Cross-section of the elastically compliant connection: 1 – channel section, size 40; 2 – flat bar 1400 × 32.

5 Conclusions

Based on the analysis of the calculation results, it can be concluded that the introduction of additional elastically compliant connections to the longitudinal sides of the support frame, as well as the use of refined designs of the support frame and the finite-stiffness cables, ensures high load-bearing capacity and stiffness of the considered long-span suspended roof system. At the same time, the forces and displacements arising in them under imposed loads remain within permissible limits.

Disclosure of Interests. The authors have no competing interests to declare that are relevant to the content of this article.

References

1. Shymanovskyi O, Shalinskyi V, Chaban N (2013) Механика висячих систем [mechanics of suspended systems]. Stal, Kyiv
2. Baran W, Jędraszak B (2007) Solution of the hyperboloid shell in the ruled parameterization. In: 53rd scientific conference of the committee of civil and water engineering of the polish academy of sciences and the committee of science of the polish association of engineers and technicians, Krynica 2007, conference materials, vol 2. Białystok University of Technology Publishing House, Białystok, pp 431–438

3. Bielak S (1976) Rectified shells. Scientific Papers of the Higher School of Engineering in Opole, Construction series 3:19/1976, Opole
4. Bielak S (1995) Nonlinear theory of shells: Part II. Higher School of Engineering in Opole, Studies and monographs 83, Opole
5. Flugge W (1950) Das Relaxationsverfahren in der Schalenstatik. In: Federhofer-Girkmann-Festschrift, Wien
6. Girkmann K (1957) Surface girders. introduction to the elastostatics of discs, plates, shells and sheet metals. Arkady, Warsaw
7. Reichhart A (2002) Geometrical and structural shaping of corrugated sheet metal shells. Publishing House of Rzeszów University of Technology, Rzeszów
8. (1997) Справочник, Сварные строительные конструкции, том 2, Типы конструкций [Handbook, Welded Building Structures, vol. 2, Types of Structures]. E.O. Paton Institute of Electric Welding, Kyiv
9. ДСТУ Б В.1.2-3:2006 (2006) Прогини і переміщення. Вимоги проектування [DSTU B V.1.2-3:2006, deflections and displacements—Design requirements]. Ministry of Construction of Ukraine, Kyiv

Diagnostics and Preservation of Built Heritage

New Possibilities for Diagnosing the Structure of Historic Buildings Using Ultrasound

Joanna Drobiec[1(✉)] and Łukasz Drobiec[2]

[1] Institute of Civil Engineering, University of Zielona Góra, Zielona Góra, Poland
j.drobiec@ib.uz.zgora.pl

[2] Faculty of Civil Engineering, Silesian University of Technology, Gliwice, Poland

Abstract. Wood and brick are among the most common materials found in historic buildings. Their computational analyses require determining the strength of these materials. Destructive testing undoubtedly provides the most reliable results, but sampling historic buildings is often difficult and, in many cases, impossible. Determining the strength of materials can be accomplished using non-destructive methods. However, this requires both destructive and non-destructive testing to correlate material strength with the non-destructively measured parameter. The article describes the potential applications of ultrasonic methods to determine the strength of wood and brick. Two devices emitting transverse and longitudinal ultrasonic waves were used in the tests. The study was conducted on nearly 64 wood beams and over 125 bricks obtained from historic buildings throughout Poland. After non-destructive testing, the elements were destructively tested: wooden beams were tested for bending to determine the wood class, and bricks were tested for compression to determine the strength of the masonry elements.

Keywords: Historic Buildings · Wood · Masonry Units · Non-Destructive Testing · Ultrasonic Methods

1 Introduction

Historic buildings can be constructed in various ways, but they often feature brick and wood. When conducting computational analyses of historic buildings, it is essential to understand their mechanical parameters, including material strength [1, 5, 9, 23]. Inadequate knowledge may lead to serious consequences, even leading to damage to the monument [18, 19]. The simplest way to determine a material's strength is to extract samples from the structure and test them in a testing machine. Unfortunately, in historic buildings, this procedure is usually not possible due to conservation restrictions [1, 7, 16]. Therefore, non-destructive testing is increasingly used to determine the material strength of historical objects [4, 15, 17, 21, 22]. However, to determine the strength of the tested material, it is necessary to first obtain samples, perform non-destructive testing, then destructive testing, and finally determine the correlation between the non-destructive parameter being tested and the strength of the tested material. Therefore, the biggest challenge is obtaining sufficient historical material for both non-destructive and

Z. Perkowski et al. (Eds.): ECCE 2026, LNCE 929, pp. 181–192, 2027.
https://doi.org/10.1007/978-3-032-39515-3_14

destructive testing. Such material can be sourced from buildings undergoing renovation or demolition. This usually requires permission from the relevant Office for the Protection of Historical Monuments.

Non-destructive sclerometric or ultrasonic testing can be used to determine the strength of a material [2, 3, 6]. This paper describes the results of ultrasonic and destructive testing. 64 wooden beams and 125 ceramic bricks recovered from historic buildings were examined. Firstly, non-destructive testing was performed, followed by destructive testing, and finally, formulas describing the material strength as a function of ultrasonic wave velocity were developed. These formulas can be used to determine the strength of wood and brick in existing historic buildings.

2 Devices Used in Research

Two ultrasonic devices were used in the study: the UK 1401 Surfer (Fig. 1a) and the UK 1410 Pulsar (Fig. 1b). The former emits ultrasonic transverse waves and measures their flow velocity in the tested material, while the latter emits longitudinal waves and also measures their velocity. Both devices are equipped with exponential ultrasonic probes, so no additional acoustic coupling is required. The UK 1401 Surfer has a fixed probe spacing (15 cm). The UK 1410 Pulsar has an arbitrary probe spacing, and the device requires the probe spacing value to be entered before measurement. In both devices, the frequency of ultrasonic waves is 50 kHz.

The UK 1401 Surfer test is linear and near-surface, and the device emits and analyzes the ultrasonic transverse wave (Fig. 2a), while the UK 1410 Pulsar test is also linear, but covers the entire thickness of the element and the device emits and analyzes the velocity of the ultrasonic longitudinal wave (Fig. 2b). Wood is an anisotropic material and is often subject to biological corrosion in historic buildings [8, 14, 20, 24]. When built into a structure, it is usually accessible from two sides; therefore, the decision was made to test wooden elements with two devices. Ceramic bricks usually have a more uniform structure than wood, and in walls, access to the brick surface is usually from one side. Therefore, the bricks were only tested with the UK 1401 Surfer device. The tests were performed on the brick faces. The transverse wave tests made using the UK 1401 Surfer apply more to near-surface areas (provided that effectively measurable transverse waves reach the receiving head before the significantly higher-amplitude Reyleigh waves).

3 Testing of Wooden Elements

3.1 Research Material

For the study, 75 historic wooden beam elements were obtained from Poland. The elements dated back to the last 500 years: the oldest from the early 15th century, and the youngest from the first half of the 20th century. 64 coniferous beams (spruce, pine, and fir) were selected for the study. The elements came from roof trusses, ceilings, and timber-framed walls. A view of the elements is shown in Fig. 3. Both destructive and non-destructive tests were conducted in an air-dry state, with the wood's moisture content of 12 ± 3%, determined by the electrical resistance method.

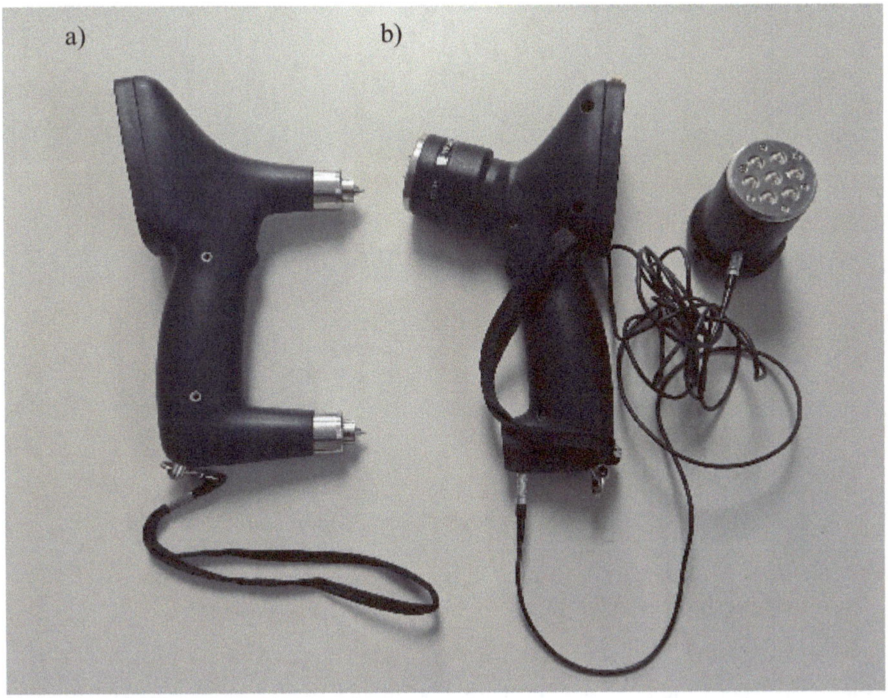

Fig. 1. Devices used in the research: a) UK 1401 Surfer, b) UK 1410 Pulsar.

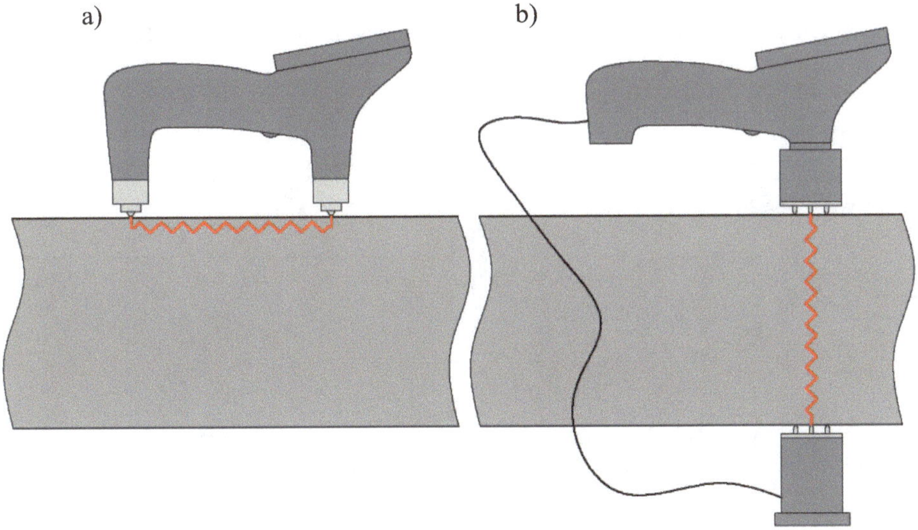

Fig. 2. Research area in devices used in tests: a) UK 1401 Surfer, b) UK 1410 Pulsar.

Fig. 3. View of the beams obtained for research.

3.2 Non-destructive Tests

Wooden beams were tested using UK 1401 Surfer and UK 1410 Pulsar ultrasonic devices. The elements were tested on two perpendicular planes, with 5 evenly spaced measurement points along the length of the beam on each plane. The measurement point spacing was L/4–30 cm (the extreme measurement points were located 15 cm from the edge of the element), where L denotes the length of the element being tested. A total of 640 measurement points were tested. The measurement results were averaged for each beam. A view of the devices during testing is shown in Fig. 4.

a) b)

Fig. 4. Testing of wooden beams with devices: a) UK 1401 Surfer, b) UK 1410 Pulsar.

3.3 Destructive Tests

After completing non-destructive testing, each beam was subjected to destructive testing at a specially prepared test stand. The beams were tested for bending, as the aim of the study was to determine the strength of the wood. Wood class depends on the characteristic bending strength of the wood. The tests were conducted in accordance with EN 380 [10], EN 384 [11], and EN 408 [12]. The tests were conducted in a single cycle until the beam failed. The load was applied using a 196 kN actuator attached to a 1000 kN steel frame anchored to the laboratory's strong floor. During the tests, force was measured using a 200 kN dynamometer, displacements were measured using a linear variable differential transformer (LVDT) at 5 points (at the end supports and every quarter of the beam span), and non-contact optical measurements were taken on the lateral surface of the tested element using the Aramis system. Figure 5 shows a view of the beam during testing. The results of beam deflection tests recorded by LVDTs and an optical system are described in detail in the doctoral thesis [8].

Fig. 5. View of a wooden beam during bending tests.

3.4 Correlation of Destructive and Non-destructive Test Results

Figures 6 and 7 show the correlations between the destructive and non-destructive ultrasonic testing results. In the case of non-destructive testing, the graphs show the average result of the tests at 10 points. Both graphs show trend lines (blue) and the lower envelope of the test results (orange). A second-degree polynomial function was used to describe the trend line and the lower envelope of the longitudinal wave test results. An exponential function was used to describe the lower envelope of the transverse wave test results.

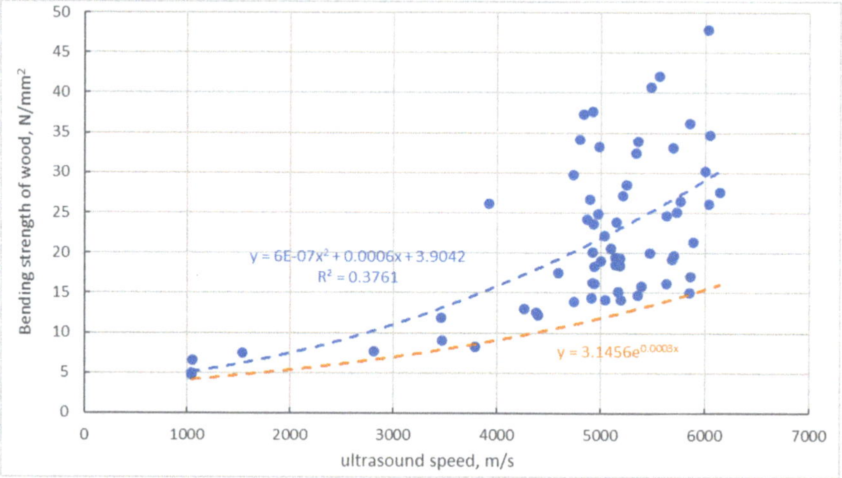

Fig. 6. Correlation of the destructive and ultrasonic (transverse wave) test results for softwood (Color figure online).

The average function (trend line) describing the bending strength of the softwood (f_m), depending on the velocity of the ultrasonic transverse wave (C_T), has the form:

$$f_m = 6 \bullet 10^{-7}\ C_T^2 + 0.0006\ C_T + 3.9042. \tag{1}$$

The lower envelope of the softwood test results is described by the equation:

$$f_m = 3.1456\ e^{0.00027\ C_T}. \tag{2}$$

The average function (trend line) describing the bending strength of the softwood (f_m), depending on the velocity of the ultrasonic longitudinal wave (C_L), has the form:

$$f_m = -5 \bullet 10^{-7}\ C_L^2 + 0.0164 C_L - 1.6165. \tag{3}$$

The lower envelope of the softwood test results is described by the equation:

$$f_m = 7 \bullet 10^{-6} C_L^2 - 0.0071\ C_L + 6.6402. \tag{4}$$

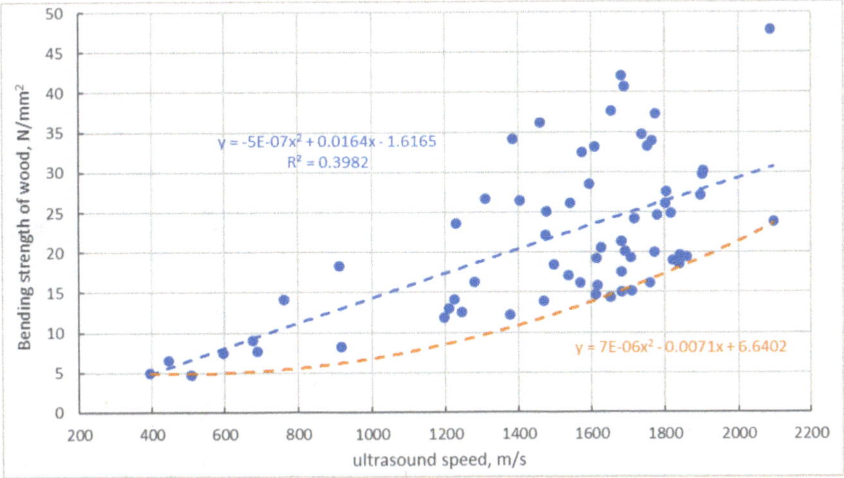

Fig. 7. Correlation of the destructive and ultrasonic (longitudinal wave) test results for softwood (Color figure online).

4 Testing of Bricks

4.1 Research Material

For the study, 125 historical clay bricks were obtained from Poland. The bricks dated back to the last 600 years: the oldest from the early 14th century, and the youngest from the first half of the 20th century. The elements originated from the demolition and renovation of historic buildings. It is important to note that the permission to collect samples for testing was obtained from the relevant Conservator of Monuments. Both destructive and non-destructive tests were conducted in an air-dry state, with moisture content measured using the electrical resistance method on the brick surfaces within the range of $2 \pm 1.5\%$.

The bricks obtained for testing had varied shapes, uneven edges, and were covered with mortar fragments (Fig. 8). The EN 772-1 [13] standard assumes testing of entire or cut-out sections of masonry units. According to this standard, masonry units should have flat surfaces (tolerance of 0.1 mm per 100 mm), which can be achieved by grinding or leveling the surface with a cement-sand mortar. Due to the need to remove hardened mortar from the sample surfaces, it was decided to grind the units and locally trim them if there were any uneven edges or gaps. Figure 9 shows examples of cut units prepared for testing.

4.2 Non-destructive Tests

Before destructive testing, the brick samples were subjected to non-destructive testing using a UK1401 Surfer device. Each brick was tested on its face surfaces, as these, together with its header surfaces, are typically visible on the wall surface. Testing could not be performed on the header surfaces because the spacing between the device's heads

a) b)

Fig. 8. Brick samples before cutting and grinding.

Fig. 9. Cut and polished bricks prepared for testing.

is 15 cm, which exceeds the width of the brick head. An example photograph of the test is shown in Fig. 10.

Fig. 10. Non-destructive ultrasonic testing of bricks.

4.3 Destructive Tests

The aim of the research was to determine the compressive strength of the brick and, consequently, the brick class. The masonry units were aligned with the head joint of a

300-t testing machine after cleaning the head surface. The load was applied at a rate that ensured sample failure within at least 1 min of the start of loading. Efforts were made to follow the recommendations given in Table 2 of the EN 772-1 standard [11]. The load was always applied to the top and bottom surfaces of the specimens. The maximum load achieved during testing of each masonry unit was recorded. An example view of the brick after testing is shown in Fig. 11.

Fig. 11. View of a brick sample after testing in a testing machine.

4.4 Correlation of Destructive and Non-destructive Test Results

Figure 12 shows the correlation between the results of destructive and non-destructive ultrasonic testing of bricks. The graph shows trend lines (blue) and the lower envelope of the test results (orange). A second-degree polynomial is used to describe the trend lines and the lower envelope.

The average function (trend line) describing the compressive strength of the brick (f_B), depending on the velocity of the ultrasonic transverse wave (C_T) has the form:

$$f_B = 9 \bullet 10^{-7}\, C_T^2 + 0.0078\, C_T - 0.8964. \tag{5}$$

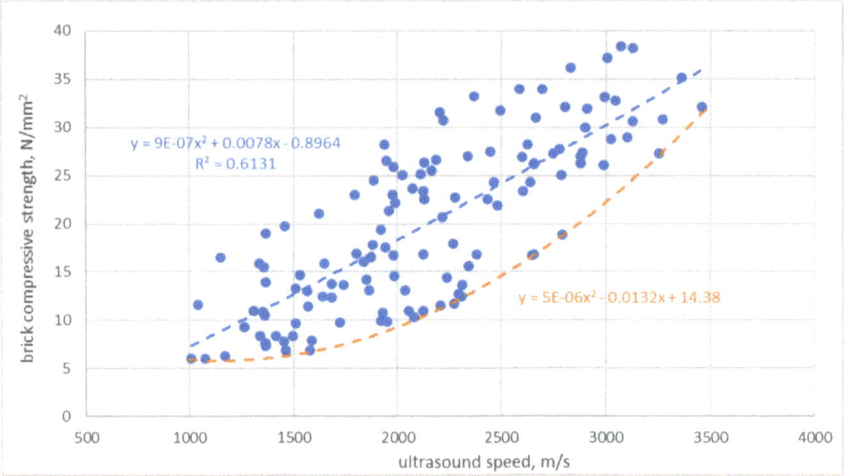

Fig. 12. Correlation of the destructive and ultrasonic (transverse wave) test results for bricks (Color figure online).

The lower envelope of the brick test results is described by the equation:

$$f_B = 5.3 \bullet 10^{-6}\ C_T^2 - 0.0132\ C_T + 14.38. \qquad (6)$$

The strength f_B given in Formulas (5) and (6) refers to the test result for entire masonry units. This strength was not converted to the normalized compressive strength of masonry unit f_B in accordance with the EN 772-1 standard [13].

5 Summary

This paper describes the results of non-destructive ultrasonic and destructive testing of historic wood and clay bricks. Based on the obtained test results, correlations between ultrasonic wave velocity and the strength of the tested material were developed. Formulas describing the strength of wood and brick as a function of ultrasonic wave velocity were adopted for the mean value of the tests and the lower envelope of the results. The curves limiting the strength from below are close to the characteristic strength, i.e., the 5% strength quantile. Presented formulas allow for non-destructive estimation of the strength of historic wood and masonry elements without the need for sampling. It is recommended to use the lower envelope of the test results, which will provide a reliable estimate. The use of trend lines is only possible when the tested elements are in good technical condition. The obtained results directly relate to strength estimates for the materials with the moisture content at which the authors tested them.

References

1. Baronio G, Binda L, Tedeschi C, Tiraboschi C (2003) Characterisation of the materials used in the construction of the Noto Cathedral. Constr Build Mater 17(8):557–571

2. Borosnyoi-Crawley D (2023) Non-destructive strength estimation of vintage clay bricks based on rebound hardness in architectural heritage buildings. J Build Eng 80:108055
3. Cabané A, Cuadros-Rojas E, Roca P, Pelà L (2025) Non-destructive evaluation of compressive strength in solid clay bricks using the rebound hammer test. Int J Archit Heritage:1–13. https://doi.org/10.1080/15583058.2025.2592139
4. Camassa D, Castellano A, Fraddosio A, Piccioni MD (2019) Improvements of the ultrasonic tomography for applications to historical masonry constructions. In: Aguilar R, Torrealva D, Moreira S, Pando MA, Ramos LF (eds) Structural analysis of historical constructions, RILEM bookseries, vol 18. Springer
5. Chen L, Kong Q, Xiong Q, Xiong H (2025) Research and application of non-destructive testing technology for timber structures: a review. Structures 80:109790
6. Dackermann U et al (2014) In situ assessment of structural timber using stress-wave measurements. Mater Struct 47:787–803
7. Drobiec J, Nowogońska B (2023) Restrictions and new possibilities of technical and conservation diagnostics of wood in heritage buildings. Civ Environ Eng Rep 33(2):106–116
8. Drobiec J (2025) Analysis of the possibilities of using non-destructive and semi-non-destructive methods to determine the strength of historic wood. Doctoral dissertation, University of Zielona Góra
9. Drobiec Ł, Grzyb K, Zając J (2021) Analysis of reasons for the structural collapse of historic buildings. Sustainability 13:10058
10. EN 380:1993 (1993) Timber structures—test methods—general principles for static load testing. European Committee for Standardization (CEN), Brussels, Belgium
11. EN 384:2016 (2016) Structural timber—determination of characteristic values of mechanical properties and density. European Committee for Standardization (CEN), Brussels, Belgium
12. EN 408:2010+A1:2012 (2012) Timber structures—structural timber and glued laminated timber—determination of some physical and mechanical properties. European Committee for Standardization (CEN), Brussels, Belgium
13. EN 772-1:2011+A1:2015 (2015) Methods of test for masonry units—part 1: determination of compressive strength. European Committee for Standardization (CEN), Brussels, Belgium
14. Espinosa L, Prieto F, Brancheriau L, Lasaygues P (2019) Effect of wood anisotropy in ultrasonic wave propagation: a ray-tracing approach. Ultrasonics 91:242–251
15. Grzyb K, Drobiec Ł, Blazy J, Zając J (2022) The use of NDT diagnostic methods and calculations in assessing the masonry tower crowned with the steel dome. Materials 15:7196
16. Grzyb K, Drobiec Ł, Zając J, Drobiec K (2025) Preliminary assessment of structural masonry damage in Malbork castle. Case Stud Constr Mater 22:e04166
17. Niemz P, Mannes D (2012) Non-destructive testing of wood and wood-based materials. J Cult Herit 13S:S26–S34
18. Nowogońska B, Drobiec J (2025) Prevention of failure of heritage buildings through diagnostic testing using the example of roof structure elements from the 15th century. Eng Fail Anal 178:109709
19. Nowogońska B (2020) Consequences of improper renovation decisions in a 17th century half-timbered building. Sci Rev Eng Environ Sci 29(4):557–566
20. Perlin LP, de Andrade Pinto RC, do Valle Â (2019) Ultrasonic tomography in wood with anisotropy consideration. Constr Build Mater 229:116958
21. Schabowicz K (2019) Non-destructive testing of materials in civil engineering. Materials 12:3237
22. Tejedor B, Lucchi E, Bienvenido-Huertas D, Nardi I (2022) Non-Destructive Techniques (NDT) for the diagnosis of heritage buildings: traditional procedures and futures perspectives. Energ Buildings 263:112029
23. Witzany J, Čejka T, Sýkora M, Holický M (2016) Strength assessment of historic brick masonry. J Civ Eng Manag 22(2):224–233

24. Zielińska M, Rucka M (2023) Assessment of wooden beams from historical buildings using ultrasonic transmission tomography. Int J Archit Heritage 17(1):249–261

Research on the Structure, Humidity, and Salinity of the Brick Walls of the Medieval Refectory

Anna Hoła(✉)

Wrocław University of Science and Technology, Wrocław, Poland
anna.hola@pwr.edu.pl

Abstract. The paper presents the result of research on the structure, moisture content, and salinity of the brick walls of the refectory in a historic Cistercian monastery complex, built at the turn of the 13th and 14th centuries in Pomerania, Poland. The wall structure was examined using videoscopy and mercury porosimetry, while mass moisture content was assessed with a non-destructive dielectric method (GANN UNI 2 meter), supplemented by gravimetric measurements. The type and concentration of salts in the masonry were determined using qualitative and semi-quantitative methods with chemical reagents and indicator papers.

The results revealed a layered, heterogeneous wall structure, with mass moisture content significantly exceeding the permissible 3% along the entire perimeter. In the surface zone, the wall was heavily damp, while in the middle zone, it was wet. Furthermore, sulfates, chlorides, and nitrates were identified, with concentrations in many areas exceeding values considered safe for brick walls. A thorough analysis of the findings revealed that the primary cause of the excessive moisture was attributed to the free capillary rise of water from the ground. This discovery served as the foundation for recommendations on repair and renovation measures.

Keywords: Medieval Refectory · Brick Wall · Structure Analysis · Moisture and Salinity Testing · Case Study

1 Introduction

The preservation of material cultural heritage in the form of architectural monuments is in the broader public interest and plays an important role in sustainable development. Properly conducted renovations of buildings of high historical value foster the development of local communities, contribute to the preservation of cultural identity, and enhance the tourist attractiveness of regions; they also support the circular economy. However, a necessary condition for such interventions is that renovation works are consistently based on reliable research results.

In historic buildings constructed of ceramic brick, a common problem is excessive moisture and salt content in the walls [e.g., 1–3]. According to the classification adopted in the extant literature, excessive moisture is defined as exceeding the permissible mass

Z. Perkowski et al. (Eds.): ECCE 2026, LNCE 929, pp. 193–206, 2027.
https://doi.org/10.1007/978-3-032-39515-3_15

moisture level of 3% [4], while excessive salinity is defined as exceeding the limit concentration of salts harmful to walls, which is 0.15% [4, 5].

One of the main causes of excessive dampness and salinity of walls is prolonged, direct contact between brick walls lacking moisture protection and the adjacent ground. It should be noted that until the beginning of the 20th century, horizontal and vertical moisture barriers were not used. To limit the effects of ground moisture, buildings were constructed on stone foundations, or underground drainage systems were installed in their immediate vicinity in order to locally lower the groundwater level. Over time, as building density increased, soil and groundwater conditions changed, and existing drainage systems were often disrupted or damaged. This facilitated uncontrolled moisture ingress from the ground, together with the harmful salts dissolved in it, into brick walls and their upward capillary transport, leading to progressive dampness and salt accumulation [6].

In addition, due to water evaporation from excessively damp walls through exposed wall surfaces, salt concentration increases in near-surface capillaries, leading to their gradual crystallization. This process is accompanied by the sorption of moisture from the air by salt crystals, further increasing the wall's moisture content. The crystals also exert crystallization pressure on the capillary walls [7], which, once the tensile strength of the walls is exceeded, leads to deterioration of the brick-and-mortar structure.

Moisture-induced deterioration of masonry progresses from the wall surface toward its interior and is often intensified by frost damage [8]. The consequences of these processes include a gradual reduction in wall thickness, deterioration of mechanical properties that may compromise structural safety, degradation of thermal performance, and the development of mould fungi on the interior wall surface, among other effects.

Effective mitigation of excessive dampness and salt contamination in brick masonry of historic buildings, as well as the prevention of the associated destructive processes, requires comprehensive investigations aimed at determining both the extent and the causes of moisture and salt presence [9]. It is also essential to identify the wall structure, including material homogeneity, stratification, and the presence of voids or cavities [10], as well as the physical properties of the ceramic bricks used in its construction, which account for approximately 75–80% of the wall volume. Of particular importance are parameters such as total porosity, water absorption, and pore size distribution [11, 12]. This knowledge is crucial for making appropriate repair and conservation decisions, including selecting suitable repair materials for excessively damp, salt-contaminated historic masonry.

The aim of this study is to present, through a case study, the scope of investigation and the research methods used to provide a reliable assessment of the moisture condition of historic brick masonry dating from the medieval period. No such investigations had previously been carried out in the building under study, and earlier conservation efforts were limited to partial replacement of deteriorated lime mortar in near-surface zones and the removal of salt efflorescence and mould from the wall surface. The study also aims to present the results of the investigations and, based on their analysis, to identify the causes of excessive dampness and salinity of the examined walls. The knowledge obtained provides a basis for planning effective conservation measures and contributes to a deeper understanding of the degradation processes affecting historic architecture. For this reason, it may also be of significant cognitive value for other researchers investigating brick structures from the same historical period.

2 Research Object

The object of the study was a refectory located on the ground floor within a historic post-Cistercian monastic complex, erected at the turn of the 13th and 14th centuries in Pomerania, Poland [13]. The post-monastic complex was entered into the register of historic monuments in 1984. It forms part of a larger Cistercian-cathedral complex, which in 2014 was designated a Historic Monument by decision of the President of the Republic of Poland, one of the highest forms of heritage protection in the country.

The refectory serves a representative function and is also open to visitors (see Fig. 1). It has the form of an elongated rectangle and is covered by a ribbed cross vault supported by massive walls with a thickness of nearly 200 cm (Fig. 2). The walls were constructed of solid ceramic bricks with dimensions of 320 mm × 120 mm × 90 mm and lime mortar. Clear signs of moisture and frost-related deterioration are visible on the wall surfaces, including salt crystallization, mould growth, and brick degradation (Fig. 3).

Fig. 1. General view of the refectory.

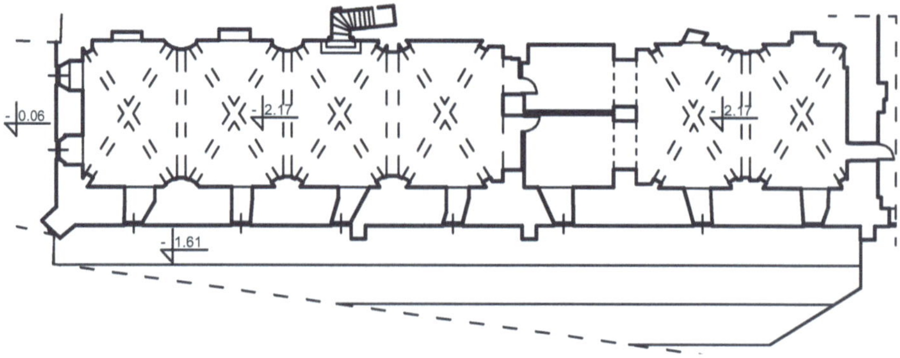

Fig. 2. Plan of the refectory.

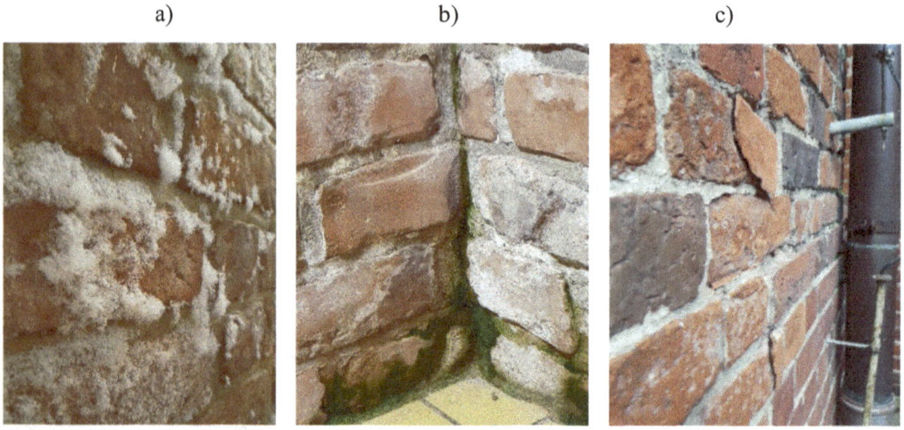

Fig. 3. Close-ups of wall sections: interior surfaces with salt efflorescence (a) and mold (b), and an exterior section with frost-damaged bricks (c).

3 Description of the Conducted Research

Prior to planning the scope of restoration works, it was deemed necessary to conduct comprehensive investigations to assess the condition of the brick walls of the refectory, with particular emphasis on their macro and microscopic structures, as well as the degree and causes of dampness and salinity.

3.1 Structure Analysis

The macroscopic structure of the brick walls was examined at three randomly selected locations using a videoscopic method with a Videoinspector 3D device equipped with digital image recording functionality. The procedure involved drilling a 14 mm diameter hole in the wall and gradually deepening it. After each drilling stage, a microcamera was inserted into the hole to observe the wall structure.

Selected physical properties of the bricks incorporated into the refectory walls were also determined, including water absorption by mass, bulk density, true density, total porosity, and capillary pore size distribution. These analyses were carried out on three samples collected from fragments of bricks loosely detached from the walls.

Initially, water absorption by mass was determined using the gravimetric method. The collected material was dried in a laboratory oven at 105 ± 5 °C until a constant mass was reached. After drying, the samples were weighed, then immersed in distilled water at room temperature until a stable water-saturated mass was achieved. After removal from water, the samples were gently dried with a paper towel to remove surface water, and their mass was recorded. Water absorption by mass was calculated using the formula:

$$W_m = \frac{m_{sat} - m_{dry}}{m_{dry}} \times 100\% \tag{1}$$

where:

W_m – water absorption by mass [%],
m_{dry} – mass of the dry sample [g],
m_{sat} – mass of the saturated sample [g].

Other physical properties were determined using mercury porosimetry with a Micromeritics AutoPore IV 9510 device. Samples for porosimetric analysis were prepared by crushing brick fragments into 2–3 mm granules, which were then homogenized. Six representative samples of approximately 2 g each were taken from the granules. Each sample was initially air-dried at 110 °C for 1 h and then degassed in the penetrometer (25 °C, 30 min, 50 μm Hg). Next, the penetrometer with the sample was filled with mercury at 0.01 MPa. Based on the known mercury volume in the empty penetrometer and in the penetrometer with the sample, bulk density was calculated as:

$$\rho_a = \frac{m_{dry}}{V_t} \tag{2}$$

where:

ρ_a – bulk density [g/cm^3],
m_{dry} – mass of the dry sample [g],
V_t – sample volume (difference between the mercury volume in the empty penetrometer and in the penetrometer with the sample) [cm^3].

Mercury pressure was then gradually increased to 414 MPa to allow penetration into the sample's porous structure (intrusion stage). At this stage, true density was calculated as:

$$\rho_s = \frac{m_{dry}}{V_{solid}} \tag{3}$$

where:

ρ_s – true density [g/cm^3],
m_{dry} – mass of the dry sample [g],

V_{solid} – volume of the solid part of the sample (volume of material excluding all pores determined by mercury intrusion) [cm^3].

The volume of open pores was determined after completion of the intrusion stage by measuring the change in mercury level in the penetrometer tube. Total (open) porosity was then calculated as:

$$n = \frac{V_p}{V_t} \times 100\% \quad (4)$$

where:

n – total (open) porosity [%],
V_p – volume of open pores in the sample [cm^3],
V_t – total volume of the sample [cm^3].

Finally, mercury pressure was gradually reduced to 0.1 MPa, resulting in partial extrusion of mercury from the pore structure (extrusion stage). The volumetric proportions of pores of different widths were determined from changes in the mercury level in the penetrometer tube; according to Washburn's equation, the applied pressure range of 0.01 MPa to 414 MPa corresponded to pore diameters from 150 μm to 3.5 nm.

3.2 Moisture Tests

The mass moisture content of the brick walls was determined using both the non-destructive dielectric method and the destructive gravimetric method, following a methodology previously developed by the author [8].

The non-destructive tests were performed using the dielectric method. Measurements were conducted with a GANN UNI 2 meter equipped with an active B50 spherical probe, which allows the assessment of the average moisture content of the brick wall to a depth of approximately 60 mm from the surface. Since the meter provides dimensionless readings, it was calibrated using small brick samples from the near-surface zone of the wall, whose moisture content had been determined gravimetrically under laboratory conditions. A total of 30 samples were collected. After sampling, the bricks were weighed, sealed in airtight containers, and transported to the laboratory, where they were dried to constant mass at 105 ± 5 °C. Their mass moisture content was then calculated according to Eq. (1).

The 30 samples yielded 30 pairs of results (dimensionless meter readings x and gravimetrically determined mass moisture U_m), allowing the determination of a correlation representative of the meter and the bricks from the investigated walls. This correlation was described using an exponential function:

$$U_m = f(x) = 0.1667\, e^{0.0355\, x} \quad (5)$$

where:

U_m – mass moisture content [%],
x – meter reading [–].

The obtained coefficient of determination $R^2 = 0.79$ indicates a good fit of the correlation to the experimental data. This relationship was subsequently used to estimate U_m at all points measured with the GANN UNI 2 m.

Dielectric measurements were performed both inside the refectory and on the exterior of the building, along the perimeter of the walls, at two heights above the floor/ground level. This allowed the creation of profiles of mass moisture distribution U_m along the height and length of the walls. Profiles of moisture distribution across the wall thickness were prepared based on gravimetric measurements of samples collected from the wall.

3.3 Salinity Tests

The type and concentration of water-soluble salts harmful to masonry, present in the wall as chlorides, sulfates, and nitrates, were determined using qualitative and semi-quantitative methods with appropriate chemical reagents and indicator strips. The same 30 brick samples used for gravimetric determination of mass moisture content were employed for salinity analysis.

The previously dried material was ground into a fine powder using a mortar. To prepare the stock solution, 5 g of the powdered material was placed in a beaker, to which 50 ml of distilled water was added. After mixing and allowing the solid material to settle, the solution was filtered through a funnel with filter paper. The resulting clear solution constituted the stock solution, from which approximately 5 g aliquots were taken for subsequent analyses.

4 Results and Analysis

4.1 Structure of the Walls

The investigation of the macroscopic structure revealed that the brick walls of the refectory have a layered construction. The facing layers—external, with a thickness of approximately 45 cm, and internal, with a thickness of approximately 35 cm—were made of solid ceramic brick, whereas the central layer, approximately 145 cm thick, consists of stone (see Fig. 4). No air voids were identified at the interfaces between the individual layers.

In turn, the results of the investigations of the physical properties of the bricks incorporated into the masonry, carried out using gravimetric methods and mercury porosimetry, are presented in Tables 1 and 2 and in Fig. 5.

These studies showed that both the bulk density and the true density of the bricks used in the construction of the refectory walls are typical for this type of material. The water absorption by mass of the examined solid brick was 20.26%, which falls within the upper range of values characteristic of historic bricks. The total porosity is also high, averaging 26.60%, indicating a relatively high pore content in the material. These parameters demonstrate that the brick exhibits a considerable capacity for water absorption and, consequently, a high susceptibility to dampness and salinity. The results of these investigations are consistent with the organoleptic assessment and with the moisture measurements of the refectory walls, which are discussed in the subsequent

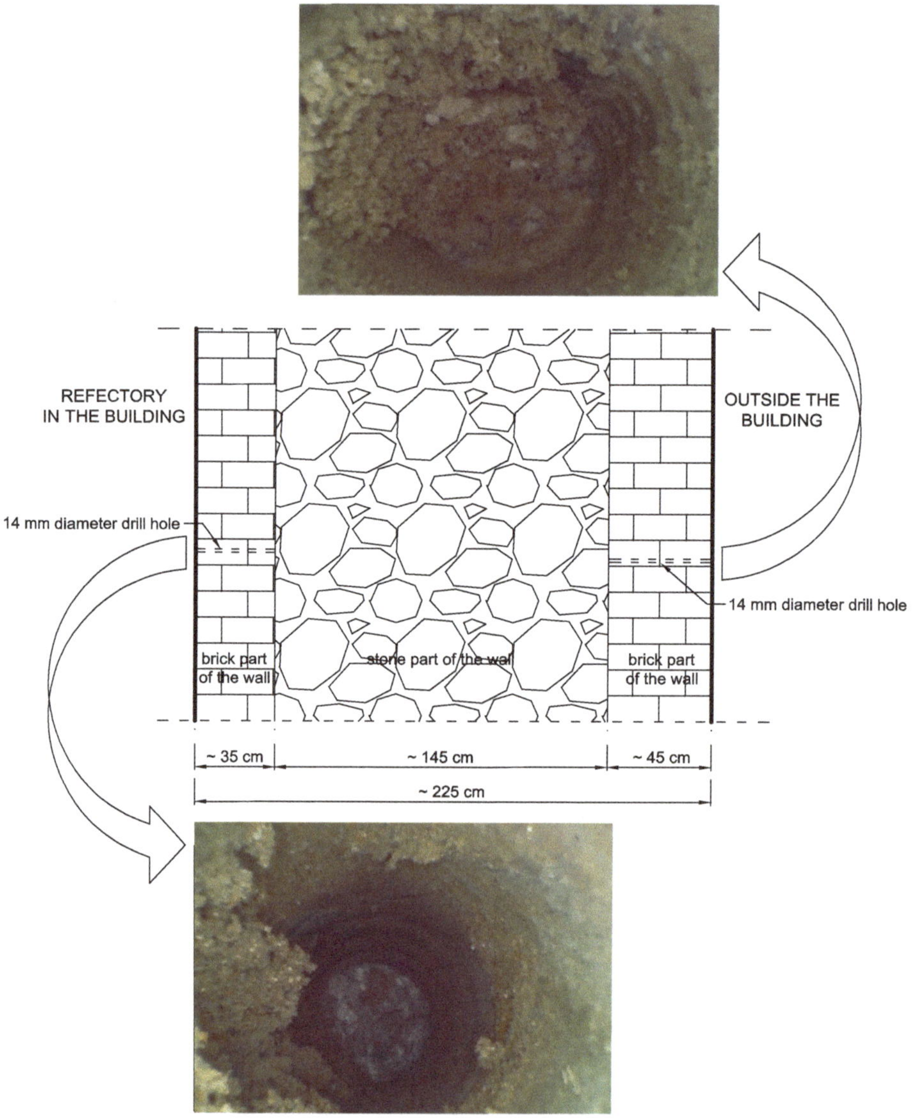

Fig. 4. Cross-section of the layered wall of the refectory and images obtained using a videoscope in the drilled holes.

part of the article. In the analyzed samples, pores with diameters smaller than 1 μm and in the 1–3 μm range constitute the largest proportion (>60%). According to [14, 15], pore sizes in this range are unfavorable for the durability of ceramic materials, as they promote the accumulation of water in fine capillaries, thereby increasing the material's susceptibility to frost damage and degradation under adverse atmospheric conditions. By contrast, pores with diameters greater than 3 μm, which account for nearly 40% of

Table 1. Summary of the average values of selected physical parameters characterizing the ceramic bricks incorporated into the refectory walls.

Bulk density ρ_a [g/cm^3]	True density ρ_s [g/cm^3]	Water absorption by mass W_m [%]	Total porosity n [%]	Median pore diameter [µm]
1.800	2.454	20.26	26.60	1.68

Table 2. Average percentage distribution of pore sizes in samples taken from the brick walls of the refectory according to the defined ranges.

Adopted pore size range [µm]	Proportion of pores within the given size [%]
100–200	5.2
6–100	23.6
3–6	10.8
1–3	24.3
<1	36.1

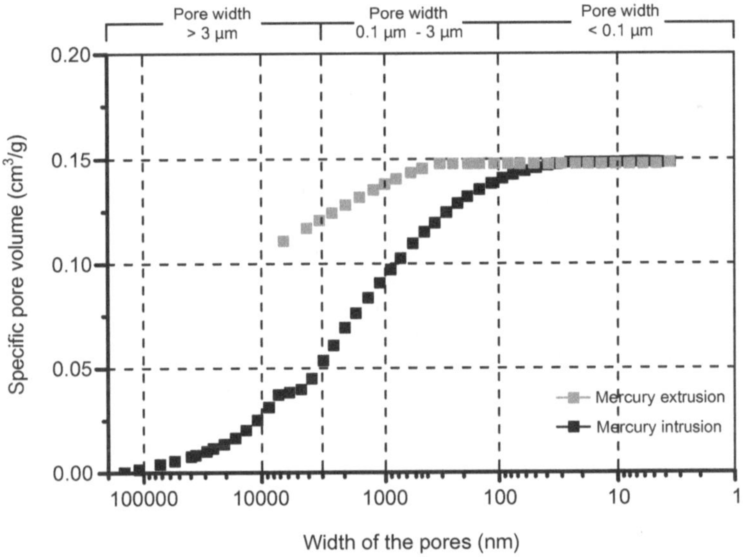

Fig. 5. Distribution of pore sizes in the examined ceramic brick within the range of 200 µm to 3 nm [9].

the pores in the examined ceramic material, are considered beneficial for its durability according to [14, 15].

4.2 Moisture Content of the Walls

The mass moisture content U_m was measured at 170 points located at different heights above the floor/ground level and at various depths within the wall. The results showed that the brick walls of the refectory clearly exceeded the permissible 3% mass moisture content along the entire perimeter.

In the near-surface zone, on both the exterior and interior sides of the room, the brick wall is highly damp and wet (average $U_m = 8.10\%$, maximum $U_m = 18.93\%$). Within the deeper sections of the walls, on both the external and internal sides, the moisture content remains high (>12%), with an average $U_m = 13.69\%$ and a maximum $U_m = 16.93\%$. The high level of moisture persists up to a height of approximately 150 cm above the floor level. Example profiles of the mass moisture distribution across the wall thickness and along the wall height of the refectory are shown in Figs. 6 and 7, respectively.

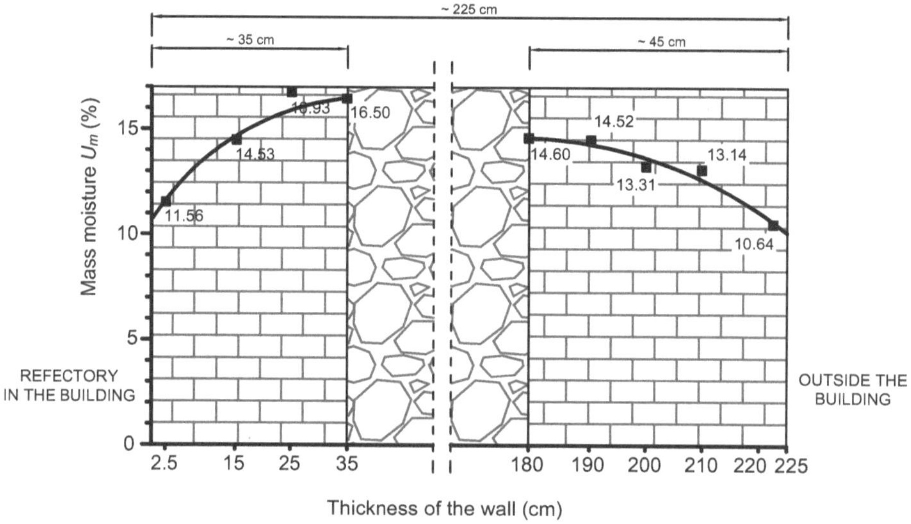

Fig. 6. Example profile of the mass moisture content U_m across the thickness of the brick wall at a height of 80 cm above the floor of the refectory.

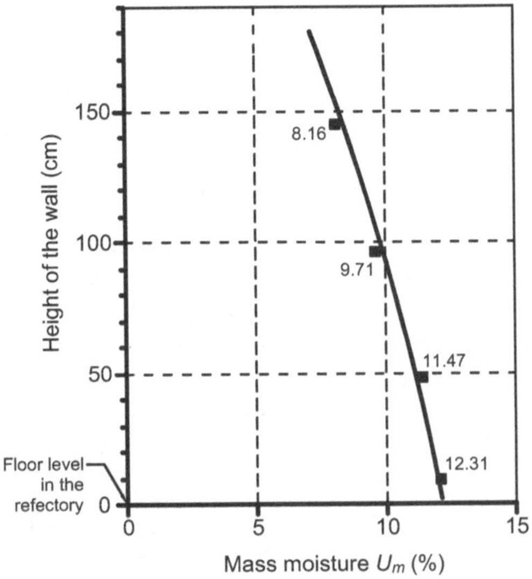

Fig. 7. Example profile of the mass moisture content U_m along the height of the brick wall.

4.3 Salinity of the Walls

The qualitative and semi-quantitative investigations of the types and concentrations of salts showed that the brick walls of the refectory contain harmful salts, including sulfates, chlorides, and nitrates. The mass concentration of sulfates exceeds the permissible safe level of 0.15% for brick masonry in many locations.

The concentration of sulfates ranges from 0.20% to 1.60% (average 0.395%), which, according to the classification in [16], qualifies as high (>1.50%), medium (0.50–1.50%), or low (<0.5%), depending on the location. Chloride concentrations vary from 0.017% to 0.10% (average 0.046%), which should be considered a low level of salinity (<0.20%). Nitrates are present in the walls only sporadically, with concentrations ranging from 0.00% to 0.25% (average 0.015%), and are classified as low (<0.10%) or, in some locations, medium (0.10–0.30%).

4.4 Causes of Dampness and Salinity and Key Recommendations

Based on the analysis of the research results, the main cause of excessive dampness and salinity in the brick walls of the medieval refectory was identified as unrestricted capillary rise of groundwater from the foundation, caused by the absence of horizontal damp-proofing in the walls.

For the brick wall on the interior side of the refectory, the dampness process is further enhanced by moisture absorbed from the air by salts that have crystallized on the wall surface and within the near-surface pores. This process is promoted by insufficient ventilation in the refectory and a relative air humidity maintained at approximately 80%. For the brick wall forming the external part of the walls, dampness is additionally enhanced by lateral capillary rise from the ground due to the lack of vertical damp-proofing.

The research results provided the basis for formulating recommendations to develop a detailed scope of restoration and conservation work for the brick walls of the refectory. The most important restoration and conservation recommendations are as follows:

- Installation of a horizontal damp-proof barrier in the brick walls on the interior side of the refectory by drilling injection holes just above the floor level and filling them with a silane-based injection cream.
- Installation of a horizontal damp-proof barrier in the brick walls on the exterior side, after excavation, at the same level as on the interior side, by drilling injection holes and filling them with a silane-based injection cream.
- Construction of a continuous vertical damp-proof barrier using a mineral-based sealing micro-mortar on the surface of the excavated external wall.
- Desalination of the wall using desalinating compresses.
- Removal of mould efflorescence and disinfection of the affected wall surfaces with a fungicidal chemical agent.
- Filling voids with new restoration ceramic bricks, handmade and characterized by total porosity and pore size distribution similar to those of historic bricks, using lime mortar appropriate for the restoration of historic masonry.

5 Conclusions

1. The investigation of the wall structure of the medieval refectory, with a thickness exceeding 200 cm, using a videoscopic method, revealed a clearly layered construction. The external and internal layers, approximately 45 cm and 35 cm thick, respectively, were made of solid ceramic bricks, whereas the central layer consists of stone infill. Supplementary investigations of the physical properties of the bricks used to construct the walls, carried out using the gravimetric method and mercury porosimetry, showed a high total porosity of 26.60% and a relatively large proportion of pores (~40%) with diameters exceeding 3 μm, which are considered beneficial for the durability of the bricks. The results obtained expand understanding of medieval wall-construction techniques and the material properties of the bricks used in northern Poland.
2. Mass moisture content measurements of the brick walls of the refectory, performed using both non-destructive and destructive methods, indicated significant dampness. Both the external and internal sides of the refectory walls are highly damp or wet in the near-surface zone up to approximately 150 cm above the floor, with average mass moisture content U_m values of 8.10% and 9.20%, respectively. In the central zone, the walls remain wet, with an average mass moisture content U_m of 13.69%. Furthermore, the presence of harmful salts, particularly sulfates, was confirmed, with maximum concentrations exceeding 1.60% in some locations within the walls.
3. An important outcome of the study was the determination of a correlation for the investigated refectory, described by an exponential function, between the readings x of the applied dielectric meter and the mass moisture content U_m of the walls. This relationship, presented in the article, can serve as a useful tool for non-destructive assessment of moisture in brick walls in other medieval buildings constructed using similar techniques.

4. The main cause of excessive dampness and salinity in the brick walls of the refectory was identified as unrestricted capillary rise of groundwater from the foundation due to the absence of horizontal damp-proofing. This process is further intensified by additional factors: in the external wall, by lateral capillary rise from the ground, and in the brick wall on the interior side of the refectory, by moisture sorption from the air by salts crystallized on the wall surface and in its near-surface zone.
5. The results presented in this article are not only of cognitive value but may also be used in comparative studies conducted by other researchers investigating brick masonry structures dating from the medieval period.

References

1. Franzoni E, Berk B, Bassi M, Marrone C (2023) An integrated approach to the monitoring of rising damp in historic brick masonry. Constr Build Mater 370:130631. https://doi.org/10.1016/j.conbuildmat.2023.130631
2. Nowogońska B, Korentz J (2025) Increased moisture in brick pillars caused by improper repairs – a case study. In: Branco JM, Lima DF, Moscoso Y, Silva SM (eds) Moisture in buildings. ICMB 2025, Lecture notes in civil engineering, vol 776. Springer. https://doi.org/10.1007/978-3-032-09054-6_18
3. Hoła A, Matkowski Z, Hoła J (2017) Analysis of the moisture content of masonry walls in historical buildings using the basement of a medieval town hall as an example. Procedia Eng 172:363–368. https://doi.org/10.1016/j.proeng.2017.02.041
4. Rokiel M (2019) Hydroizolacje w budownictwie (waterproofing in construction). Dom Wydawniczy MEDIUM, Warszawa
5. Adamowski J, Hoła J, Matkowski J (2005) Problemy renowacyjne zawilgoconych monumentalnych obiektów barokowych (renovation problems of damp monumental baroque buildings). Renowacje i Zabytki 1(13):130–138
6. Kubik J (2000) Przepływ wilgoci w materiałach budowlanych (moisture flows in building materials). Oficyna Wydawnicza Politechniki Opolskiej, Opole
7. Gentilini C, Franzoni E, Bandini S, Nobile L (2012) Effect of salt crystallisation on the shear behaviour of masonry walls: an experimental study. Constr Build Mater 37:181–189. https://doi.org/10.1016/j.conbuildmat.2012.07.086
8. Hoła J (2018) Degradacja budynków zabytkowych wskutek długotrwałego zawilgocenia – wybrane problemy (the degradation of historic buildings due to excessive moisture – selected issues). Budownictwo i Architektura 17(1):133–148. https://doi.org/10.24358/Bud-Arch_18_171_17
9. Hoła A (2020) Methodology for the in situ testing of the moisture content of brick walls: an example of application. Arch Civ Mech Eng 20:114. https://doi.org/10.1007/s43452-020-00120-3
10. Karczmarczyk S (2020) Rola norm projektowych w rzeczoznawstwie murów zabytkowych. In: Goszczyńska B (ed) XVI Konferencja Naukowo-Techniczna "Warsztat Pracy Rzeczoznawcy Budowlanego". Zarząd Główny PZITB, Warszawa, pp 323–342
11. Wesołowska M, Kaczmarek A, Hoła J (2021) The influence of external environmental conditions on properties of ceramic building materials with waste material additives. Materials 14(11):2982. https://doi.org/10.3390/ma14112982
12. Hoła A, Kierzek K, Żak A (2024) Preliminary microstructural research of clay solid bricks taken from historic buildings located in Poland. Arch Civ Mech Eng 24:159. https://doi.org/10.1007/s43452-024-00972-z

13. Łużyniecka E (2014) Pelplin i Doberan. Architektura opactw cysterskich spokrewnionych filiacyjnie (Pelplin and Doberan: the architecture of related Cistercian abbeys). Oficyna Wydawnicza Politechniki Wrocławskiej, Wrocław
14. Maage M (1984) Frost resistance and pore size distribution in bricks. Mater Struct 17:345–350. https://doi.org/10.1007/BF02478706
15. Stryszewska T, Kańka S (2019) Forms of damage of bricks subjects to cyclic freezing and thawing in actual conditions. Materials 12(7):1165. https://doi.org/10.3390/ma12071165
16. WTA (1991) Merkblatt 2-2-91: Ergaenzungen zum Merkblatt 2-2-91 – Sanierputzsysteme. WTA

Reconstruction and Adaptive Operation of Historic UNESCO Hydrotechnical Infrastructure: From Aqueducts and Qanats to Tajchy

Lucia Nováková(✉)

Department of Classical Archaeology, Trnava University, Trnava, Slovakia
lucia.novakova@truni.sk

Abstract. Historic hydrotechnical infrastructures based on gravity-driven conveyance, water storage, and controlled distribution—most notably qanats, Roman aqueducts, and the Slovak tajchy reservoir systems—represent long-lived socio-technical systems whose performance depends on the interaction between engineering design, operational practice, and governance. This chapter presents a comparative analysis of these three system types within the institutional context of UNESCO World Heritage, focusing on the relationship between Outstanding Universal Value (OUV), authenticity, integrity, and long-term management requirements. The study is based on a structured narrative review of technical and historical scholarship, UNESCO and ICOMOS nomination and evaluation documents, and site management plans, with particular emphasis on the tajchy of the Banská Štiavnica mining district as an integrated network of reservoirs, collection ditches, and driving channels developed to supply hydraulic power for mining operations. Across all systems, sustained functionality has relied on robust engineering concepts, appropriate material solutions, regular maintenance, and governance mechanisms capable of coordinating labor, allocating water, and managing operational risks. Contemporary challenges are examined through a risk-informed perspective addressing climate variability and extremes, groundwater depletion, sedimentation processes, urban pressures, and water-quality constraints. The chapter concludes by outlining criteria-based comparability among the systems and identifying implications for integrating historic hydrotechnical infrastructure into sustainable, risk-aware regional water management strategies.

Keywords: Hydrotechnical Heritage · Gravity-Driven Water Systems · Adaptive Management

1 Introduction

Historic water infrastructure is rarely just a work of engineering. They are also institutions, landscapes, and cultural practices. Qanats, Roman aqueducts, and the Slovak tajchy system each emerged under different hydroclimatic constraints and socio-economic demands, yet all share a core logic: the use of terrain and gravity to secure reliable

Z. Perkowski et al. (Eds.): ECCE 2026, LNCE 929, pp. 207–219, 2027.
https://doi.org/10.1007/978-3-032-39515-3_16

water services over long time horizons. Their endurance is partly technical—embedded in design choices, material performance, and maintenance access—but also sociopolitical, depending on institutions that coordinate upkeep and regulate distribution and access [2, 3, 11]. The UNESCO World Heritage framework provides a governance setting that shapes how these infrastructures are documented, assessed, protected, and communicated. The 1972 Convention establishes international obligations for the identification and protection of heritage of Outstanding Universal Value (OUV), and foregrounds the need for effective protection and management measures [42]. Within this framework, hydrotechnical heritage poses distinctive challenges: system integrity may depend on linear elements (channels, ditches, tunnels), hidden components (qanat galleries), and broader catchment conditions (reservoir inflows, groundwater recharge), as well as on visually prominent structures such as arcaded aqueduct bridges [8, 32].

This article compares three system types that exemplify different combinations of capture, conveyance, storage, and distribution:

1. Qanats as subsurface conveyance and groundwater-tapping infrastructures, typically maintained through vertical access shafts [10, 46];
2. Roman aqueducts as long-distance conveyance systems supplying urban centers and associated water cultures [8, 22];
3. Slovak tajchy as networks of surface reservoirs with collection and driving channels developed to stabilize water supply and provide hydraulic power for mining and processing, particularly in the Banská Štiavnica region [5, 24].

The objective is not to establish a universal typology, but to offer a coherent comparative account that remains technically credible and governance-aware. Particular attention is given to adaptive operation, understood here as the practical system of decisions and practices through which infrastructure is sustained under changing conditions—technological, environmental, institutional, and economic. This includes routine inspection and maintenance, negotiated allocation or service rules, repairs and selective modernization, and risk management under climate variability and extremes [39, 40]. A further goal is to present the Slovak tajchy in a manner that is analytically comparable to qanats and aqueducts without collapsing their differences. Comparability is treated as conditional and criteria-based, rather than as an assertion of equivalence. This is especially important in a heritage setting, where claims of uniqueness or exceptionality can inadvertently conflate cultural significance with technical performance, thereby obscuring operational constraints such as dam safety, water quality, and catchment governance [4, 18].

2 Materials and Methods: Evidence Base and Comparative Framework

2.1 Study Design and Evidence Sources

The article is structured as a narrative review with comparative document analysis, using three categories of evidence. First, it draws on specialist research addressing historic water supply and hydraulic engineering, long-term operation, and patterns of reuse and repair [2, 16, 28]. Second, it uses UNESCO/ICOMOS nomination and evaluation documentation as formalized accounts of heritage significance, integrity conditions, and management expectations, including materials related to the Historic City of Yazd and Las Médulas [14, 15], as well as UNESCO World Heritage Centre documentation on relevant tentative list nominations involving aqueduct-related cultural landscapes and ensembles [43–45]. Third, it draws on Slovak management plans for the Banská Štiavnica property and its technical monuments, treated as governance instruments that identify protection needs, threats, and operational challenges [32, 33].

Sources were included when they provided at least one of the following:

- technically grounded description of system components and operating principles;
- evidence of long-term use, repair, or transformation;
- discussion of governance, stakeholder participation, and heritage management;
- identification of threats and risk drivers, including climate-related pressures;
- empirical evidence relevant to safety, sedimentation, or water quality constraints.

2.2 Comparative Variables and Unit of Analysis

The unit of analysis is the hydrotechnical system, not merely a single structure. Comparison is organized across six variables that remain meaningful for all three system types:

(1) functional chain (capture–conveyance–storage–distribution–use);
(2) structural logic and materiality (dominant construction types and typical vulnerabilities);
(3) maintenance regime (access, periodicity, repair traditions);
(4) governance and allocation (institutional arrangements enabling operation and upkeep);
(5) heritage attributes and integrity conditions (including hidden and linear components);
(6) risk drivers and operational thresholds (climate variability, groundwater decline, sedimentation, and safety constraints).

This framework allows comparison without forcing uniformity. For example, the maintenance regime is present in all three system types, but its technical form differs: qanat maintenance is structured through shaft access and tunnel cleaning; aqueduct maintenance through channel clearing and structural repairs; tajchy maintenance through embankment surveillance, outlet/spillway function, and channel upkeep [10, 24, 30, 39].

3 System Typologies and UNESCO-Framed Heritage Significance

3.1 Qanats: Subsurface Gravity Systems and Their Landscape Expressions

Qanats are subsurface galleries that tap groundwater—often via a deeper mother well—and convey it downslope by gravity to points of use. Vertical shafts serve multiple roles: removing excavated material during construction, providing ventilation, and later providing access for inspection and maintenance [10]. The regular pattern of shafts can form a persistent surface signature, readable in landscape morphology and aerial imagery.

As a technological strategy, qanats are designed to minimize losses and energy use by avoiding continuous mechanical pumping, provided that groundwater levels remain sufficient and maintenance is sustained [46]. Historical scholarship emphasizes the long-standing role of qanats in Middle Eastern water cultures and notes their concentration in Iran [11, 26, 29].

In contemporary terms, qanat sustainability is increasingly constrained by aquifer-level dynamics and competition with pumped wells. Mechanized abstraction can accelerate groundwater decline and destabilize qanat hydraulics and structures, contributing to drying, collapses, and service loss [25]. These processes highlight that qanats are not solely heritage objects; they remain entangled with groundwater governance and land-use pressures. UNESCO inscription provides an additional layer of institutionalization for selected qanat-related heritage. The ICOMOS evaluation for the Historic City of Yazd frames water management as a key enabling condition for urban form and continuity in an arid context, including the ongoing relevance of qanat-based supply systems [14]. From a system perspective, heritage significance is therefore linked not only to tangible elements such as shafts and channels, but also to the operational knowledge and governance practices through which water is allocated and infrastructure maintained, a point echoed in broader discussions of local involvement and traditional practices in the World Heritage system [29, 41].

3.2 Aqueducts: Long-Distance Conveyance Systems and Their Urban Expressions

Roman aqueducts were designed to deliver large volumes of water over long distances using controlled gradients. Their construction combined underground channels with above-ground arcades where topography required (Fig. 1). The technical effectiveness of these systems depended on surveying, careful gradient control, and a capacity for repair and adaptation over time [8, 13, 29]. Aqueducts supported urban water cultures—public baths, fountains, and civic water distribution—that structured social life and urban identity [2, 27]. The diversity of Roman and post-Roman aqueduct technologies includes features such as inverted siphons, which required particular engineering solutions and presented distinct vulnerability profiles compared to open channels and arcades [17]. Regional studies in Iberia document both construction histories and patterns of abandonment, underscoring that aqueduct lifecycles were shaped by shifting political economies, urban growth, and changing maintenance capacity [22, 39].

Post-Roman trajectories further complicate any simple decline narrative. Evidence from Spain shows that aqueducts could remain functional through reuse, repair, and selective rebuilding, producing hybrid material histories rather than a single original state [39]. Elsewhere, reassessments of systems such as Caesarea emphasize that archaeological interpretation can revise assumptions about hydraulic configuration, reinforcing the need for cautious technical claims in heritage discourse [34]. In a different register, Thessaloniki's water-supply history illustrates how successive infrastructures accumulate and transform over centuries, often intersecting with earlier hydraulic traces rather than simply replacing them [47]. UNESCO documentation also highlights how aqueduct-related heritage is often embedded within broader ensembles and landscapes rather than treated as isolated monuments. Tentative list nominations in France emphasize networks of

Fig. 1. Arcaded remains of the Roman aqueduct at Olba, preserved within a rocky valley landscape near the ancient urban area. Photo: Erik Hrnčiarik.

ancient cities, territories, and associated infrastructures, including aqueducts [44], while documentation relating to Split similarly points to the significance of integrated heritage ensembles in which infrastructural elements can be relevant to interpretation and management [43]. Las Médulas provides a further example where hydraulic works are integral to an extractive landscape; the ICOMOS evaluation records debates over environmental transformation alongside recognition of heritage value [15].

3.3 Slovak Tajchy: Reservoir Networks and Their Mining Landscape Expressions

The Slovak tajchy are artificial surface reservoirs connected by systems of collection and driving channels (Fig. 2). They were developed to address limited surface flows, hydrogeological constraints, and energy demands in mining districts, particularly in the Banská Štiavnica ore region. Stored water could be delivered to power dewatering, ore processing, and metallurgical operations, forming an integrated hydraulic-energy system [5, 24]. Historical accounts emphasize the sophistication of the Banská Štiavnica system and its integration with broader mining infrastructure. The network's scale is described in the mid-20th century literature as comprising dozens of reservoirs and extensive channel alignments, with a reduced number surviving today [12, 24].

Comparable Slovak waterworks beyond Banská Štiavnica, including the Turčekovský water supply (15th century) and the Špaňodolinský water supply (1639), illustrate that engineered water conveyance was also part of a wider regional tradition [48]. The tajchy were shaped by notable engineers such as Jozef Karol Hell, Matej Kornel Hell, and Samuel Mikovíny, whose contributions are associated with innovations in pumping and system design and with the broader modernization of mining operations [5,

Fig. 2. Klinger: an 18th-century tajch, showing the embankment and the inlet section of the supply channel that fed water into the reservoir for hydraulic power in mining operations. Photo: Lucia Nováková.

48]. Their significance is therefore simultaneously technical, economic, and cultural: the system both enabled mining productivity and became embedded in a distinctive montane landscape, later recognized as a heritage landscape [19].

Within the UNESCO framing, the tajchy are interpreted as technical monuments integrated with the historic core of Banská Štiavnica and its cultural landscape, reinforcing their relevance beyond engineering history alone [20, 24]. Archaeological and heritage research on mining towns also highlights the methodological and interpretive challenges of investigating complex industrial-urban systems whose key attributes span built fabric, subsurface works, and landscape infrastructures [21]. Management planning documents similarly emphasize that system integrity can be undermined by gaps in legal protection and ownership clarity affecting linear components such as channels and ditches [32].

4 Adaptive Operation Over the Long Term: Maintenance, Repair, and Transformation

Long-term operation of historic hydrotechnical systems has depended on maintenance regimes and on the ability to integrate repairs and selective modernization without undermining system coherence. While technological design enables performance, it does not guarantee endurance: continuity depends on institutional arrangements capable of mobilizing labor, resources, and expertise across generations [2, 41].

In qanat systems, maintenance is facilitated by the vertical shafts that enable cleaning, stabilization of tunnels, and management of sediment and collapse risks. This maintenance infrastructure is inseparable from the social institutions that allocate water and coordinate upkeep; where those institutions weaken, physical degradation accelerates [11, 25]. Qanat design is often presented as sustainable in a narrow energy sense, yet its operational viability in modern settings is increasingly conditioned by groundwater governance and by competition with pumped wells that change aquifer dynamics [25, 46].

Roman aqueduct maintenance historically involved routine channel cleaning, repair of cracks and structural elements, and occasional rebuilding of vulnerable segments. In post-Roman contexts, reuse and repair could sustain function even as political and economic settings changed [23]. Evidence from Iberia emphasizes functioning aqueducts in later periods as products of compromise, repair practices, and adaptive reconfiguration [16]. Urban water systems could also be reinterpreted and redesigned in later centuries: the restoration of the Acqua Vergine (Aqua Virgo) in sixteenth-century Rome illustrates how restoration can draw simultaneously on precedent and experiment in order to reconcile hydraulic function with urban representation and institutional objectives [37].

For the Slovak tajchy, long-term operation requires attention to reservoirs, embankments, outlets, spillways, and channel networks. Historical accounts record repeated interventions: embankments were raised to increase storage, seepage and stability were managed through repairs, and system elements were reconfigured as mining demands shifted [5, 24]. Some reservoirs were reduced in operational relevance or taken out of active use when new capacities were built or when industrial energy regimes changed. The nineteenth-century spread of steam-powered pumping and later electrification progressively reduced dependence on hydraulic power from stored water, transforming the tajchy from core mining infrastructure into multipurpose resources for water supply, fire protection, and, eventually, recreation (Fig. 3) [24, 48].

Fig. 3. Červená Studňa: a historic tajch traditionally used for ice skating during the winter season. Photo: Lucia Nováková.

Adaptive operation in contemporary heritage settings also includes governance and stakeholder participation. Heritage sites face management challenges linked to funding, regulatory complexity, and competing uses—issues discussed for World Heritage settings in different national contexts [1]. Local participation can strengthen stewardship and legitimacy, but it must be integrated into coherent management strategies and institutional responsibilities rather than treated as an automatic benefit [35, 41]. Tourism-oriented initiatives can likewise generate both opportunities and pressures: studies of World Heritage tourism emphasize the need to balance community interests, visitor expectations, and conservation requirements [36, 49]. In the specific case of aqueduct-related attractions, stakeholder involvement has been argued to shape prospects for tourism development and interpretation, as discussed for the Cherchell aqueducts [6].

Finally, the legitimacy of interventions—especially those described as reconstruction—requires careful framing in heritage doctrine. Reconstruction debates from the Venice Charter tradition to later doctrinal developments stress the need to avoid conjectural rebuilding that could undermine heritage values [38]. In hydrotechnical contexts, this implies that technically necessary repairs (e.g., stabilizing an embankment, rehabilitating an outlet) should be distinguished from speculative reconstruction of lost forms, and that interventions should remain compatible with the evidentiary basis and the integrity conditions of the heritage system.

5 Climate, Risk, Safety, and Water Quality: A Risk-Informed Perspective

Historic water infrastructures are exposed to multiple interacting hazards: hydroclimatic variability, extreme events, seismicity, pollution, changing land use, and governance constraints. A risk-informed perspective distinguishes hazard (what can go wrong), exposure (what is at stake), vulnerability (how susceptible the system is), and impacts (human, environmental, and heritage losses). Climate change amplifies this perspective by altering precipitation regimes and increasing the likelihood or intensity of extremes, which can stress both physical fabric and institutional capacity [40].

For qanats, a central contemporary hazard is groundwater decline. Mechanized abstraction and aquifer depletion can reduce or eliminate qanat flow and can destabilize subsurface galleries, raising the likelihood of collapse and long-term system failure [25]. Exposure is concentrated among communities and agricultural systems that rely on qanat-fed water, while vulnerability is shaped by the weakening of traditional institutions and by regulatory or economic structures that privilege short-term pumped abstraction. This coupling of hydrology and governance reinforces the importance of long-term monitoring and coordinated allocation as part of ongoing operation [11, 41].

For aqueducts, risks include earthquakes, floods, and long-term material deterioration, particularly in urban environments where pollution and development pressures can accelerate decay and complicate interventions [8, 31]. Studies of specific structures, such as inverted siphons, underscore that vulnerability can be linked to both original design constraints and later modifications, making systematic inspection and documentation important for both safety and conservation (Fig. 4) [17]. Roman dam scholarship also highlights how hydraulic infrastructures have historically required careful evaluation of structural behavior and safety, offering interpretive parallels for contemporary risk assessment [7].

For the Slovak tajchy, the risk profile is strongly shaped by their character as historic reservoir infrastructures with embankments and spillways [30]. Climate variability and extremes can generate rapid inflows and overflow risks; land-use changes in catchments

Fig. 4. Aspendos aqueduct with the inverted siphon system, where pressure-based conveyance imposed high structural and maintenance demands. Photo: Erik Hrnčiarik.

can increase sediment yield and degrade retention capacity; and long-term sedimentation can reduce operational flexibility and ecological quality. Empirical work on historical reservoirs near Banská Štiavnica documents changing retention characteristics, illustrating that sedimentation and hydrological shifts can materially affect system performance [18]. Management and funding constraints further condition vulnerability: rehabilitation and routine maintenance require sustained institutional commitment, and gaps in the legal and ownership status of linear elements can undermine system-wide integrity [33, 34].

Water quality is a cross-cutting constraint that shapes which contemporary functions are feasible. Where reservoirs or channels are used or considered for potable supply, water treatment, catchment protection, and monitoring become decisive. Slovak water-treatment scholarship emphasizes that future potable-water viability depends on treatment capacity and planning rather than on generalized assumptions about clean sources [4]. Where systems serve recreation and tourism, operational goals shift toward stable access, safe bathing conditions, and landscape amenities, but these goals must remain compatible with safety and conservation needs and with the ecological dynamics of reservoirs and catchments [36, 49]. Risk management in heritage infrastructures is therefore not a separate technical add-on; it is integral to responsible stewardship. This is visible in the historic record as well: accounts of the tajchy describe how breaches and near-failures shaped subsequent repairs and operational decisions [24], while contemporary discussion of Slovak restoration and financing highlights the persistent challenge of sustaining preventive maintenance and coordinated governance over time (Fig. 5) [9].

Fig. 5. While the reservoir of the Beliansky tajch has undergone recent reconstruction, most supply and collection ditches are currently non-operational, largely because of modern disturbances. Photo: Lucia Nováková.

6 Conclusions

Qanats, Roman aqueducts, and the Slovak tajchy illustrate different but comparable ways in which societies have managed water through gravity-driven infrastructures. Their comparability is strongest when it is framed through shared variables: system integration across the water-service chain, institutionalized maintenance, governance of allocation and access, and landscape-level effects that bind technical function to cultural meaning [8, 10, 19]. At the same time, differences remain decisive for both operation and heritage management. Qanats are tightly coupled to groundwater levels and subterranean stability; aqueducts balance conveyance engineering with urban embedding and structural vulnerabilities such as siphons; tajchy are storage-centered and entail dam-safety and sediment-management obligations that intensify under climate variability [17, 18, 25].

Within the UNESCO World Heritage governance context, these systems underscore the importance of treating integrity as a property of connected infrastructures and their enabling conditions, not merely of isolated monuments. Linear and hidden components—qanat galleries and shafts, aqueduct corridors and channels, tajchy ditches and outlets—can be structurally and legally fragile, yet they may be critical to sustaining the attributes through which heritage value is expressed and interpreted [32, 42]. UNESCO/ICOMOS documentation also shows that hydrotechnical heritage often appears within broader ensembles and cultural landscapes, as illustrated by the evaluation of Las Médulas and the nomination of Yazd [14, 15], and by tentative list dossiers that frame aqueducts within wider territorial and urban systems [43, 44].

Adaptive operation emerges as a central concept for contemporary stewardship because it links conservation values to operational practices. In practical terms, adaptive operation involves: regular inspection and preventive maintenance; clear governance and responsibility allocation; monitoring of key variables (groundwater levels, inflows, sedimentation, seepage, water quality); and decision rules that define safety thresholds and acceptable trade-offs among uses. These requirements do not diminish heritage value; they are essential to sustaining it under modern pressures, including climate stressors and tourism-driven expectations [36, 40, 49].

For the Slovak tajchy, the main implication is that long-term sustainability depends on coupling heritage-sensitive conservation with the operational responsibilities of historic reservoir infrastructures. This includes maintaining embankment safety and spillway capacity, addressing sedimentation and catchment erosion, and differentiating feasible functions (potable supply, recreation, fire protection, ecological services) based on water quality and treatment capacity [4, 18]. The same logic applies, in different technical forms, to qanats and aqueducts, where infrastructure endurance depends on institutionalized maintenance and on governance able to manage resource competition and development pressures [16, 25, 41].

In conclusion, these hydrotechnical systems should be understood as living infrastructures with layered histories rather than as static relics. Their technical and cultural significance is sustained through ongoing operational decisions—decisions that increasingly must be risk-aware, climate-attentive, and institutionally grounded if heritage value is to remain legible and meaningful for future generations.

References

1. Al-Tokhais A, Thapa B (2020) Management issues and challenges of UNESCO world heritage sites in Saudi Arabia. J Herit Tour 15(1):103–110
2. Angelakis A et al (2012) Evolution of water supply throughout the millennia. IWA Publishing, London
3. Barghouth JM, Al-Sa'ed RMY (2009) Sustainability of ancient water supply facilities in Jerusalem. Sustainability 1(4):1106–1119
4. Barloková D et al (2018) Úprava vody a jej budúcnosť v podmienkach Slovenska. In: Sborník příspěvků VODA ZLÍN 2018. Moravská vodárenská, Olomouc, p 184
5. Bárta V (1968) Banská Štiavnica: Sprievodca Mestom a Okolím. Stredoslovenské vydavateľstvo, Banská Bystrica
6. Chennaoui Y (2024) The importance of stakeholder involvement in promoting the Cherchell aqueducts as tourist attractions: current situation and prospects. ARCHive-SR 8(2):1–11
7. Díaz Marta M (1984) Comentarios el estudio conjunto sobre la presa romana de Consuegra. Rev Obras Públicas 3223:191–205
8. Fernandez Casado C (1968) Los acueductos romanos. Inf Constr 20(197):53–82
9. Galková H (2024) Havária na tajchoch Veľká Vodárenská a Klinger (Február 1977). Etnologické rozpravy 31(1):29–34
10. Ghasemi H et al (2013) Application of traditional architectural structure as sustainable approach to mitigation of shortage water supply in desert regions. Acad J Sci 2:125–132
11. Goblot H (1979) Les qanats: Une technique d'acquisition de l'eau. Éditions de l'École des Hautes Études en Sciences Sociales, Paris
12. Grega V, Vozár J (1964) Banská Štiavnica. Stredoslovenské vydavateľstvo, Banská Bystrica

13. Hauck G (1986) Structural design of the Pont du Gard. J Struct Eng 112(1):105–120
14. ICOMOS (2026) Historic city of Yazd. Report on the nomination for the world heritage list (1544-ICOMOS-2148-en). ICOMOS. https://whc.unesco.org/en/list/1544/documents. Last accessed 15 Jan 2026
15. ICOMOS (2026) The Las Médulas Roman gold-mining area. Report on the nomination for the world heritage list (803-ICOMOS-949-en). ICOMOS. https://whc.unesco.org/en/list/803/documents/. Last accessed 15 Jan 2026
16. Martínez Jiménez J (2012) Reuse, repair and reconstruction: functioning aqueducts in post-Roman Spain. In: Jervis B, Kyle A (eds) Make-do and mend: archaeologies of compromise, repair and reuse. Archaeopress, Oxford, pp 27–39
17. Kessener PM (2000) The aqueduct at Aspendos and its inverted siphon. Am J Archaeol 13:105–132
18. Kubinský D et al (2015) Changes in retention characteristics of historical artificial water reservoirs near Banská Štiavnica, Slovakia. Open Geosci 1:880–887
19. Labuda J (2022) Človek a montánna krajina neskorého stredoveku a novoveku: príklad Banskej Štiavnice. Archaeologia historica 47(2):719–730
20. Labuda J (2000) Pozoruhodné nálezy zo Starého mesta v Banskej Štiavnici. Archaeologia historica 25(1):7–24
21. Labuda J (2007) Špecifiká archeologického výskumu banského mesta. Archaeologia historica 32(1):103–112
22. Leather GM (2002) Roman aqueducts in Iberia. Sullom Side Books, Sullom Side
23. Leigh S (1997) The "reservoir" of Hadrian in Athens. J Rom Archaeol 10:279–290
24. Lichner M et al (2002) Banská Štiavnica: Svedectvo času. Štúdio Harmony, Banská Bystrica
25. Maghrebi M et al (2022) Anthropogenic decline of ancient, sustainable water systems: Qanats. Groundwater 61(1):e13248
26. Moosavi SAA (2006) Qanat invention puzzle. In: Angelakis AN, Koutsoyiannis D (eds) Proceedings of the 1st IWA international symposium on water and wastewater technologies in ancient civilizations. National Agricultural Research Foundation, Heraklion, pp 535–540
27. Müller-Wiener W (1977) Bildlexikon zur Topographie Istanbuls: Byzantion, Konstantinupolis, Istanbul bis zum Beginn des 17. Jahrhunderts. Wasmuth, Tübingen
28. Nováková L (2024) Cesty vody – ako historické akvadukty menili krajinu. In: Guzdek A (ed) Prostor pro život: Cesty. Proceedings of the 2nd international conference Krajina Sídla Památky 2024, Brno, Czech Republic. Vysoké učení technické v Brně, Brno, pp 220–227
29. Nováková L (2024). Suppl. Sustainability of the UNESCO World Heritage Sites) UNESCO water structures: heritage, innovation and sustainable use. Cult Soc Econ Polit 4(2):27–45. https://doi.org/10.2478/csep-2024-0009
30. Nováková L (2025) Akvadukty, kanáty a tajchy: udržateľnosť v adaptácii vodných stavieb UNESCO. In: Guzdek A (ed) Prostor pro život: voda. Proceedings of the 3rd international conference Krajina Sídla Památky 2025, Brno, Czech Republic. Vysoké učení technické v Brně, Brno, pp 298–311
31. Ordax SA, Pizarro Gómez FJ (2006) Monumentos artísticos de Extremadura. Junta de Extremadura, Mérida
32. Paučulová Ľ et al (2010) Manažment plán lokality UNESCO Historické mesto Banská Štiavnica a technické pamiatky okolia: 1. etapa – podklady a analýzy. Ministerstvo kultúry SR, Bratislava–Banská Štiavnica
33. Paučulová Ľ et al (2010) Manažment plán lokality UNESCO: Historické mesto Banská Štiavnica a technické pamiatky okolia, 2. etapa edn. Ministerstvo kultúry SR, Bratislava–Banská Štiavnica
34. Porath Y (2002) The water-supply to Caesarea: a re-assessment. In: Amit D, Patrich J, Hirschfeld Y (eds) The aqueducts of Israel. Cambridge University Press, Cambridge, pp 104–129

35. Poulios I (2014) Discussing strategy in heritage conservation: living heritage approach as an example of strategic innovation. J Cult Heritage Manage Sustainable Dev 4(1):16–34
36. Richards G (2018) Cultural tourism: a review of recent research and trends. J Hosp Tour Manag 36:12–21
37. Rinne KW (2007) Between precedent and experiment: the restoration of the Acqua Vergine (1560–1570). In: Roberts L, Schaffer S, Dear P (eds) The mindful hand: inquiry and invention from the late renaissance to early industrialization. Edita, Amsterdam, pp 95–115
38. Román A (2002) Reconstruction—from the Venice charter to the charter of cracow 2000. In: Estrategias relativas al patrimonio cultural mundial. La salvaguarda en un mundo globalizado. Principios, prácticas y perspectivas. Proceedings of the 13th ICOMOS general assembly and scientific symposium. Comité Nacional Español del ICOMOS, Madrid, pp 117–119
39. Sánchez López E, Martínez Jiménez J (2016) Los acueductos de Hispania: Construcción y abandono. Fundación Juanelo Turriano, Madrid
40. Seekamp E, Jo E (2020) Resilience and transformation of heritage sites to accommodate for loss and learning in a changing climate. Clim Chang 162(1):41–55
41. Sullivan S (2004) Local involvement and traditional practices in the world heritage system. In: Linking universal and local values, vol 49. UNESCO World Heritage Centre, Paris, pp 49–55
42. UNESCO (2026) Convention concerning the protection of the world cultural and natural heritage. UNESCO. https://whc.unesco.org/archive/convention-en.pdf. Last accessed 15 Jan 2026
43. UNESCO (2026) Diocletian's palace and the historical nucleus of split (extension). Report on the tentative list nomination (ref. no. 2015). UNESCO. https://whc.unesco.org/en/tentativelists/2015/. Last accessed 15 Jan 2026
44. UNESCO (2026) Les villes antiques de la Narbonnaise et leur territoire: Nîmes, Arles, Glanum, aqueducs, via Domitia. Report on the tentative list nomination (ref. no. 1658). UNESCO. https://whc.unesco.org/en/tentativelists/1658/. Last accessed 15 Jan 2026
45. UNESCO (2026) The Great Wall of Gorgan. Report on the tentative list nomination (ref. no. 6199). UNESCO. https://whc.unesco.org/en/tentativelists/6199/. Last accessed 15 Jan 2026
46. Wessels JI (2016) Qanats of Syria. In: Angelakis AN, Chiotis E, Eslamian S, Weingartner H (eds) Underground aqueducts handbook. CRC Press, Boca Raton, pp 229–240
47. Yannopoulos S et al (2017) Evolution of water supply infrastructures of Thessaloniki city, Hellas, through centuries. Eur Water 58:465–472
48. Zámora P et al (2003) Dejiny baníctva na Slovensku, vol 1. Grafička agentúra Bratislava, Košice
49. Zbuchea A (2020) World heritage sites, local communities and tourists. PLURAL 8(2):77–90

Stabilisation and Correction of Tower Leaning Through Controlled Ground Weakening: A Case Study of the Dansker Tower at Kowalewo Pomorskie Castle

Tomasz Szkuta(✉)

Toruń Branch, Polish Association of Civil Engineers and Technicians (PZITB), Szeroka 34, 87-100 Toruń, Poland
tomszkuta@gmail.com

Abstract. The dansker tower of the castle in Kowalewo Pomorskie is a valuable remnant of medieval defensive and sanitary architecture whose condition has deteriorated due to long-term differential settlement of the foundation subsoil. The preserved fragment exhibits a noticeable inclination, and geodetic monitoring indicates that deformation remains active, posing a risk to both the structure and its surroundings.

This paper evaluates the feasibility of stabilising and correcting the tower's inclination through controlled ground weakening, considering the monument's historic character, geotechnical conditions, and monitoring results. The study is based on archival records, technical and geotechnical documentation, and deformation measurements collected prior to intervention.

The analysis identified unfavourable groundwater conditions and non-uniform subsoil deformability as the main causes of inclination. A stabilisation concept is proposed based on a gradual, controlled reduction of soil bearing capacity at the foundation level, combined with structural protection and continuous monitoring.

The proposed method is minimally invasive and suited to historic structures where conventional strengthening or straightening techniques are difficult to apply. It may provide a basis for further design and conservation work.

Keywords: Dansker Tower · Kowalewo Pomorskie · Historic Structure · Structural Inclination · Controlled Ground Weakening · Structural Stabilisation

1 Introduction

The dansker tower of the castle in Kowalewo Pomorskie is one of the few preserved relics of the former Teutonic complex (Fig. 1) and an important testimony to medieval technical culture in the Chełmno Land [4, 5]. Originally serving sanitary and communication purposes, it was situated outside the principal defensive perimeter and connected to the convent house via a gallery (Fig. 2). Despite its advanced state of destruction and the loss of a substantial part of its original fabric, the preserved fragment (Fig. 3) remains an important element of the historical landscape and a subject of interest to researchers in architecture, archaeology, and conservation [4, 5].

Z. Perkowski et al. (Eds.): ECCE 2026, LNCE 929, pp. 220–233, 2027.
https://doi.org/10.1007/978-3-032-39515-3_17

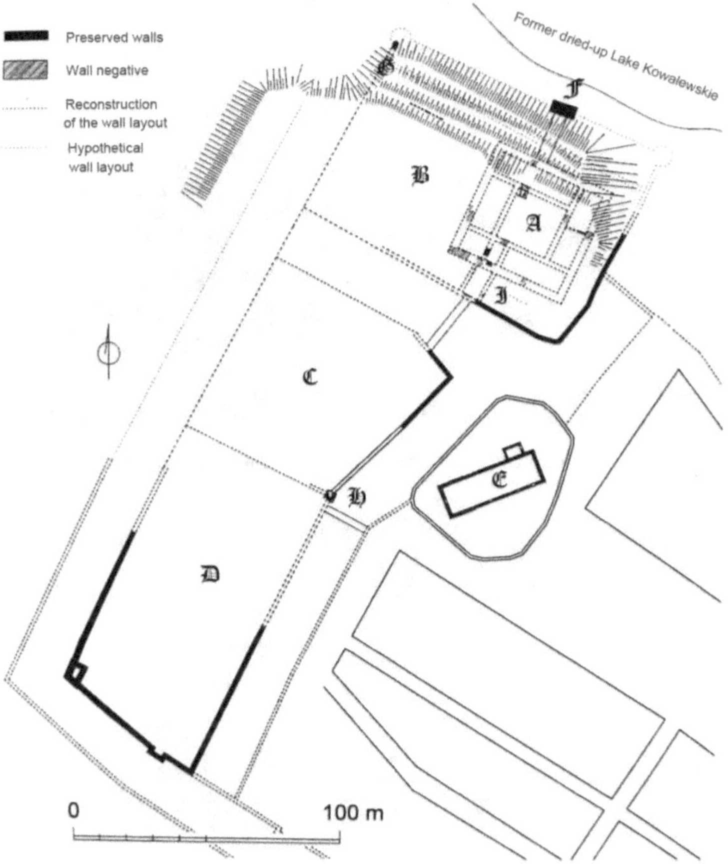

Fig. 1. Reconstruction of the castle plan. The letter F marks the dansker tower [4].

In recent years, particular attention has been drawn to the progressive inclination of the preserved structure. This deformation results from the long-term influence of unfavourable groundwater conditions, historical site transformations, and gradual deterioration of the masonry. Because of the monument's historical character and high structural sensitivity, the choice of intervention method must reconcile structural safety requirements with the principle of the minimum possible interference with the historic fabric.

The stabilisation and rectification of leaning masonry structures is a particularly complex problem, especially when historic buildings founded in difficult ground conditions are concerned. In such cases, conventional methods of foundation strengthening or forceful straightening may lead to secondary damage, loss of authenticity of the historic substance, or an unacceptable level of execution risk. For this reason, staged solutions based on detailed geotechnical investigation, continuous monitoring, and gradual control of subsoil response are very important [1–3].

Fig. 2. Probable appearance of the castle based on research [4].

Fig. 3. Current state: a single tower against the background of the no longer existing castle and lake [4].

Flagship examples reported in the literature include the intervention carried out on the Leaning Tower of Pisa [1, 2] and corrective measures applied to other historic structures founded on subsoil susceptible to differential settlement [3]. In these cases, the combination of geotechnical action, long-term displacement monitoring, and a cautious approach to the extent of inclination correction proved to be essential.

For the dansker tower in Kowalewo Pomorskie, a particularly promising approach appears to be the method of controlled ground weakening applied on the side opposite to the direction of tilt. This method enables predictable, limited corrective displacements without imposing abrupt actions on the structure. Its successful use, however, depends on proper recognition of the object's geometry, subsoil stratification, deformation dynamics, and conservation constraints.

The aim of this paper is to assess the feasibility of stabilising and controlling the correction of the inclination of the dansker tower of the castle in Kowalewo Pomorskie by means of controlled ground weakening. The paper presents the structural characteristics, the geotechnical analysis, the geodetic monitoring results, and a proposed intervention concept tailored to the monument's specific character.

2 Characteristics of the Structure and Its State of Preservation

The dansker tower of the castle in Kowalewo Pomorskie is a remnant of a medieval sanitary tower functionally related to the Teutonic castle complex. The structure was identified during archaeological and architectural investigations conducted in 2013–2015 (Fig. 4), thereby enabling the reconstruction of the spatial layout of the high castle and its communication and service facilities [4, 5]. These investigations confirmed the dansker's important role in the complex's sanitary and defensive system.

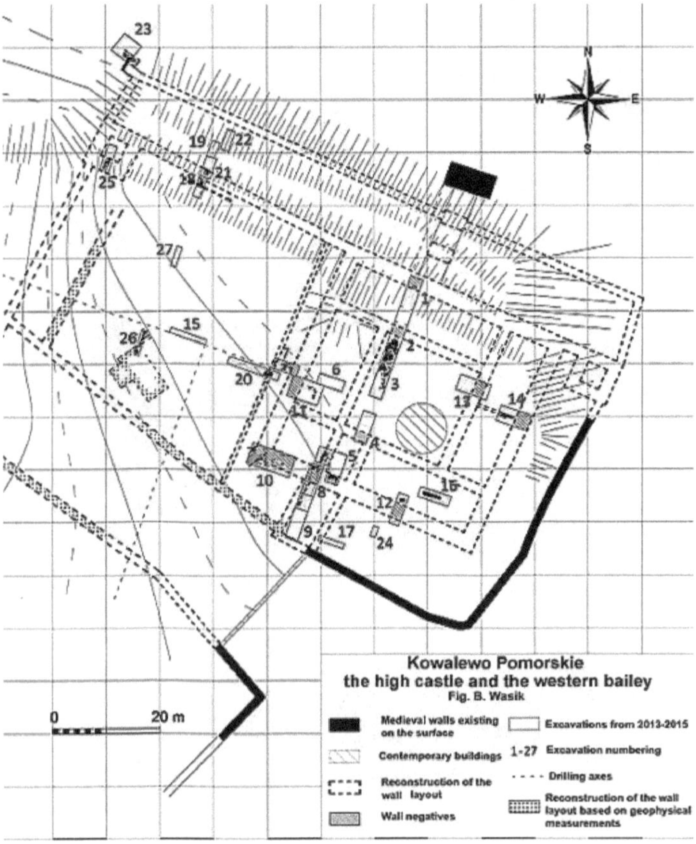

Fig. 4. Spatial layout of the high castle [4].

The site on which the structure stood is clearly sloping (Fig. 5) and, in the past, adjoined a lake that served as a natural receiver of waste (Fig. 2). Such a location was consistent with the typical arrangement of sanitary towers in Teutonic castles. At present, only a free-standing masonry relic has survived, representing a fragment of the former tower (Figs. 3 and 6).

Fig. 5. Photograph from the beginning of the twentieth century: the dansker tower and the shape of the slope (without vegetation).

Fig. 6. Contemporary photograph: present vegetation stabilising the slope.

The preserved part of the object was probably erected in the first half of the fourteenth century in solid brick bonded with lime mortar and founded on a stone foundation approximately 1.0–1.2 m thick. According to the available data, the structure has a plan of approximately 4.05 m × 8.75 m, while the height of the preserved part is about 14.5 m.

The above-ground structure still shows characteristic features of the historic brickwork, including the use of overburnt bricks, as well as relics of original architectural solutions.

The structure's state of preservation must be assessed as poor. Numerous defects in the masonry are visible, including mortar losses, cracking, and traces of long-term moisture action. The brick surface shows flaking, joint erosion, and local effects of salt crystallisation. The most alarming phenomenon is the significant deviation of the structure from the vertical, caused by differential ground settlement. This process is not merely historical but remains active.

From a conservation point of view, the structure requires measures to ensure structural safety while preserving as much of the authentic historic substance as possible. This dual perspective—engineering and conservation—determines the selection of analytical and intervention methods.

3 Source Materials and Basis of Analysis

The study was based on archival materials, technical assessments, geological and geotechnical documentation, geodetic observations, and records concerning the condition of the structure and its surroundings. The combined analysis of these data made it possible to determine both the present technical condition of the dansker and the initial conditions required to plan stabilisation measures.

The analysis used, in particular, the architectural, historical, and conservation documentation concerning the castle in Kowalewo Pomorskie [4, 5], technical assessments of the ruins of the dansker tower [6], geodetic reports prepared in 2023–2024 [7], and geological and geotechnical documentation prepared for the foundation zone of the structure [8]. Information resulting from the programme of planned conservation and protective works was also taken into account [9].

The adopted analytical approach consisted of comparing data on the object's geometry and material condition with the results of geotechnical investigations and the analysis of deformation dynamics. This made it possible to identify the principal factors responsible for the tower inclination and to assess the feasibility of corrective measures based on controlled interaction with the subsoil.

The study has a diagnostic and conceptual character. Its purpose is not to present a full execution design, but rather to justify and organise the technical premises that lead to the selection of a particular method for stabilisation and inclination correction. The presented considerations may, however, serve as a basis for further design work.

4 Site Conditions and Geotechnical Background

The dansker tower is founded in an area with complex groundwater conditions that have a decisive influence on its stability. The terrain surrounding the structure lies at elevations of approximately 86.0–88.0 m above sea level and shows clear topographic variation. On one side, the tower is adjacent to a slope, while on the northern side, there is an area transformed by infrastructure works, including underground utility networks and historically altered water conditions. In the past, this area was associated with a lake (Fig. 2) that later dried out and was filled in.

The geotechnical investigation included boreholes ranging from 6.0 to 10.5 m in depth, as well as dynamic soundings, enabling recognition of ground stratification (Figs. 7 and 8). On this basis, layers with varying bearing capacity and deformability were identified, which are of fundamental importance for the behaviour of the foundation. Particularly unfavourable was the occurrence of weaker and more settlement-prone soils in the part of the foundation zone towards which the deformation developed.

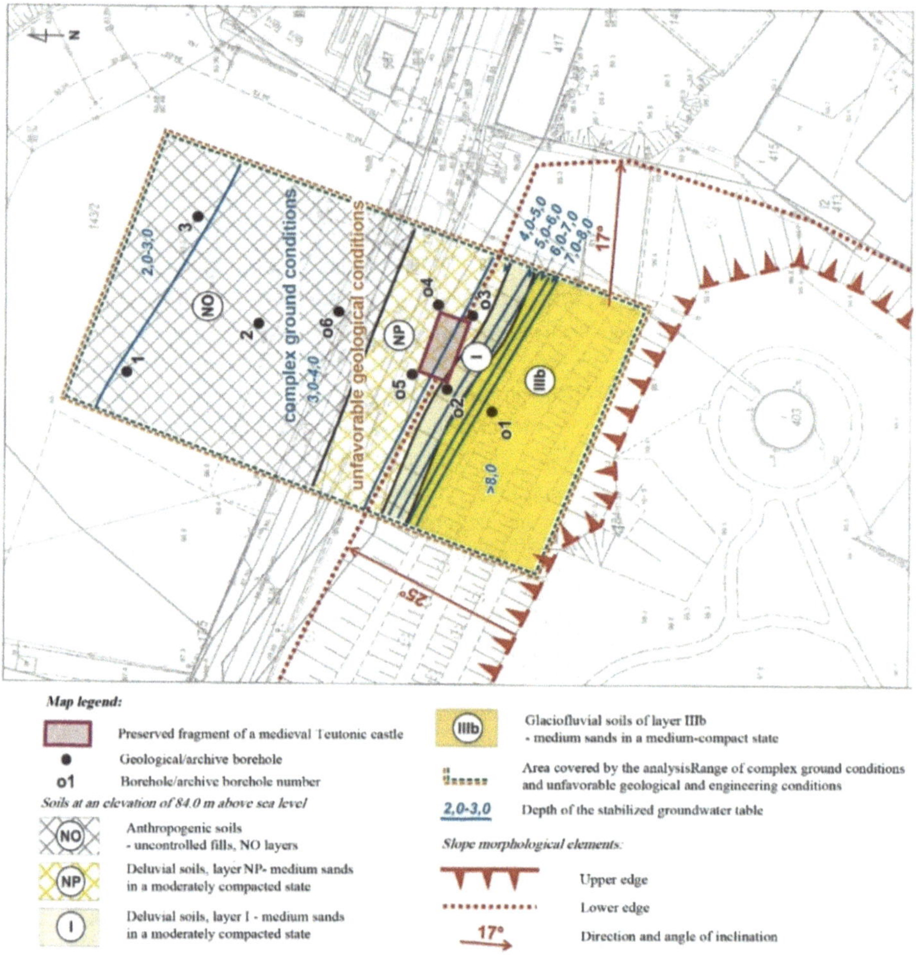

Fig. 7. Geological map.

The investigation results indicate that the arrangement of the subsoil layers is unfavourable for long-term structural stability. Although there are no clear indications of an active landslide, the possibility of secondary soil displacements must be considered, especially under changing moisture conditions, dewatering, or local earthworks in the vicinity. Under such conditions, the subsoil may respond unevenly, leading to further development of the inclination.

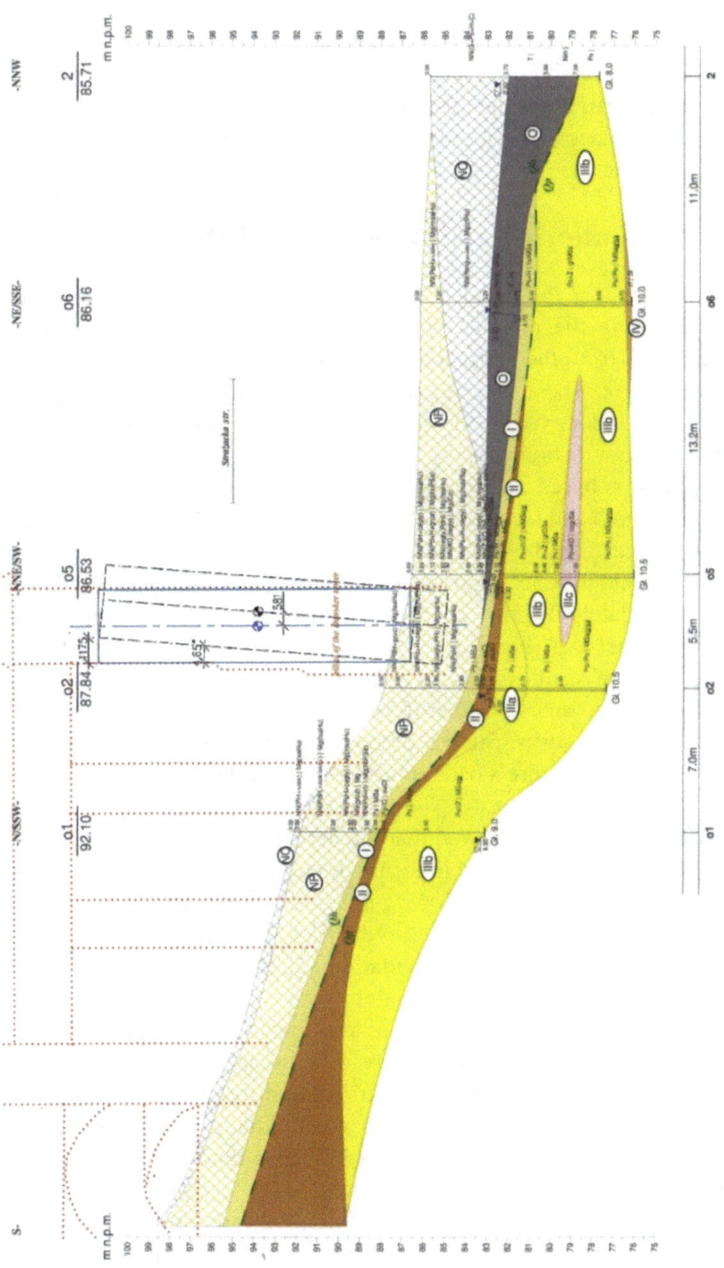

Fig. 8. Geological cross-section with the location of the dansker tower.

An important conclusion of the geotechnical investigations is that the ground conditions permit the controlled induction of limited displacements through staged reduction of bearing capacity in a selected zone. This feature supports the consideration of controlled ground weakening as a potentially effective, minimally invasive intervention technique.

It should be emphasised that the geotechnical conditions not only explain the origin of the observed deformations, but also define the limits of safe remedial action. For this reason, all execution decisions should be closely linked to the ongoing interpretation of monitoring results and to the possibility of staged correction of the planned works.

5 Geodetic Monitoring and Assessment of Deformation Dynamics

Geodetic monitoring of the dansker tower control points (Fig. 9) was carried out in the period preceding the planned stabilisation intervention to determine the actual degree of deviation from the vertical and to establish whether the deformation was stabilised or progressive. These measurements were of fundamental importance in assessing structural risk and served as the reference basis for the proposed stabilisation and inclination-correction technology.

The analysis was based on measurement series from July, October, and November 2023, and from March and October 2024 [7]. Reports from these monitoring campaigns document an increase in the structure's deviation from the vertical: in the southern direction, from 4.39° to 4.65°, and in the western direction, by 0.27°. For a structural height of approximately 14.5 m, this corresponds to a displacement of about 1.18 m relative to the vertical axis (Fig. 10).

The comparison of results from successive dates enabled assessment of changes in the object's geometry over time and identification of deformation tendencies in the period preceding the planned works. Such a monitoring time range is particularly important because it reflects the structure's behaviour under natural service conditions, without the influence of remedial measures.

The measurement results showed that the deformation process is active. Successive monitoring campaigns confirmed that the inclination is not solely the result of historical settlement processes, but is still developing. Based on the recorded data, the rate of change was estimated at approximately 3.5 cm/year. For a historic structure with weakened material integrity and unfavourable foundation conditions, this value must be regarded as significant from the point of view of safety.

The interpretation of the geodetic results is consistent with the geotechnical investigation. The observed displacements may be associated with long-term differential settlement of the subsoil, further influenced by hydrological changes in the area and earthworks carried out in the vicinity. This means that geodetic monitoring does not serve merely a documentary purpose but constitutes a fundamental diagnostic tool that confirms the necessity of stabilisation measures.

From the perspective of the planned intervention, the monitoring results serve another important function: they define the initial state against which the effectiveness of successive stages of inclination correction can be assessed. In the case of such a sensitive structure, systematic control of displacements is a prerequisite for safe execution and for limiting the risk of secondary damage.

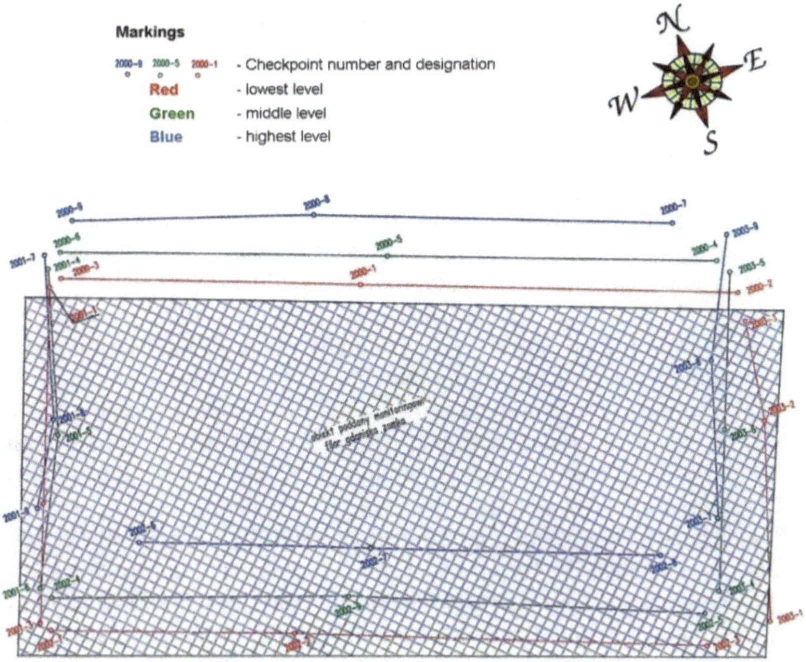

Fig. 9. Arrangement of control points.

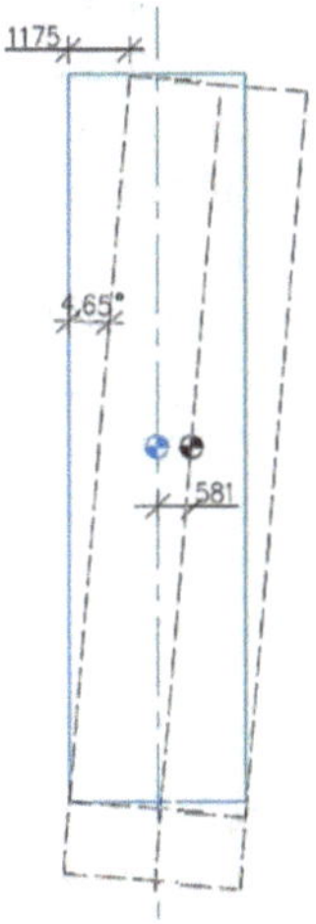

Fig. 10. Inclination of the dansker tower.

6 Assumptions of the Stabilisation and Inclination Correction Technology

The primary objective of the proposed intervention is to improve the structure's stability and reduce its inclination while maintaining the highest possible level of structural and conservation safety. Because of the historical character of the monument, it is not advisable to use abrupt methods that directly induce large structural displacements. A staged approach, in which geometric changes are induced gradually and under full measurement control, appears more appropriate.

In the analysed case, the concept is based on the use of controlled ground weakening at the foundation level. Its essence lies in gradually reducing the bearing capacity of the soil on a suitably selected side of the foundation so as to induce small, predictable displacements leading to correction of the inclination. This method requires very careful preparation and cannot be considered as an isolated geotechnical treatment. It must be combined with simultaneous structural protection, stabilisation of the surroundings, and continuous assessment of the structural response.

The first stage should involve providing structural safeguards to reduce the risk of uncontrolled displacements. For this purpose, the use of an anchoring system and temporary strengthening elements is considered, intended to absorb part of the actions during the work stages. These safeguards should be designed so that they do not cause permanent damage to the historic fabric and can be adapted to the current on-site conditions.

At the same time, slope stability and the preparation of the working zone must be ensured. Depending on the detailed site conditions, this may require retaining elements, limiting suffosion processes, and partially removing soil in the immediate surroundings of the object. The extent of such measures should result from the current assessment of the influence of earthworks on the foundation conditions.

Another important component of the technology is the installation of a strengthening ring that cooperates with the existing foundation. This solution is intended to rationalise load transfer and to create more favourable conditions for safe control of the inclination correction process. Such works should be carried out in stages on short sections to avoid local ground loosening or undesirable stress concentrations.

Only after the necessary protections have been implemented and the foundation zone prepared can the actual process of ground weakening commence. This operation should be performed gradually by drilling a limited number of boreholes with controlled geometric parameters and in a strictly defined sequence. Each stage of ground weakening must be preceded and followed by geodetic control, while subsequent actions should depend on the structure's actual response.

In practice, this means that the inclination correction technology cannot be rigidly closed. It should rather be treated as an adaptive procedure in which design assumptions are continuously verified against measurement results and site observations. Such an approach is particularly important for historic structures, for which the safety margin is often limited, and the influence of local material and geotechnical heterogeneities may be difficult to fully predict at the preliminary stage.

7 Proposed Execution Technology

The proposed technology is based on the assumption that reducing the inclination of the dansker tower can be achieved by inducing controlled, limited corrective displacements of the foundation. Unlike forceful straightening methods, this approach leverages the properties of the subsoil and enables more uniform control of the entire structure's response.

In the first place, a temporary structural protection system is envisaged. Such a system may include cable anchors fixed outside the zone of potential displacement, together with cooperating steel and retaining elements limiting the risk of a sudden increase in deformation. Detailed solutions should take into account: the geometry of the structure, the bearing capacity of the anchorage zone, and the possibility of safe installation in a historic environment.

Subsequently, the immediate surroundings of the structure, including the slope and the technological access zone, should be stabilised. The aim of these measures is to reduce the risk of secondary slips, suffosion, and uncontrolled loosening of the ground. Depending on local conditions, this may include temporary or permanent retaining elements, partial reduction of excess soil, and execution solutions allowing safe drilling operations.

An important stage is the execution of a peripheral element cooperating with the existing foundation. In the considered concept, a reinforced concrete ring beam located below ground level, connected to the existing foundation and constructed in stages, is envisaged. Such an element is intended to improve the cooperation of the foundation zone, reduce the risk of local damage, and create more predictable conditions for further corrective measures.

The principal phase of inclination correction consists of gradually weakening the ground by drilling boreholes of limited diameter and controlled depth within a designated zone. The boreholes should be carried out symmetrically with respect to the structure's axis, with quantitative limits per time unit, so that the structure's response can be continuously assessed. Too rapid a pace of work could result in loss of control over the deformation process; therefore, staging and strict technological discipline are of key importance.

During the execution of works, continuous geodetic monitoring and ongoing observation of the structure and subsoil conditions are indispensable. Assessment should concern not only changes in the value of inclination, but also possible new cracks, changes in the width of existing fractures, local settlements, and the response of the surrounding ground. These results must form the basis for decisions on continuation, modification, or suspension of subsequent stages of the works.

After completion of the main intervention phase, post-execution monitoring is necessary. Its aim is to confirm that the obtained structural state is stable and that no secondary deformation occurs due to stress redistribution or changes in groundwater conditions. In the case of the dansker tower, as a structure of high historical value and limited safety reserve, this stage should be treated as an integral part of the entire technology.

8 Discussion

The presented concept of stabilisation and inclination correction of the dansker tower through controlled ground weakening arises from the need to reconcile two fundamental requirements: structural safety and the preservation of the historic substance. In the case of masonry structures with advanced material deterioration, all methods that involve direct forceful action on the structure increase the risk of generating new damage. This concerns particularly those buildings whose material arrangement is non-homogeneous and whose internal cohesion has been weakened by long-term environmental exposure.

An important advantage of the proposed approach is its potentially minimally invasive character. The inclination correction is not achieved by imposing a large one-time displacement, but by controlled use of the deformability of the subsoil. This creates the possibility of transferring changes to the structure more gently, thus reducing the risk of sudden cracking or loss of fragments of historic material. Such a solution appears especially appropriate for an object that no longer performs its original utility function but must be preserved as a relic of high cultural and cognitive value.

At the same time, it must be emphasised that the effectiveness of this method depends on a very strict execution regime. Any underestimation of ground conditions, too rapid a pace of weakening, or insufficient protection of the working zone could lead to effects contrary to those intended. For this reason, the method requires not only good preliminary documentation but also readiness to continuously adjust assumptions based on observations made during the planned works.

A limitation of the analysed concept is also the fact that it is based on diagnostic data and preliminary recognition rather than on full execution and long-term observation of realised effects. This means that the presented solution should be regarded as a technically justified intervention concept that requires further refinement at the execution design stage. Nevertheless, even at the present stage, it may be indicated as one of the most rational among minimally invasive options.

From a conservation point of view, it is important that the proposed technology allows the relic to be preserved in situ and limits interference with the historic structure. This aspect distinguishes it from more radical solutions that might require partial dismantling, rebuilding, or intensive foundation intervention. In the case of the Kowalewo Pomorskie dansker, the preservation of the authentic material arrangement and the structure's historical location is important not only technically but also culturally and cognitively.

9 Conclusions

The performed analysis confirms that the preserved fragment of the dansker tower of the castle in Kowalewo Pomorskie is in a state of structural risk due to long-term differential settlement of the subsoil and progressive degradation of the masonry.

The geotechnical investigation demonstrated the presence of unfavourable groundwater conditions that promote further deformation. The arrangement of the subsoil layers and their differentiated deformability should be regarded as the principal cause of the observed inclination of the structure.

The geodetic monitoring results confirmed that the deformation process is active. The recorded rate of change and the scale of the inclination indicate the need to undertake protective and stabilising measures within a relatively short time.

Controlled ground weakening may constitute a rational basis for a stabilisation and inclination correction technology, provided that it is strictly combined with structural protection, stabilisation of the surroundings, and continuous geodetic and technical monitoring.

The proposed solution aligns well with the specific character of a historic structure, as it enables limiting interference with the historical fabric and preserving the relic *in situ*. For this reason, it should be regarded as particularly valuable in the context of cultural heritage protection.

The presented concept requires further development at the execution design stage, including refinement of technological parameters, safety criteria, and the detailed monitoring programme during and after the completion of works.

References

1. Burland JB, Jamiolkowski M, Viggiani C (2003) The stabilisation of the leaning tower of Pisa. Soils Found 43(5):63–80
2. Jamiolkowski MB (2001) The leaning tower of Pisa: end of an odyssey. In: Proceedings of the 15th international conference on soil mechanics and geotechnical engineering, vol 4, Istanbul, Turkey, pp 2979–2996
3. Ovando-Shelley E, Tamez Gonzalez E, Santoyo Villa E (1997) Options for correcting differential settlements in Mexico city's metropolitan cathedral. In: Proceedings of the 14th international conference on soil mechanics and foundation engineering, vol 4, Hamburg, Germany, pp 1437–1442
4. Wasik B (2020) Układ przestrzenny i architektura zamku w Kowalewie Pomorskim w świetle badań z lat 2013–2016. In: Stare i nowe dziedzictwo Torunia, Bydgoszczy i regionu (II), Studia i materiały z dziedzictwa kulturowego Torunia i regionu, vol 3, Toruń, pp 27–40. [317–323] (CC BY-ND)
5. Wasik B, Wiewióra M (2017) Zamek w Kowalewie Pomorskim. Dzieje warowni w świetle wyników badań archeologiczno-architektonicznych. Acta Universitatis Nicolai Copernici: Archeologia 35:129–158
6. Szkuta T (2025) Technical expertise of the ruins the dansker tower of the castle in Kowalewo Pomorskie. Unpublished technical report. PZITB, Toruń Branch, Toruń, Poland
7. Markowski T, Geomiar UG (2024) Geodetic monitoring reports for the dansker tower of the castle in Kowalewo Pomorskie. Unpublished documentation, Golub-Dobrzyń, Poland
8. Szczuczko TT, Geolit SC (2024) Geological and geotechnical documentation for the foundation area of the dansker tower of the castle in Kowalewo Pomorskie. Unpublished documentation, Toruń, Poland
9. Rubnikowicz-Góźdź H, Witkowska K, Rdesińska M (2024) Programme of conservation and protective works for the dansker tower of the castle in Kowalewo Pomorskie. Unpublished documentation, Poland

Social Spatial Justice, Spatial Planning

Early Twentieth Century Urban Buildings in Portugal: Main Sanitary Requirements for Dwellings and Their Impacts in Working-Class Housing

Anabela M. Moreira(✉)

TECHN&ART – Technology, Restoration and Arts Enhancement Center, Polytechnic University of Tomar, Estrada da Serra, Quinta do Contador, Tomar 2300-313, Portugal
anamoreira@ipt.pt

Abstract. The labor and social changes arising from the first industrial revolution brought about a number of profound transformations in European societies. The accumulation of people in urban centers and the insufficient sanitary conditions of existing buildings promoted the proliferation of serious contagious diseases, for which there were no medical treatments, and which were reflected in high mortality rates. Compared with what happened in the countries that pioneered the industrial revolution, in Portugal, the industrialization processes took longer, and the problems resulting from growing urban agglomeration were exacerbated sometime later. Nevertheless, the need to regulate the sanitary and hygienic conditions of the buildings erected in Portuguese urban centers quickly became apparent. Thus, in 1903, the *Regulamento de Salubridade das Edificações Urbanas* came into force, establishing different rules for housing. The paper aims to present the main regulatory requirements for the healthiness of Portuguese buildings in the early twentieth century. The analysis in this study focused on affordable housing for the working class. The study undertaken was based on documentary research of coeval sources, namely the regulatory decree of 1903 and written publications in the areas of architecture and construction. The study aims to contribute to the understanding of the main causes of insalubrity in Portuguese working-class housing, the main sanitary requirements for dwellings, and the efforts made in Portugal to disseminate and implement the regulations in affordable housing.

Keywords: Building Heritage · History of Construction · Sanitary and Hygienic Housing Conditions

1 Introduction

The affordable housing crisis, which currently poses a number of challenges for Portugal and other European countries, has also been felt in the past. With the advent of the Industrial Revolution, the nineteenth century witnessed a major expansion of Europe's urban population [1, 2]. Urban areas attracted large numbers of rural people to work in the growing industries [3, 4]. Some industrialized cities increased their populations

Z. Perkowski et al. (Eds.): ECCE 2026, LNCE 929, pp. 237–252, 2027.
https://doi.org/10.1007/978-3-032-39515-3_18

exponentially over a few decades, such as the city of Mulhouse in France, which tripled its population in five decades [5 pp. 170]. These cities were polluted, overcrowded, unhealthy, and infested with contagious diseases. With no decent housing to accommodate the workers, living conditions in most of these new urban areas were miserable [6, 7]. These circumstances triggered widespread movements across Europe [5 pp. 164–169], [7, 8 pp. 135] and posed an enormous challenge to policymakers and the construction industry, despite the different pace of industrial and urban development of the various European parties.

Bairoch [9 pp. 259] explains the pace of industrial development of European countries in terms of four groups, the first of which includes the countries that were the first to be affected by the Industrial Revolution, and the fourth of which includes, among others, Portugal, as industrialization had made very little progress before 1914. Cordeiro [10] pp. 315–319 presents a paper in which he analyzes some factors that may partly explain Portugal's industrial backwardness compared to other countries: the lack of financing and the difficulty for entrepreneurs to obtain credit from banks, but also the limited use of energy from the steam engine [10], due to the lack of natural resources: coal and other main raw materials [11 pp. 427]. Antunes [4] adds factors related to the political instability that agitated Portugal at that time.

The answer to the problem of unhealthy housing began to emerge in England in the 1830s, when the first legislative measures were taken [12 pp. 144]. In the 1840s, technical reports, carried out separately by English, French, and Belgian hygienists, revealed the miserable accommodation of the working classes in the growing European cities. Several congresses at the time on health and social conditions led to recommendations for well-ventilated, bright rooms and adequate water and sewage systems in buildings [5 pp. 166–169]. A recent study by Harris and Helgertz [13] discusses the contribution of improved public health and domestic hygiene to declining mortality rates in different parts of Europe during the late nineteenth and early twentieth century. From another perspective, Timmins [6] examines the evolution of the configuration and housing conditions of the working class in Lancashire during the nineteenth century, through back-to-back housing, one of the defining features of industrial towns in Britain during the Industrial Revolution period.

In the mid-nineteenth century, Henry Roberts (1803–1876) in England and Emile Muller (1823–1889) in France developed housing models to improve the living conditions of the working classes. These models quickly spread throughout Europe thanks to their presentations at the Great Exhibitions (London 1851, and Paris 1867) and at various international congresses on welfare [5, 14]. In England, an attempt was made in 1861 to solve this threat to public health, and successive laws were published. These laws were unified in 1890 by the Housing of the Working Classes Act.

In nineteenth-century England, one of the most innovative responses to the problem of providing fair living conditions for working-class housing was granted by private initiative, namely the model dwellings companies [7]. Morris [7] explains that these companies, which were developed by a group of market institutions, played a more significant role than local government agencies in providing low-cost housing for the working classes in London at that time. The author also demonstrates that investors were offered rates of return comparable to those available from other domestic investment opportunities.

Although there are various studies on the Portuguese working-class neighborhoods and their buildings of the early twentieth century, very little research has been conducted on the sanitary regulation of urban buildings, their application, and their impacts, particularly in housing in this category. This paper focuses on the building requirements set out in the Portuguese regulation, 'Regulamento de Salubridade das Edificações Urbanas' (which can be translated as 'Health regulations for urban buildings') (RSEU) [15], which came into force in 1903, and examines the housing conditions of the poorest classes in order to find out whether this regulation was able to provide an effective response to improve their living conditions in the years following their publication. To this end, the study was based on an analysis of the hygiene conditions described by some hygienists and technicians of the time, as well as the efforts made in Portugal to disseminate and implement the RSEU. This regulation emerged after some attempts made in the last two decades of the nineteenth century, through proposed laws for affordable houses for the working class, but they had no practical effect [16].

This paper is structured as follows: the methodology is set out in Sect. 2; in Sect. 3, the context of housing before the publication of the health regulation is presented through a review of the literature; in Sect. 4, the main construction requirements of the health regulation are presented; in Sect. 5, the texts on affordable housing published in the journal between 1900 and 1913 are discussed; and in Sect. 6 the conclusions are presented.

2 Methodology

The preparation of this study was based on documentary research conducted through consultation of primary and secondary sources available in Portuguese digital libraries. Primary sources were coeval regulations and reports, and secondary sources were press articles and papers on the same subject. Table 1 lists the sources consulted and available in Portuguese digital repositories. This research began in 2023 and was conducted again between July and December 2025.

Table 1. Analyzed sources.

Title	Type	Year
'A Construcção Moderna' [17]	Periodical	1900–1919
'Regulamento Geral de Saúde e Beneficiência Pública' [18]	Law	1901
'Regulamento de Salubridade das Edificações Urbanas', RSEU [15]	Law	1903
'Habitações populares' [12]	Report	1909

The study of the Portuguese regulations published and in force during the considered period provides an understanding of the constructive solutions that were intended to be implemented in the buildings. The original legislative documents relating to sanitary requirements for buildings were consulted via official digital publications and repositories. A set of economic and financial reports carried out during that period by Matta [12] was also considered.

The analysis was carried out through coeval texts issued in the Portuguese periodical '*A Construcção Moderna*' (ACM) [17], which can be translated as 'Modern Construction', published fortnightly between 1900 and 1919. This periodical was the first Portuguese journal to focus on architecture and construction subjects, e.g., published technical texts on building materials and systems, announced events in the field of architecture and construction, presented opinion articles, reproduced articles published in other journals of the time, and plan sets. The analysis for this study focused on the topic of affordable housing texts published between 1900 and 1913, i.e., the beginning of its publication and ten years after the entry into force of the regulation on sanitary requirements for Portuguese urban buildings, RSEU [15].

Initially, a survey of articles related to affordable housing for the working class was conducted. It was found that not all texts were published in the same issue; longer texts (e.g., the RSEU text) were distributed in different issues of the periodical. It was also noted that the published texts were diverse, including news about international congresses and competitions related to affordable housing; presentations of texts issued at conferences; transcriptions of texts published in the national and foreign press; transcriptions of national and foreign laws; opinion articles; technical reports; and drawings.

The present study focused on technical texts and reports (e.g., those addressing poor construction practices and their potential impact on housing conditions) as well as regulations. This approach allows us to understand ACM's [17] role in disseminating laws and constructive solutions adopted in other countries, as well as publicizing national studies conducted at the time. Additionally, it sheds light on the evolution of the working-class housing problem during that period by examining whether living conditions improved and, if so, whether these improvements were significant.

3 The Background of Portuguese Regulation

In 1833, epidemic crises spread to the most densely populated areas of Lisbon, which the public authorities linked to poor sanitary conditions in residential buildings [11 pp. 429], [19 pp. 509], [20 pp. 253]. In doing so, the public authorities sought to establish urban building regulations, such as wider streets with water drainage, as well as the construction of sewerage and water supply networks that covered the entire city. With the decree of April 19, 1834, the city of Lisbon took over the responsibility of cleaning and lighting, as well as the construction of the water supply and sewage systems [20 pp. 253–254].

The end of the nineteenth century witnessed a series of economic and social transformations driven by the development of Portuguese industry [21, 22, 23]. The advent of the Industrial Revolution led to an increase in urban population due to the migration of people from rural areas to Lisbon, which, along with Oporto, was one of the main Portuguese urban centers where industrial development took place [22 pp. 40–41]. The increase in population facilitated the spread of infectious diseases such as tuberculosis, which affected different social classes [19 pp. 509].

The literature review shows the publication of different studies that allow for contextualizing Portuguese industry in the nineteenth century, and the living conditions of the working population in Portugal's two largest cities, Lisbon and Oporto. Cordeiro [10] presents a study in which he characterizes the existing industry in the city of Oporto in the second half of the nineteenth century. The author states that urban growth of Oporto did not follow the growth model seen in most industrialized centers, because the industrial structure of the city was made up of small units largely installed in the households of the industrialists themselves. This author's work reveals that the overcrowding conditions in urban spaces and buildings, as described in major industrialized European cities like London, did not have the same impact in Portugal's second-largest city, Oporto.

Alcântara [24] presents a study that identifies the location of the main industries in the city of Lisbon, based on the Portuguese industrial survey carried out in 1890 and complemented by other sources. The study reveals that the existing industries in Lisbon were dispersed both geographically and in terms of their diversity, and identifies a variety of production areas (textiles, metallurgy, ceramics, and tobacco) that were developed using the working classes. This work defines the typologies of the urban agglomerations formed by the working class in Lisbon during that period.

Despite the smaller size of the Portuguese industry compared to some of the main European industrial cities, the Portuguese state has shown concern to reduce its harmful consequences. Pinto [25] puts into context the nineteenth-century Portuguese regulations that established conditions for the operation of industrial units, with the aim of minimizing the effects on neighboring populations. The populations most affected by the growth of the industrial units that proliferated in urban centers were those with fewer economic resources, such as the workers in the industry itself.

Pereira [19] presents a description of the two main types of urban organization designed to accommodate the working class that flocked to the city of Lisbon to work in industry from the 1870s onwards: 'pátios' and 'vilas operárias'. Pátio (in singular form, or pátios in plural) can be translated as 'courtyard'. The Portuguese term 'vila operária' (in singular form, or 'vilas operárias' in plural) can be translated as 'working-class neighborhoods'.

The pátios were basement dwellings, most of which lacked drainage, sinks, or water supply, and were located in areas without sunlight and exposed to continuous dampness [19 pp. 511]. The pátios were the initiative of dynamic landlords who built squat houses to rent out to workers at the back of their buildings, but also made use of unhealthy basements, convents, and abandoned palaces that they bought at public auctions [19 pp. 511–512], [23 pp. 27]. The pátios developed in the old quarters of Lisbon and also on its outskirts. As Antunes [4] noted, the patios are not an original nineteenth-century urban form. In Lisbon, similar courtyards had existed since at least the Roman and Moorish periods. Nevertheless, this urban form became widespread in the nineteenth century, particularly among the poorest populations. During Lisbon's demographic growth in the second half of the nineteenth century, the working classes faced a housing crisis. In response, real estate entrepreneurs seized the opportunity to address this issue by adding more floors to pre-existing buildings and converting cellars, mansards, old convents, and unoccupied palaces into small dwellings.

The vilas operárias, which came after the pátios, consisted of groups of buildings built by private initiative specifically to house low-income families in a concentrated manner. The vilas operárias had some improvements in formal and constructive characteristics when compared to the pátios [19 pp. 511], [21 pp. 70], [16 pp. 2], for example, they had slightly larger areas and included drainage sinks [23 pp. 26–27]. One of the characteristics of vilas operárias was to maximize the amount of land available for construction, so that their developers could build as many houses as possible and optimize their earnings [19 pp. 513–514]. Vieira [23] identifies some real estate developers of pátios and vilas operárias who operated in Lisbon at the end of the nineteenth century [23 pp. 26–30].

Teixeira [21] refers that in the city of Oporto the working classes were housed in old buildings with poor sanitary conditions, and from 1850 onwards in new housings buildings known as ilhas. The Portuguese word 'ilha' (in singular form, or 'ilhas' in plural) can be translated as 'island'. The ilhas consisted of rows of small single-story houses, usually no larger than 16 m^2, built in the backyards of old bourgeois homes. Most of the ilhas had no water supply, the toilets were shared by all the inhabitants and whole families lived in these small and unhygienic houses [21 pp. 67–68].

Vieira [20] uses documentation from the Lisbon Municipal Archive as well as existing buildings in the city to compare the spaces occupied by the working class with those of the upper class. The study highlights the precarious housing conditions of the working class between 1890 and 1910 [20 pp. 319–395].

As Antunes [4] points out, although the ilhas in Oporto and the pátios and vilas operárias in Lisbon are both linked to working-class housing, it is difficult to identify similarities between the developments in Portugal, and particularly in Lisbon, and those in other European cities due to the complexity, diversity, and uniqueness of the industrial housing heritage in each location.

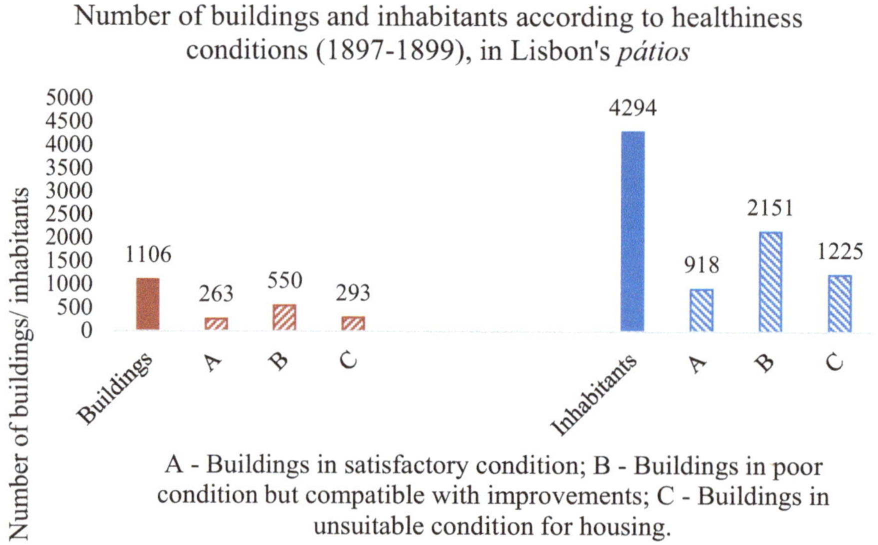

Fig. 1. Number of buildings (on the left) and inhabitants (on the right) according to the report published in [17] (i.e., ACM no. 88, 1 March 1903, pp. 29) and carried out by the Portuguese Council, '*Conselho de Melhoramentos Sanitários*', between 1897 and 1899.

In 1903, the journal ACM [17] presented the results of a report drawn up by the engineer Serra Prado for the Portuguese Council, the Conselho de Melhoramentos Sanitários (this Council was created by Decree on October 24, 1901 with the aim of issuing opinions on the healthiness conditions of buildings; Conselho de Melhoramentos Sanitários can be translated as 'Sanitation Improvement Council') on Lisbon's pátios where the working class lived. In this report, pátios are described as filthy and miserably inhabited, not being exempt from various unhealthy inconveniences without fresh air, light, or possible cleanliness [17] (i.e., ACM n. 88 1 March 1903, pp. 29). According to what the journal published, the study was carried out in 18 parishes, where 4,294 people lived in 1,106 buildings (Fig. 1). The graph in Fig. 1 shows that, according to the published report, only 263 dwellings had satisfactory living conditions, representing 918 people or 21% of the inhabitants surveyed.

4 The Portuguese Sanitary Regulation for Urban Buildings

As Teixeira [21] points out, in the last decades of the nineteenth century, the poor sanitary conditions of many urban dwellings, the spread of epidemics, and the resulting high mortality rates forced the municipal authorities to confront and solve the serious problems in Lisbon and Oporto. Some of the measures adopted included municipal codes requiring building permits, specific rules for the construction of septic tanks in new buildings, or the inspection of buildings by municipal inspectors [21 pp. 73]. Ildefonso et al. [16] present the initiatives and legislative practices undertaken in Portugal between 1883 and the early years of the twentieth century to address these housing problems [16 pp. 4–5].

In 1901, the Regulamento Geral de Saúde e Beneficiência Pública [18] (which can be translated as 'General Regulation on Health and Public Welfare') established that the health of places and dwellings was a competence of the health services, and two years later, the Portuguese regulation on sanitary conditions, RSEU [15], came into force. These regulations defined the sanitary and environmental conditions of new housing and required that these buildings be inspected by the municipal health authorities.

The RSEU [15] was published on February 14, 1903. The regulation was approved by the Portuguese Council, Conselho Superior de Obras Públicas e Minas e de Hygiene Pública (which can be translated as 'Superior Council of Public Works, Mines, and Public Health') and signed by the President of Conselho de Ministros (which can be translated as 'Ministry Council'), by the Ministry and Secretary of the Estado dos Negócios do Reino (which can be translated as 'State of the Kingdom's business') and by the Ministry and Secretary of the Estado dos Negócios das Obras Públicas, Comércio e Indústria (which can be translated as 'State of Public Works, Trade, and Industry'). This regulation was intended to standardize the health conditions of residential and non-residential buildings, i.e., churches, theaters, workshops, schools, and asylums. It was in force until 1951, when it was replaced by an updated regulation, Regulamento Geral das Edificações Urbanas. (which can be translated as General Regulation for Urban Buildings).

The regulation contains 60 articles and is divided into three chapters, according to the framework shown in Table 2: the first chapter deals with the health of the soil, the second with buildings, and the third with the general provisions.

Table 2. Framework of the 1903 Portuguese sanitary regulation [15].

Chapter	Section
1. Healthiness of land	
2. Healthiness of buildings	– Potable water tanks. – Drop pipes. – Siphons. – Ventilation pipes. – Latrines and basins. – Urinals and other drainage systems. – Cesspools. – Animal accommodations.
3. General provisions	

The construction requirements are defined for general buildings, with specific provisions for other types of use, particularly regarding ventilation conditions, in response to the concerns of the hygiene movement of the time. Table 3 shows the main constructive requirements and corresponding building objectives established in [15].

Chapter 1, dedicated to the healthiness of the land, imposes restrictions on the construction of buildings on land that was damp or contaminated by harmful substances, next to cemeteries, and on land where water from domestic use or harmful industries had been deposited.

Chapter 2, relating to the healthiness of buildings, set out specific requirements, as summarized in Table 2. The regulation laid down rules for the height of building facades, which depended on the width of the streets ([15], article 5), and defined the minimum height of the floors so that there was no water seepage and that there were no harmful fumes. These dwellings had to be equipped with latrines and had to ensure ventilation and lighting from the inside through windows ([15], article 18). The foundations of the buildings should include impermeable materials, at least 0.15 m thick to prevent rising damp ([15], article 10), and the first floor of the dwellings had to be covered with a waterproof layer or an air gap of 0.60 m, which had to communicate with the outside ([15], article 10). The staircases to the upper floors were to be lighted and ventilated and provide a smooth ascent ([15], article 9). In order to ensure ventilation and lighting in residential buildings, windows should cover at least 10% of the floor area ([15], article 11). In the bedrooms, the minimum volume allowed was 25 m^3 per person ([15], article 13).

For the disposal of domestic waste, each dwelling should be equipped with a sink and a latrine with an adequate drainage system ([15], articles 42 and 43). The sink was intended for domestic waste and should preferably be located outside the house, although it could also be installed on an outside wall and near a window ([15], article 43). The latrine could be placed outside or inside the house, provided that in the latter case it was installed next to a window with a connection to the outside. The latrine had to be adequately drained by a water tank with a flush toilet ([15], article 42).

Table 3. Main goals and corresponding constructive requirements, settled in [15].

Goal	Constructive requirements
• Prevent rising damp	– Walls and floor waterproofed.
• Ensure the ventilation and lighting of spaces	– '*Patios*': with a minimum area of $30m^2$ if the height of the building was less than 18 m, or 40 m^2 if the height was more than 18 m (article 19).
	– Interior staircases: installation of fans and skylights (article 9).
	– Bedrooms: mandatory window (article 13) with a minimum surface area of 10% of the floor area (article 11).
• Ensure rainwater drainage around the building	– Proper construction of roofs (article 15).
	– Roof drainage systems (articles 16,17].
	– Drainage piping systems (articles 25,27–33, 35).
• Ensure the disposal of domestic wastewater	– Installation of glazed ceramic or limestone sinks, cast in a single piece (article 43).
	– The sink had to be laid on masonry with a layer of hydraulic cement (article 43).
• Ensure sewage drainage systems	– Sewage drainage systems (articles 25–31,34).
	– Siphons (articles 36–38).
	– Ventilation pipes (articles 39–41).
• Smoke extraction	– Chimneys with non-combustible materials, with rounded corners (article 14).

Construction requirements were also laid down to ensure the proper drainage of rainwater from the roof, which included the proper construction of the roof and the drainage system with gutters and downpipes.

Chapter 3 of the regulation requires new buildings to comply with the sanitary conditions laid down in the text and obliges local councils to draw up sanitary regulations adapted to local conditions, giving them powers to inspect the healthiness of buildings. The general provisions explain the fines to be applied in the event of non-compliance by building owners. For Lisbon and Oporto (Portugal's most populous cities), the regulation established the requirement of a license issued by the town hall for new buildings and for alterations in existing buildings.

5 Analysis and Discussion

An analysis of the texts published in ACM [17] shows that health in working-class housing was a recurring theme in the journal.

Before 1903, the regulation came into force, the main articles on the topic called for legislation on the sanitary conditions of buildings [17] (i.e., ACM, no. 100, 1 July 1903, pp. 127; no. 102, 20 July 1903, pp. 142–143). Between 1900 and 1903, different articles were published on the health requirements for buildings, particularly on the importance of ventilation, natural light, sanitation, and the orientation of facades. In 1903, the year in which the regulation came into force, the journal published its complete text, divided into six parts and printed in six issues [17] (i.e., ACM, no. 103, 1 August 1903, pp. 152; no. 110, 10 October 1903, pp. 208; no. 111, 20 October 1903, pp. 216; no. 112 10 November 1903, pp. 224; no. 113, 10 November 1903, pp. 232; no. 115, 1 December 1903, pp. 248).

Figure 2 shows the number of articles published between 1903 and 1913 on the subject of sanitary conditions of working-class housing or affordable houses. Some of these articles were divided into several parts and published in different issues of the journal. Between 1903 and 1908, there was a steady increase in the number of articles on the subject, with the latter numbering 9 (Fig. 2).

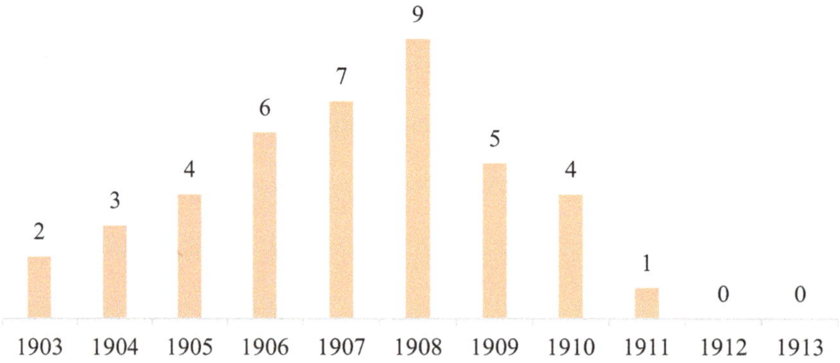

Fig. 2. Articles published in [17] on affordable housing, between 1903 and 1913.

In May 1904, the journal ACM [17] published an article by Motopa & Triglypho describing the living conditions of the poorest classes in the city of Oporto, i.e., houses without sewers, rooms without light in narrow streets, some of which only received sunlight once a year [17] (i.e., ACM, no. 132, 20 May 1904, pp. 94). This report refers to the ilhas and seems similar to descriptions of other industrialized European cities in terms of the living conditions of the lower social classes living in dense urban areas [1, 5, 26]. In general, buildings for the working class used cheap and low-quality construction materials [19, 21], which was reflected in the poor condition of the dwellings: water leaks, high levels of moisture, poor ventilation, scarce natural light, and the absence of sanitation, in short, small and unhealthy spaces.

In 1905, the journal wrote, in the article 'Habitações Económicas' (which can be translated as 'Affordable Housings') that, despite Lisbon's good climatic conditions compared to other European cities, the mortality rate was over 24 per thousand, compared to 18 in London and 20 in Paris [17] (i.e. ACM, no. 169, 20 June 1905, pp. 102). The excess of deaths was attributed to the unhealthy housing conditions in Lisbon: poorly ventilated houses without light [17] (i.e., ACM, no. 169, 20 June 1905, pp. 102). The essay is in line with the hygienic currents of the time, which warned of the direct link between poor housing conditions in dense urban areas and the spread of epidemics.

After 1905, the articles published in the journal on the subject of affordable housing focused mainly on general sanitary requirements for urban areas and residential buildings, on the dissemination of foreign legislation, on the explanation of proposals for financing problems, and on the presentation of Portuguese affordable houses, showing the contribution and efforts to solve this complex problem. The articles published in 1906 focus mainly on the dissemination of the conclusions of international congresses on health requirements for urban areas and residential buildings, and on the announcement of the winning projects in the Rothschild French competition, for the construction of affordable housing; while those published in 1907 translate and describe the legislation on affordable housing that came into force in France. This French law was enacted in April 1907 and amended the law of November 30, 1894, on housing; it refers to the creation of commissions for the protection of affordable housing and social welfare, and the establishment of conditions for financial benefits for buildings intended for this type of lodging. The protection commissions thus created confirmed the healthiness of the houses that could receive the benefits provided for by the law. The dissemination of legislative proposals aimed at promoting the financing of affordable housing was a recurring theme in the texts of the journal between 1907 and 1908.

In 1908, the journal transcribed a study carried out by Augusto Montenegro, President of the Conselho de Melhoramentos Sanitários (which can be translated as 'Sanitary Improvements Council'), entitled 'O Problema da Habitação' (which can be translated as 'The Housing Problem'). The essay was published in three parts and reflected the hygienist concerns of the time: reiterating the link between the spread of tuberculosis and the lack of conditions in unhealthy houses; promoting the responsibility of the state in the creation of building societies; and recommending the use of a health label on houses recognized as being in a satisfactory sanitary condition [17] (i.e., ACM, no. 264, 10 July 1908; no. 265, 20 July 1908; and no. 266, 1 August 1908). In the same year, the journal published a three-part article on a proposed law on the financing of affordable housing, which envisaged the creation of building societies and technical and advisory committees with functions that seem to resemble those of 1907 French law [17] (i.e., ACM, no. 272, 1 October 1908; no. 273, 10 October 1908; and no. 274, 20 October 1908).

In 1909, the economic and financial study 'Habitações Populares' [12] (which can be translated as 'Low-incoming Housing') was published, dedicated to the construction of low-income housing. This study presented a legislative overview of the main initiatives to solve the problem of affordable housing in different European countries (i.e., England, France, Germany, Switzerland), but also in Brazil and in the United States of America [12 pp. 143–168]. The study also mentioned some private initiatives to build affordable housing in Portugal between 1880 and 1900 [12], in parallel with what was happening in other industrialized countries. In Portugal, private initiatives through societies for the construction of affordable housing for the working class included housing for sale and rent in new buildings and the rebuilding of old palaces. The housing typologies varied between rooms for two occupants, dormitories, and two-story buildings with backyards.

The latter, which accommodated four occupants, included a living room, two bedrooms, and a kitchen [12, pp. 129–132]. In the same year, four texts were published in ACM [17], two of which divided into four different issues, summarized the laws applied by other countries (England, Belgium, France, Italy, Germany and Austria) to affordable housing, namely in the categories of new houses, houses to be built and general housing matters [17] (i.e., ACM, no. 284, 1 February 1909; no. 286, 20 February 1909; no. 289, 20 March 1909; no. 290, 1 April 1909; and no. 293, 1 May 1909). The March issue features a model of an affordable single-family house (Fig. 3) located in Lisbon and built by a private developer and building contractor, Domingos Sarzedello [17] (i.e., ACM, no. 287, 1 March 1909, pp. 177–178). The houses built by Mr. Sarzedello had five to six rooms, a basement, a yard or a small garden, and plumbing for water and gas. As illustrated in Fig. 3, the architectural plans depict a two-story dwelling with a basement and a garden.

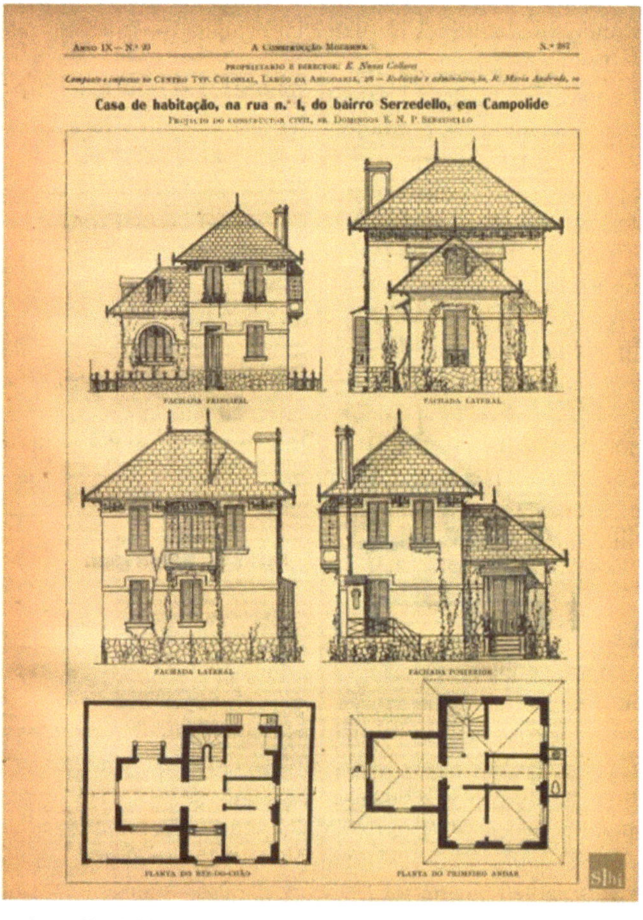

Fig. 3. Plan set of an affordable house built by Domingos Serzedello and published in [17] (i.e., ACM no. 287, 1 March 1909, pp. 177).

In 1910, the journal published two articles on affordable housing: one on the French law [17] (i.e., ACM, no. 318, 10 February 1910, pp. 138–139) and the other on the importance of healthiness requirements which was divided into 5 parts [17] (i.e., ACM, no. 322, 20 March 1910, pp. 174–176; no. 323, 1 April 1910, pp. 183–184; no. 324, 10 April 1910, pp. 191–192; no. 325, 1 April 1910, pp. 198–200; no. 326, 1 May 1910, pp. 206–207). The latter discusses the advantages of healthy housing for workers' health and consequently for their productivity, and also refers to legislation in some European countries.

In 1911, the journal ACM [17] was renamed 'A Construcção Moderna e as Artes do Metal' (which can be translated as 'Modern Construction and Metal Arts'). In May, an article was published dealing with the construction requirements of affordable housing, the organization of the neighborhood in which it is located, and its financial solution. This text was divided into two issues [17] (i.e., ACM, no. 345, 6 May 1911, pp. 66–67; no. 346, 20 May 1910, pp. 74–75).

In the following years, there were no publications in ACM [17] on the specific topic of affordable housing. Between 1900 and 1913, the journal published several essays on good building practices and on the main building problems. However, the articles that referred to solutions to technical problems in buildings, such as inadequate ventilation, dampness in the walls, or sanitation, were presented for housing in general and not specifically for affordable housing.

The study by Matta [12] confirmed the existing problems in the housing conditions of the working classes. Despite the 1903 regulation and some private promoter initiatives, the consulted coeval sources show that, in general, there was no significant improvement in the urban housing conditions of the Portuguese poorest classes in the years following its entry into force. These findings converge with those of the studies by [19, 20 pp. 381–382, 21].

6 Conclusions

This paper outlines the primary regulatory requirements for the health and safety of Portuguese buildings in the early twentieth century.

In the late nineteenth and early twentieth centuries, tuberculosis and other infectious diseases spread in overcrowded environments with poor sanitary conditions, such as the housing intended for the working class, who came from rural areas and flocked to Portuguese urban centers with some industrial dynamics. The modest salaries of most workers allowed them only to pay rents for poor-quality, small, and unhealthy housing. In Oporto, the ilhas were built by small merchants and artisans who invested their meager capital there. The small investments made by these developers only allowed them to build small and poor dwellings, which nevertheless provided what the low-income working classes could afford. The ilhas and the pátios suited the limited investment capacity of their builders, for whom these forms of housing represented a safe and profitable investment for their small savings. However, the poor sanitary conditions of many of these dwellings led to the spread of serious infectious diseases, resulting in high mortality rates.

Faced with this serious public health problem, the Portuguese state, like other countries with more advanced industries, was concerned to reduce its harmful effects.

The Portuguese health regulation for urban buildings, RSEU [15], came into force in 1903. The regulation laid down different requirements for various parts of the building to ensure the health and well-being of its occupants, such as an adequate drainage system, the waterproofing of walls and roofs, or ventilation and lightning through a minimum area of windows.

For this study, coeval sources were sought to analyze the living conditions of the working class that settled in Lisbon and Oporto. This analysis was carried out through the descriptions given by hygienists and technicians of the time in the Portuguese press, namely in 'A Construcção Moderna' [17], a journal dedicated to architecture and construction. Between 1903 and 1913, the journal published about 40 articles on affordable housing for the working classes, thereby helping to raise awareness of this complex issue at the time. The texts analyzed show the different approaches adopted by the journal: reporting on studies and surveys, describing and summarizing foreign legislation, and presenting proposals for financing or disseminating good practices in the promotion of affordable housing.

In the two decades leading up to the implementation of the Portuguese regulation of 1903, several affordable housing bills were proposed but not approved. Private initiatives for affordable housing construction also emerged during this period, as outlined in Matta's 1909 study [12]. Despite the health regulation coming into force in 1903, and the efforts of the journal to disseminate information about this public health issue and highlight its importance, it can be concluded that the living conditions of the working classes remained poor until at least 1910. However, the literature review also revealed some minor improvements in housing conditions for the working class, particularly in Lisbon.

This introductory study is expected to provide a basis for further research from other perspectives. For example, it could compare the constructive requirements established by Portuguese regulations with those of other European countries. It could also investigate the extent to which construction solutions from other countries (i.e., building materials and systems) have been adopted or adapted for use in Portuguese housing projects.

References

1. Cullen F (2011) The provision of working-and lower-middle-class housing in late nineteenth-century urban Ireland. Proc R Ir Acad 111C(1):217–251. https://doi.org/10.1353/ria.2011.0015
2. Grabowski M (2019) The 18th and 19th century industrialization process as the main aspect of city creation and its impact on contemporary city structures: the case of Lodz. IOP Conf Ser 471:082046. https://doi.org/10.1088/1757-899x/471/8/082046
3. Andrews RM (1995) The development of the residential suburb in Britain 1850–1970. Urbani Izziv 28–29:15–27. https://doi.org/10.5379/urbani-izziv-en-1995-28-29-002
4. Antunes G (2025) Industrial housing clusters in nineteenth-century Lisbon: finding spatial patterns. Urban Hist. 52:159–174. https://doi.org/10.1017/s0963926823000433
5. Van Praet C (2019) The opposite of Dante's Hell? the transfer of ideas for social housing at international congresses in the 1850s– 1860s. In: Spiegeleer C (ed) The civilising offensive: social and educational reform in 19th century Belgium. De Gruyter, pp 163–188. https://doi.org/10.1515/9783110581546-007

6. Timmins G (2013) Housing industrial workers during the 19th century: back-to-back housing in textile Lancashire. Ind Archaeol Rev 35(2):111–127. https://doi.org/10.1179/0309072813z.00000000021
7. Morris S (2001) Market solutions for social problems: working-class housing in nineteenth century London. Econ Hist Rev 54(3):525–545. https://doi.org/10.1111/1468-0289.00202
8. Fijalkow Y (2000) La notion d'insalubrité. Un processus de rationalisation 1850–1902. Rev d'Histoire du XIXe Siècle 20(21):135–156. https://doi.org/10.4000/rh19.213
9. Bairoch P (1988) Cities and economic development, from the dawn of history to the present. University of Chicago Press, Chicago
10. Cordeiro JMC (1996) Empresas e empresários portuenses na segunda metade do século XIX. Análise Social 31(136–137):313–342. https://doi.org/10.31447/AS00032573.1996136.03
11. Cascão R (1993) O crescimento demográfico: ritmos e factores. In: Mattoso J (ed) História de Portugal: o liberalismo (1807–1890). Círculo de Leitores V, Lisboa, pp 425–439
12. Da Matta C (1909) Habitações populares. Imprensa da Universidade de Coimbra, Coimbra. https://www.fd.unl.pt/Anexos/Investigacao/1234.pdf. Last accessed 25 Feb 2026
13. Harris B, Helgertz J (2019) Urban sanitation and the decline of mortality. Hist Fam 24(2):207–226. https://doi.org/10.1080/1081602x.2019.1605923
14. Kihlberg J (2021) European reform movements and the making of the international congress, 1840–1860. Int Hist Rev 43(3):488–507. https://doi.org/10.1080/07075332.2020.1808506
15. Regulamento de Salubridade das Edificações Urbanas (RSEU) (1903). Legislação régia. https://legislacaoregia.parlamento.pt/V/1/88/132/p99. Last accessed 25 Feb 2026
16. Ildefonso S et al (2019) A proto-habitação social na cidade de Lisboa: uma leitura integrada da vila operária no contexto urbano atual. Urbe Rev Bras Gestão Urbana 11. https://doi.org/10.1590/2175-3369.011.001.ao05
17. A Construcção Moderna (1900–1919) Lisboa (ACM). https://pt.revistasdeideias.net/pt-pt/a-construcao-moderna. Last accessed 05 Feb 2026
18. Regulamento Geral de Saúde e Beneficência Pública (1901) Legislação régia. https://legislacaoregia.parlamento.pt/V/1/86/133/p1070. Last accessed 05 Feb 2026
19. Pereira NT (1994) Pátios e vilas de Lisboa, 1870–1930: a promoção privada do alojamento operário. Análise Social 29(127):509–524. https://www.jstor.org/stable/41011019. Last accessed 05 Feb 2026
20. Vieira MCB (2018) História das tipologias arquitetónicas de edifícios correntes de habitação construídos na cidade de Lisboa desde início do século XVIII até à década de 1930 – Tese de Doutoramento. In: Instituto Universitário de Lisboa – ISCTE, Lisboa
21. Teixeira MC (1992) As estratégias da habitação em Portugal, 1880–1940. Análise Social 27(115):65–89. https://www.jstor.org/stable/41010899. Last accessed 05 Feb 2026
22. Alcântara A (2013) Uma geografia de Lisboa Operária em 1890. In: Paço AS, Teixeira C, Godinho P, Varela R, Pereira VB (eds) Atas do I Congresso de História do Movimento Operário e dos Movimentos Sociais em Portugal. Instituto de História Contemporânea da Faculdade de Ciências Sociais e Humanas da Universidade Nova de Lisboa, vol I, Lisboa, pp 38–52
23. Vieira C (2020) As habitações operárias em Lisboa, na segunda metade do século dezanove. Cadernos do Arquivo Municipal 2(13): 17–39. https://cadernosarquivo.cm-lisboa.pt/index.php/am/article/view/104/78. Last accessed 05 Feb 2026
24. Alcântara A (2020) A habitação das "classes laboriosas". Espaços de residência operária na Lisboa do final do século XIX. Cadernos do Arquivo Municipal 2(13):41–61. http://hdl.handle.net/10362/104684. Last accessed 05 Feb 2026

25. Pinto SMG (2018) The licensing of unhealthy, uncomfortable or dangerous establishments in the nineteenth century and the procedures' ground-plans of the Coimbra district. Bol Arq Univ Coimbra 31(1):125–162. https://doi.org/10.14195/2182-7974_31_1_5
26. Vaughan L (2018) Disease, health and housing. In: Mapping society: the spatial dimensions of social cartography. UCL Press, pp 24–60. https://doi.org/10.2307/j.ctv550dcj.7

Degrowth, Demographic Change, and Spatial Justice: A Joint Educational Model for Rural-Urban Planning

Maurizio Scarciglia[1], Anna Szczegielniak[2](✉), and Luca Trabattoni[2]

[1] Breda University of Applied Sciences, Breda, the Netherlands
[2] Faculty of Civil Engineering and Architecture, Opole University of Technology, Opole, Poland
a.szczegielniak@po.edu.pl

Abstract. This research lays the groundwork for a joint educational initiative between Breda University of Applied Sciences and Opole University of Technology, focusing on degrowth-oriented urban and spatial planning applied to rural villages in the context of European migration. As contemporary migration increasingly transcends national borders and frames Europe as a shared spatial system, comparative analysis across national contexts enables the identification of common strategies. These strategies support a research process within an international collaborative course, highlighting links between demographic change, degrowth principles, and transnational academic cooperation.

Across Europe, rural areas face challenges such as depopulation, population ageing, and declining access to essential services. Within this context, migration can offer opportunities to revitalise local economies, reinforce social networks, and enhance community resilience. From a degrowth perspective, migration is not only a demographic phenomenon but also a potential driver of sustainable development, supporting local food systems, multifunctional landscapes, and reduced reliance on energy-intensive mobility and consumption patterns. At the same time, urban areas struggle to provide affordable housing and dignified living conditions for social groups drawn by economic opportunities. The dominant growth-oriented paradigm, linked to neoliberal economic values, has contributed to environmental degradation and the climate crisis, underscoring the need for alternative visions of prosperity that respect planetary boundaries.

Building on these premises, the collaboration uses education as a vehicle for degrowth research through two design exercises, differentiated by location and scale. By aligning curricula and sharing assignments and project sites, the joint course integrates spatial scales and strengthens students' understanding of degrowth.

Keywords: Degrowth · Rural Areas · Teaching-Based Research

Z. Perkowski et al. (Eds.): ECCE 2026, LNCE 929, pp. 253–263, 2027.
https://doi.org/10.1007/978-3-032-39515-3_19

1 Introduction

Degrowth emerges from a fundamental critique of economic growth as the organising principle of contemporary societies [1]. Articulated initially by ecological economists [2], degrowth challenges the assumption that continuous expansion of production and consumption is either feasible or desirable on a finite planet. At its core, degrowth recognises that perpetual economic growth is incompatible with planetary boundaries and ecological sustainability [3, 4]. When applied to urban contexts, the degrowth critique reveals how growth imperatives shape spatial development in problematic ways. Urban planning has become subordinated to entrepreneurial governance models that prioritise real estate development and the attraction of mobile capital [5]. This growth-oriented urbanism manifests in interconnected pathologies: the financialisation of housing drives displacement; the prioritisation of economic competitiveness occurs at the expense of social reproduction; spatial expansion produces sprawl; socio-spatial inequalities are exacerbated; and metabolic demands create extractive relationships with surrounding regions [6, 7].

Post-growth economics envisions alternatives to perpetual expansion, proposing economic models that prioritise stability, sufficiency, and well-being within ecological limits [8]. Key principles include provisioning systems that ensure universal access to fundamental needs through decommodified institutions; working time reduction to distribute employment more equitably; circular and sharing economies that maximise material efficiency; and the relocalisation of production and consumption to reduce transport energy demands and strengthen community resilience [9]. In this context, the dichotomy between city and countryside raises critical questions about how the two will support each other in a future characterised by co-dependence. In the book and exhibition *Countryside, a Report*, Rem Koolhaas presents narratives of global countrysides that document possible futures of coexistence between city and nature, triggering further discussion of the need to shift habitat models away from capitalist and neoliberal frameworks [10].

Degrowth urbanism translates theoretical foundations into concrete planning principles [11, 12]. Decommodification of urban space removes land, housing, and essential services from market logic through social housing models, community land trusts, expanded public space, and rent controls [13]. Commons-based governance emphasises collective management of shared resources through participatory decision-making, transparency mechanisms, and community-enforced monitoring systems [14, 15]. Relocalisation and proximity planning reduce energy consumption and build resilience through urban agriculture, distributed energy systems, local fabrication infrastructure, neighbourhood-scale services, and prioritisation of pedestrian and bicycle infrastructure [16]. Metabolic approaches reimagine urban-rural relations, moving from extractive patterns toward reciprocal relationships through closing material loops, regionalising flows, restoring reciprocity, and creating productive landscapes [17].

Spatial justice (the fair distribution of resources, opportunities, and environmental quality across space) represents a central concern for degrowth urbanism [18]. Degrowth reframes Lefebvre's [19] conceptualisation of the "right to the city" by emphasising: the right to remain in place, resisting displacement; the right to collective determination through participatory planning processes; the right to deceleration from competitive imperatives; the right to adopt low-impact lifestyles; and the right to participate in diverse economies, including cooperative and solidarity practices [20]. For migration contexts, degrowth approaches emphasise territorial redistribution toward areas experiencing demographic decline, the right to rural residence and livelihoods, solidarity economies that enable participation outside conventional employment structures, commons-based integration through community participation, and the recognition of diverse contributions beyond wage labour [21].

Contemporary urban crises, characterised by housing unaffordability, social polarisation, and ecological degradation, represent consequences of growth-dependent development models. Financialisation has transformed housing from shelter into a financial asset, driving speculation and displacement [6]. Urban growth drives ecological degradation through carbon emissions, water overconsumption, material extraction, waste generation, and habitat destruction [22]. Rural depopulation reflects growth-dependent economic structures that systematically extract value from rural territories.

2 Methodology

In light of the above analyses, this chapter presents the teaching methodology for courses addressing urban and rural planning from a post-growth perspective, combining critical reflection on contemporary development models with practical design tools.

The course was developed within the framework of the European university KreativEU, of which Opole University of Technology (OUTech) and Breda University of Applied Sciences (BUAS) are members, and represents the outcome of a partnership between two institutions representing distinct yet complementary teaching and research traditions. The course is based on the integration of two parallel courses: *Sustainable Design and Environmental Protection* at Opole University of Technology (Architecture programme) and *Planning for Degrowth* at Breda University of Applied Sciences (Built Environment programme). This collaboration is grounded in the mutual exchange of academic and pedagogical experience. The contribution of Breda University of Applied Sciences focuses primarily on regional and urban planning scales, whereas Opole University of Technology contributes a rural and architectural perspective. The course assumes the integration of these two scales as equally important components of the educational process. A significant element of the collaboration is the confrontation of differing national contexts of Poland and the Netherlands, encompassing diverse geographical conditions, spatial planning systems, demographic processes, and socio-cultural patterns. The comparison between a highly urbanised, low-lying country (the Netherlands), requiring extensive flood protection measures [23], with a mature planning system, and a country (Poland) characterised by a diverse structure of rural areas [24], the experience of systemic transformation [25], and processes of rural depopulation [26], creates a framework for in-depth reflection on rural-urban relationships. The combined potential of both institutions enables the achievement of educational outcomes that would not be attainable through activities conducted independently by each university.

Within this comparative framework, migration constitutes a structuring theme that is embedded across both course modules. In the Dutch context, migration is primarily associated with established and diverse immigrant communities settled in both urban and peri-urban areas, where labour migration (particularly in agriculture, logistics, construction and food processing) has produced significant spatial effects, including pressure on housing stock and localised social tensions arising from rapid demographic change. In the Polish context, migration dynamics have shifted substantially in recent decades: while rural depopulation through emigration to western Europe and domestic urbanisation remains a dominant concern, Poland has simultaneously become a destination for labour migrants, particularly from Ukraine and other eastern European and central Asian countries, whose presence in smaller towns and rural areas introduces new demands on spatial planning and service provision. The course methodology requires students to engage with these differentiated migration realities by mapping the spatial footprint of migrant populations in their assigned project sites and analysing how existing spatial planning instruments address, or fail to address, the needs of these populations. This analytical layer informs the subsequent design phase.

The curriculum introduces the concept of degrowth in the context of spatial planning, urban planning, and architecture, analysing its application in the development and planning of rural and urban areas. A key component of the educational process is the development of students' ability to critically analyse how the idea of growth shapes national spatial development and generates specific spatial and social problems [27], while simultaneously understanding the theoretical foundations of these processes. The course emphasises the ability to translate abstract concepts into spatial design proposals, both urban and architectural, and to understand the relationships between regional, urban, and rural planning scales and architectural design. An important component is also raising awareness of the significance of public participation in design processes [28], as well as the role of socio-geographical context, in which the same ideas may materialise in different ways [29–33]. Complementing analytical and design competencies is the development of skills in communicating spatial concepts through drawings and presentations (Fig. 1). At the same time, the course contributes to the development of international cooperation in education and research, strengthening inter-institutional relationships, creating a foundation for the organisation of future joint courses on other topics, and establishing a sustainable network supporting the exchange of experience and best educational practices.

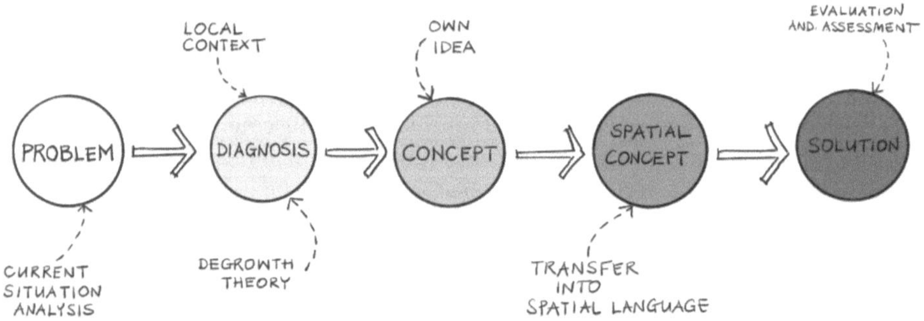

Fig. 1. Diagram of the didactic process (own elaboration).

Design assignments conducted within the course refer to specific locations in the Netherlands and Poland related to development in rural-urban areas. They are jointly developed and prepared by both partner institutions. Working with real locations enables students to attempt to solve actual spatial problems in specific places, distinguishing the assignments from abstract design exercises and encouraging deeper engagement in the design process. The scope of work includes the preparation of two presentations, a conceptual and a final one, in the form of multimedia presentations supported by oral commentary, as well as the preparation of one or more drawing boards in a specified format containing architectural and urban drawings, such as plans, sections, axonometries, and diagrams, and the construction of a physical model using any selected technique. Example design topics may include planning strategies for centres undergoing post-industrial transformation, peripheral areas struggling with depopulation, issues of densification and production in peri-urban zones, and the development of regional agroecological networks. Among these, the spatial accommodation and integration of migrant populations in rural and small-town settings constitutes a cross-cutting design challenge that students are expected to address in relation to housing typologies, shared service provision, productive landscapes, and community infrastructure.

From an organisational perspective, the course is designed in a blended learning format, combining on-site teaching with remote work. Classes at OUTech take the form of an intensive workshop week, whereas at BUAS, they operate within the structure of a traditional design studio. The programme is supplemented by regular online meetings conducted throughout the semester (Fig. 2). This organisational structure enables integration of the course between the universities and continuous coordination of the teaching process. An important element of the teaching method is group work, which fosters student integration and facilitates the exchange of ideas and experiences within each university, as well as peer learning [30–34]. At the same time, the use of online communication tools enables joint discussion and analysis of student design proposals across countries, as well as international evaluation of student work by the academic staff of both institutions, creating a space for transnational exchange of ideas and perspectives.

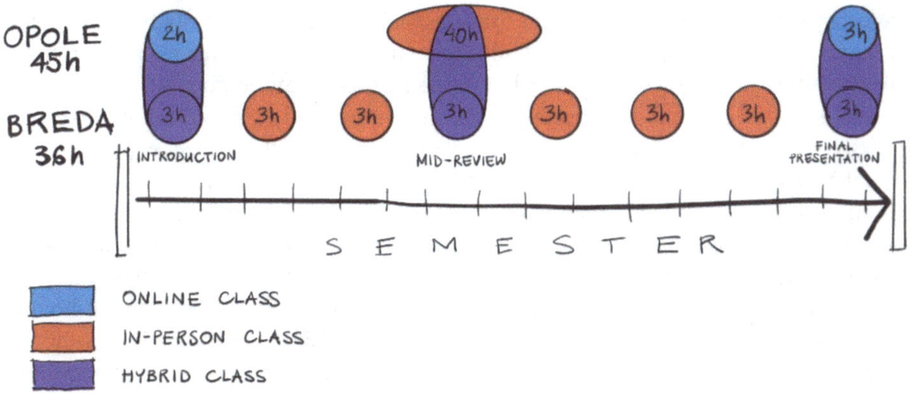

Fig. 2. Scheme of semester organisation at the two universities (own elaboration).

The teaching process includes the assessment of student work at the intermediate and final stages, conducted through multimedia presentations, design drawings, and oral commentary presented to the entire international group of students. The proposed spatial concepts are evaluated against clearly defined criteria, including the accuracy of diagnosing the spatial and organisational problem, the adequacy of spatial solutions in relation to identified conditions, and the correctness of the applied design tools. An important element of the evaluation is also the degree to which the presented concepts align with the objectives of degrowth theory or engage in its critical reinterpretation, as well as the potential positive impact of the proposed solutions on achieving sustainable development goals at local and regional scales. The substantive assessment is complemented by evaluations of presentation quality, communication clarity, and argument coherence.

3 Debate and Discussion

Within the shared design studio, it is essential to clearly define the scope of the research topic to frame the didactic work as support for an experimental research trajectory. From this perspective, adopting a multiscale approach informed by theories of degrowth and territorial sustainability establishes a general field of inquiry that can subsequently be articulated more specifically within the two design courses.

The general area of reference concerns the city-countryside relationship, understood not merely as a spatial condition but as a system of functional, social, and economic interdependencies between rural settlements and the urban contexts to which they relate. Contemporary urban theory has increasingly addressed this relationship through a critical reassessment of growth-oriented urbanisation models and the exploration of alternative spatial configurations capable of maintaining social and ecological balance [31, 32, 35, 36].

Agricultural intensification generates profits for external corporations while farmers receive declining shares of value and rural environments experience degradation [23, 33]. Resource extraction, energy generation, and tourism similarly benefit external investors while communities experience environmental and social degradation. Migration patterns within growth-dependent Europe concentrate in high-growth urban regions offering employment opportunities, while shrinking regions experience emigration and difficulty attracting newcomers [24, 34]. Across Europe, migrant populations of diverse socioeconomic backgrounds and national origins are already settled in both urban and rural areas, where they participate in labour markets, contribute to local economies, and shape the social fabric of communities. In rural contexts specifically, migrant labour has become structurally embedded in sectors such as agriculture, food processing, construction, and care work, sustaining productive activities that would otherwise face severe labour shortages due to demographic decline [35]. Yet this presence is not without spatial consequences and/or opportunities. In many rural municipalities, the existing housing stock is inadequate, leading to overcrowding and informal arrangements; in other situations, the presence of new migrant groups becomes an opportunity for repopulation of rural villages, deserted by the urbanisation process [10]. A degrowth perspective on migration must therefore engage critically with the spatial and infrastructural conditions required to make settlement sustainable and socially cohesive. This includes planning for adequate and dignified housing, ensuring equitable access to public services, supporting local economies that can absorb and benefit from diverse labour contributions, advancing cultural and social inclusivity, and fostering participatory governance mechanisms through which both established and new residents can negotiate shared spatial futures.

European responses to irregular migration flows increasingly emphasise containment through campisation, concentrating migrants in segregated facilities and restricting mobility [25, 36]. Although the majority of migrants in Europe reside outside camp structures, having obtained various forms of legal status or settled informally within existing communities, the campisation model raises specific spatial justice concerns relevant to the degrowth discourse. Degrowth approaches recognise rural settlement as potentially beneficial for migrants, receiving communities, and territorial balance, particularly in depopulating areas characterised by vacant housing, land availability, underutilised infrastructure, and social capacity for revitalisation [21]. However, camps located in rural contexts offer limited means of integration and employment, fostering isolation and loneliness among incoming populations and unnecessarily delaying their active integration into local communities through social and work activities due to asylum status recognition processes. Rural areas are also addressed in contemporary urban debates about their productive role in the future of urbanisation, envisaging scenarios of automated landscapes in which agriculture and data centres raise questions about whether these areas will remain inhabited by humans or be integrated with robotic systems [26, 37]. The role of the countryside, often positioned as a victim of urban extractivism, requires further critical examination within degrowth frameworks.

The connection between degrowth and earlier utopian imaginaries is today reinterpreted through the notion of limits and controlled development rather than through idealised spatial forms. Central to this perspective is the idea that urban systems should be socially and territorially dimensioned, not in an exclusionary sense, but as a condition for maintaining livability, cohesion, and ecological responsibility [38]. The definition of limits (social, ethical, and physical) becomes a foundational principle for settlement organisation, opposing the paradigm of unlimited expansion and hyperproduction of both goods and space [36].

This vision of a network of settlements maintaining a manageable social scale re-emerges in contemporary urban debates, from Yona Friedman's reflections on decentralised and adaptable spatial systems [39] to more recent concepts such as the "15-min city", which advocates for proximity-based urban living and functional decentralisation [16]. Similarly, critical urban scholars have highlighted the need to rethink the relationship between urban concentration and territorial equity, questioning the sustainability of dominant metropolitan models [37, 40].

From these theoretical premises, as already seen, two main interpretative perspectives emerge: From the urban perspective, the concentration of population and economic activities in major cities has strengthened their hub role while exacerbating territorial imbalances, housing stress, and functional dependencies. From a rural perspective, once anchored in self-sufficient agricultural economies, many rural settlements have become marginal and increasingly dependent on urban centres due to economic restructuring and the decline of agriculture. Within this dynamic, the issue of centralisation versus decentralisation becomes critical. As cities concentrate services and jobs, many rural areas lose autonomy and attractiveness, becoming dormitory settlements or facing abandonment, especially in disadvantaged or poorly connected regions [41, 42]. This rural crisis reflects major territorial transformations and raises questions about the future of rural regeneration and alternative living models. The focus of the joint course will be the critical reassessment of integrated urban-rural models, which gain renewed relevance. Rather than proposing idealised forms, contemporary theory and practice suggest incremental, network-based, and territorially embedded approaches capable of rebalancing urban and rural systems, fostering sustainable decentralisation, and reactivating marginal territories [16, 38].

The application of the theory discussed so far will be in the comparison between Dutch and Polish rural territories, which plays a strategic role within the research framework, as it highlights two distinct models of spatial organisation and territorial governance. The Dutch context is characterised by strong spatial planning traditions, high infrastructural integration, and an intensive management of the rural landscape, often conceived as an active component of a diffuse urban system. Rural settlements are generally well connected and functionally integrated, reducing the risk of abandonment while at times reinforcing metropolitan dependency. By contrast, the Polish rural context is marked by greater fragmentation and heterogeneity, shaped by the political and economic transformations of the late twentieth century. While many rural areas maintain closer ties to agricultural practices and local economies, they are also more exposed to depopulation, isolation, and marginalisation, particularly in less developed regions [42, 43]. This condition reveals both structural vulnerability and latent potential for experimenting with alternative pathways of territorial regeneration. A degrowth-oriented vision can reframe the city-countryside relationship by challenging unlimited urban expansion and promoting the redistribution of functions, populations, and resources across territories. It positions rural areas as active components of a polycentric system and supports decentralised, locally grounded models aimed at socio-spatial balance, ecological responsibility, and long-term resilience.

4 Conclusions

The course is scheduled to launch in the 2026/27 academic year as a joint course under the auspices of KreativEU. Its content has been developed to allow cyclical implementation in subsequent academic years within the same course, while maintaining the flexibility to change student locations. At the same time, the course's organisational structure has been designed to be replicated across other subjects and even fields of study, while maintaining a coherent organisational concept based on parallel classes at two different universities. This design constitutes a significant element of didactic innovation, enabling the sustainable integration of learning processes implemented in separate institutional contexts.

The course is based on the belief that cultural exchange is needed to address issues that, as demonstrated, cannot be reduced to local dynamics but require a broader perspective. From the perspective of a teaching-by-research process, the validity of the results is best assessed in a comparative context. This comparison is supported by both equal responses (student-student) and cultural differences (student-teacher). Building an educational system based on comparing different systems opens opportunities for incremental development in two directions: multidisciplinary and geographical. Both scenarios are part of a consortium vision (such as the KreativEU consortium, in which both the universities of Breda and Opole are already involved), where teaching and academic activities move hand in hand with experimental and research activities, drawing on multidisciplinary partners and fostering cultural exchange.

The comparison between Dutch and Polish rural areas is expected to reveal how different spatial planning traditions, land tenure systems, and infrastructural conditions shape the possibilities for degrowth-oriented interventions. In the Netherlands, where rural landscapes are intensively managed and functionally integrated into metropolitan systems, the design challenge lies in decoupling rural settlements from growth-dependent logics of efficiency and productivity, exploring models of relocalised economies instead, shared housing, and ecological restoration that reduce metropolitan dependency. In Poland, where rural fragmentation and depopulation present different structural conditions, the challenge shifts toward reactivating underutilised land and housing stock, strengthening local service networks, and designing settlement patterns that support demographic stabilisation through inclusive spatial strategies—including the integration of migrant populations as active contributors to rural regeneration. The comparative framework is intended to test whether and how these principles translate differently across national contexts, and to identify spatial typologies and planning instruments that may be transferable or require context-specific adaptation. In this sense, the course does not merely aim to apply degrowth theory illustratively but seeks to contribute, through iterative design experimentation across successive academic cycles, to the construction of an evidence base for degrowth-oriented spatial planning in European rural territories. The integration of the migration theme within this framework ensures that questions of demographic change, social inclusion, and spatial justice remain central to the design process.

Disclosure of Interests. The authors declare that they have no known competing financial interests or personal relationships that could have appeared to influence the work reported in this paper.

References

1. Kallis G, Kostakis V, Lange S, Muraca B, Paulson S, Schmelzer M (2018) Research on degrowth. Annu Rev Environ Resour 43:291–316. https://doi.org/10.1146/annurev-environ-102017-025941
2. Georgescu-Roegen N (1971) The entropy law and the economic process. Harvard University Press. ISBN: 978–0674868236
3. Raworth K (2017) Doughnut economics: Seven ways to think like a 21st-century economist. Chelsea Green Publishing, London, UK. ISBN: 978–1603586746
4. Hickel J (2020) Less is more: how degrowth will save the world. William Heinemann, London, UK. ISBN: 978–1787331387
5. Harvey D (1989) From managerialism to entrepreneurialism: the transformation in urban governance in late capitalism. Geogr Ann Ser B, Hum Geogr 71(1):3–17. https://doi.org/10.1080/04353684.1989.11879583

6. Aalbers MB (2016) The financialization of housing: a political economy approach. Routledge. ISBN: 978–1138943137
7. Swyngedouw E, Heynen NC (2004) Urban political ecology, justice, and the politics of scale. Antipode 35(5):898–918. https://doi.org/10.1111/j.1467-8330.2003.00364.x
8. Jackson T (2017) Prosperity without growth, 2nd edn. Routledge. ISBN: 978–1138935417
9. D'Alisa G, Demaria F, Kallis G (eds) (2015) Degrowth: a vocabulary for a new era. Routledge. ISBN: 978–0415704700
10. Koolhaas R, AMO (2020) Countryside: a report. TASCHEN; Guggenheim Museum Publications. ISBN: 978–3836584395
11. Xue J (2022) Urban planning and degrowth: a missing dialogue. Local Environ 27(4):404–422. https://doi.org/10.1080/13549839.2020.1867840
12. Savini F (2019) The economy that runs on waste: accumulation in the circular city. J Environ Policy Plan 21(6):675–691. https://doi.org/10.1080/1523908X.2019.1670048
13. Scheller D, Thörn H (2018) Governing 'sustainable urban development' through self-build groups and co-housing: the cases of Hamburg and gothenburg. Int J Urban Reg Res 42(5):914–933. https://doi.org/10.1111/1468-2427.12652
14. Foster S, Iaione C (2016) The city as a commons. Yale Law Policy Rev 34:281. https://papers.ssrn.com/sol3/papers.cfm?abstract_id=2653084. Last accessed 15 June 2026
15. Ostrom E (1990) Governing the commons: the evolution of institutions for collective action. Cambridge University Press, Cambridge, UK. ISBN: 978–0521405991
16. Moreno C, Allam Z, Chabaud D, Gall C, Pratlong F (2021) Introducing the "15-min city": sustainability, resilience and place identity in future post-pandemic cities. Smart Cities 4(1):93–111. https://doi.org/10.3390/smartcities4010006
17. Kennedy C, Pincetl S, Bunje P (2011) The study of urban metabolism and its applications to urban planning and design. Environ Pollut 159(8–9):1965–1973. https://doi.org/10.1016/j.envpol.2010.10.022
18. Soja EW (2010) Seeking spatial justice. University of Minnesota Press, Minneapolis, MN. ISBN: 978–0816676217
19. Lefebvre H (1996) In: Kofman E, Lebas E (eds) Writings on cities. Blackwell. (Original work published 1968)
20. Gibson-Graham JK (2006) A postcapitalist politics. University of Minnesota Press. ISBN: 978–0816641983
21. Hedberg C, do Carmo RM (2012) Translocal ruralism: mobility and connectivity in european rural spaces. Springer. https://doi.org/10.1007/978-94-007-2315-3
22. Haas W, Krausmann F, Wiedenhofer D, Heinz M (2015) How circular is the global economy? an assessment of material flows, waste production, and recycling in the European union and the world in 2005. J Ind Ecol 19(5):765–777. https://doi.org/10.1111/jiec.12244
23. Donegan KS, Adamson G, Donegan HA (2007) Indexing the contribution of household travel behaviour to sustainability. J Transp Geogr 15(4):245–261. https://doi.org/10.1016/j.jtrangeo.2006.08.004
24. Rosner A, Wesołowska M (2020) Deagrarianisation of the economic structure and the evolution of rural settlement patterns in Poland. Land 9(12):523. https://doi.org/10.3390/land9120523
25. Bąk I, Wawrzyniak K, Oesterreich M (2021) The impact of transformational changes on the socio-economic conditions of the rural population. An example of Poland. Agriculture 11(5):403. https://doi.org/10.3390/agriculture11050403
26. Wesołowska MO (2023) Depopulation villages in Poland—current status and possible transition scenarios. Rural Reg. Dev. 1(2):10006. https://doi.org/10.35534/rrd.2023.10006
27. Lessmann C (2014) Spatial inequality and development – Is there an inverted-U relationship? J Dev Econ 106:35–51. https://doi.org/10.1016/j.jdeveco.2013.08.011

28. LopezDeAsiain M, Diaz-Garcia V (2020) The importance of the participatory dimension in urban resilience improvement processes. Sustainability 12(18):7305. https://doi.org/10.3390/su12187305
29. Gong H, Hassink R (2020) Context sensitivity and economic-geographic (re)theorising. Camb J Reg Econ Soc 13(3):475–490. https://doi.org/10.1093/cjres/rsaa021
30. Hammar Chriac E (2014) Group work as an incentive for learning – students' experiences of group work. Front Psychol 5:558. https://doi.org/10.3389/fpsyg.2014.00558
31. Latouche S (2009) Farewell to growth. Polity Press. ISBN 978–0745646169
32. Harvey D (2012) Rebel cities: from the right to the city to the urban revolution. Verso Books. ISBN 978–1–84467-882-2.
33. McMichael P (2009) A food regime genealogy. J Peasant Stud 36(1):139–169. https://doi.org/10.1080/03066150902820354
34. Glorius B, Doomernik J (eds) (2020) Geographies of asylum in Europe and the role of European localities. Springer. https://doi.org/10.1007/978-3-030-25666-1
35. Ryan M (2023) Labour and skills shortages in the agro-food sector, OECD food, agriculture and fisheries papers 189. OECD Publishing, Paris. https://doi.org/10.1787/ed758aab-en
36. Kreichauf R (2018) From forced migration to forced arrival: the campization of refugee accommodation in European cities. Comp Migr Stud 6(1):7. https://doi.org/10.1186/s40878-017-0069-8
37. Bedir M, Groen L, Kuijpers M, Muñoz Sanz V, Otero Verzier M (2023) Automated landscapes. Het Nieuwe Instituut
38. Latouche S (2012) In: Grillenzoni F (ed) Limite: piccolo trattato di decrescita serena. Bollati Boringhieri. ISBN 978–8833923635
39. Friedman Y (1975) Toward a scientific architecture. MIT Press. ISBN: 978–0262010331
40. Brenner N, Schmid C (2014) The "urban age" in question. Int J Urban Reg Res 38(3):731–755. https://doi.org/10.1111/1468-2427.12115
41. Woods M (2010) Rural. Routledge, London. https://doi.org/10.4324/9780203844304
42. ESPON (2020) Shrinking rural regions in Europe. ESPON EGTC
43. Bański J (2018) Rural areas of Poland in the context of changes in population and economy. Polish Academy of Sciences

Sustainable Planning of the Inland River Port Construction – A Case Study

Adam Rak(✉) and Adrian Czubak

Opole University of Technology, Faculty of Civil Engineering and Architecture, Katowicka 48, Opole 45-051, Poland
a.rak@po.edu.pl, a.czubak@student.po.edu.pl

Abstract. The paper presents a conceptual model for planning and organizing the construction process of an inland river port, taking into account technical, logistical, environmental, and organizational conditions. The study is motivated by the growing importance of inland waterway transport within sustainable transport systems and the increasing role of inland ports as multimodal nodes integrating water, rail, and road transport. The research focuses on a planned intermodal port on the Odra River near Opole, Poland. Three conceptual design variants (A, B, and C) were developed, differing in infrastructure scale, throughput capacity, and functional scope. The analysis considers both external factors, such as hydrological, geotechnical, and environmental conditions, and internal factors related to port functionality, cargo handling organization, and technological systems. The research methodology is based on a systematic approach that integrates a literature review, analysis of the legal framework, and structural-technological solutions. A universal and modular conceptual model of the investment process is proposed, covering stages from site analysis and feasibility studies through design and construction to the integration of the port into the logistics network. The results indicate that the operational efficiency of an inland port is strongly dependent on multimodal integration, phased infrastructure development, and the incorporation of environmental and climate-related requirements into the planning process. Future research will include applying multi-criteria decision-making (MCDM) methods to support the selection of the optimal investment variant.

Keywords: Inland Port · Intermodal Transport · Multimodal Logistics Hub · Construction Process Management · Sustainability · BIM

1 Introduction

Inland waterway transport (IWT) has gained considerable attention in the context of achieving a sustainable transport system in Europe. Numerous studies in this field identify key research areas necessary for integrating inland waterway transport into intermodal supply chains. The development of inland navigation and the growing importance of river ports as multimodal nodes pose new challenges for investors, designers, and public authorities in planning, organizing, and implementing construction processes. Today, inland ports no longer function solely as cargo transshipment points; they are

Z. Perkowski et al. (Eds.): ECCE 2026, LNCE 929, pp. 264–276, 2027.
https://doi.org/10.1007/978-3-032-39515-3_20

becoming integral components of logistics systems, integrating inland waterway, rail, and road transport, thereby enabling the development of modern modal logistics centers [15, 16]. Studies [11, 14–16] present a universal conceptual model for the preparation and organization of the construction process of an inland port, integrating three key areas: organizational and logistical aspects, structural and technological solutions, and legal and environmental requirements. The model considers both external factors—such as geotechnical and hydrological conditions, water availability, environmental impacts, and climatic constraints—and internal factors, including port functionality, terminal throughput, cargo-handling systems, and automation of operational processes [1, 3]. From a structural and technological perspective, the model encompasses quay structures, storage yards, warehouses, transport systems, and IT infrastructure, ensuring the reliability of the logistics center's operation. The modular structure of the port enables infrastructure scalability to meet operational needs and local conditions, while also facilitating the optimization of logistics processes in a multimodal context [2, 12, 13]. An essential complement to this analysis is the identification of legal and environmental requirements applicable in Poland and the European Union, including construction law, environmental protection regulations, water management legislation, and sustainable development standards [17, 18, 21]. These conditions constitute both constraints and guidelines for the planning, design, and implementation phases of investment projects, including environmental decisions and environmental impact assessments [13, 16]. The proposed model enables comprehensive management of the construction process, from site analysis and feasibility studies, through design and construction, to commissioning and integration of the port into the logistics system. Such an approach allows for the creation of efficient, reliable, and sustainable logistics centers while ensuring compliance with the highest structural, technological, and operational standards, which is essential for the development of inland waterway transport in Europe and Poland [9, 10, 15].

Based on a review of scientific research concerning the construction and operation of inland intermodal ports, several key conclusions can be drawn:

a) Inland intermodal ports serve as nodes linking waterborne transport with road and rail transport. Their efficient functioning depends on strong hinterland connections and flexibility in responding to changing cargo flows. Research emphasizes the need for strategic planning of terminal locations and their connections with key supply chain nodes.
b) A critical element is the coordination of operations within the port and along inland waterways, including lock operation planning, barge queue management, and synchronization with land-based terminals. System-based models and simulation tools allow assessment of the impact of various operational scenarios on the performance of the entire transport network.
c) Inland ports can support environmentally sustainable transport strategies by reducing emissions through the substitution of road transport with waterborne transport. Studies also highlight the necessity of incorporating climate change considerations into infrastructure planning and port operations.

d) Effective port management requires detailed data on cargo flows, logistics decisions, and terminal operations. Modern decision-support systems enable cost analysis, operational efficiency assessment, and environmental impact evaluation, thereby supporting both strategic and operational planning.
e) In summary, scientific research demonstrates that inland intermodal ports are not merely transshipment points but comprehensive logistics nodes whose efficiency depends on network integration, operational coordination, business cooperation, and the implementation of sustainable transport practices.

2 Description of the Research Object

Table 1. Characteristics of selected parameters of the designed port.

Parameters	Unit	Model A	Model B	Model C
Annual throughput capacity	mln tonnes/year	1.5	3.0	5.0
Total area	ha	15	25	25
Storage capacity	thousand tonnes	100	200	250
Storage yard area	ha	5.0	7.5	10.0
Port quays	number of units	1	3	16
Quay length	m	700	700	900
Port basins	number of units	1	3	9

The study analyzes the potential location and construction of an intermodal port on the Odra River in the Opole Voivodeship, in the immediate vicinity of the city of Opole. The location of the investment results from the advantageous access to the Odra River waterway and the existing rail and road infrastructure, which creates favorable conditions for the integration of inland waterway, rail, and road transport [4, 5, 8] (see Figs. 1a, 1b and 1c). The siting of the project within the Odra valley ensures direct access to an inland waterway of both national and international importance. The Odra is part of the European transport system, connecting the southern and western regions of Poland to seaports and markets in Western Europe. The port's location in the middle course of the river enables integration of inland navigation with the existing railway and road infrastructure in the Opole region. The designed facility is intended to function as an intermodal port enabling the transshipment of bulk and general cargo, with particular emphasis on thermal coal, aggregates, and other industrial raw materials. Both direct transshipment operations (vessel-railcar, vessel-truck) and indirect operations involving temporary storage in open storage yards are envisaged. The infrastructural scope of the project includes transshipment quays adapted for the handling of barges and pushed convoys, storage yards, cargo-handling equipment, a railway siding, and an internal road network. The investment requires consideration of hydrological and environmental constraints resulting from its location within the Odra valley, in particular flood risk and the need to limit dust emissions associated with coal handling operations. The planned port is an element of the regional logistics system and may serve as a multimodal node that supports the development of inland waterway transport and improves freight transport efficiency in the region. As part of the research, three cumulative conceptual models of the port's spatial and structural solutions were developed. Characteristic spatial and technical parameters are summarized in Table 1, and the spatial layouts with short descriptions are given in Figs. 1a, 1b, and 1c.

Fig. 1. a. Conceptual Model A: construction of a single coal quay with coal transport carried out via a conveyor belt system or rail transport [8]. b. Conceptual Model B: construction of three quays: coal, universal, and container, with two port basins. The rail layout along the quays and the arrival-sidings group enable further modular expansion of the port, while the road system is planned to be connected to the regional bypass [8]. c. Conceptual Model C: the model most closely aligned with a modern approach to planning and operating a logistics center, due to the proposed wide functional scope of the port and the possibility of obtaining comprehensive results enabling the development of a complete model for planning and organizing the construction process [8].

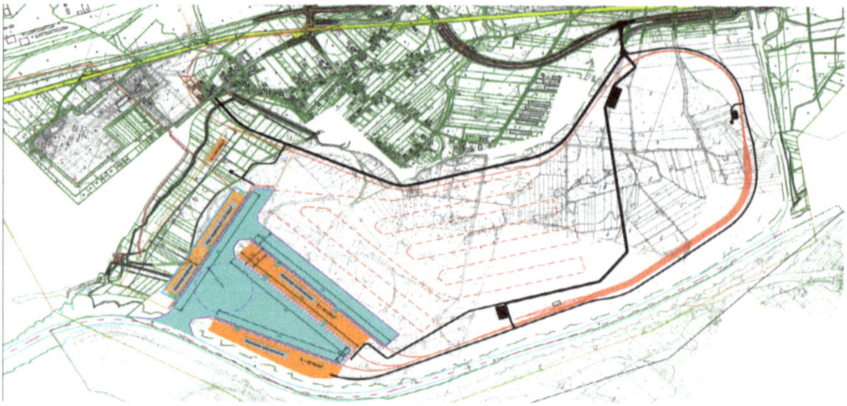

Fig. 1. (*continued*)

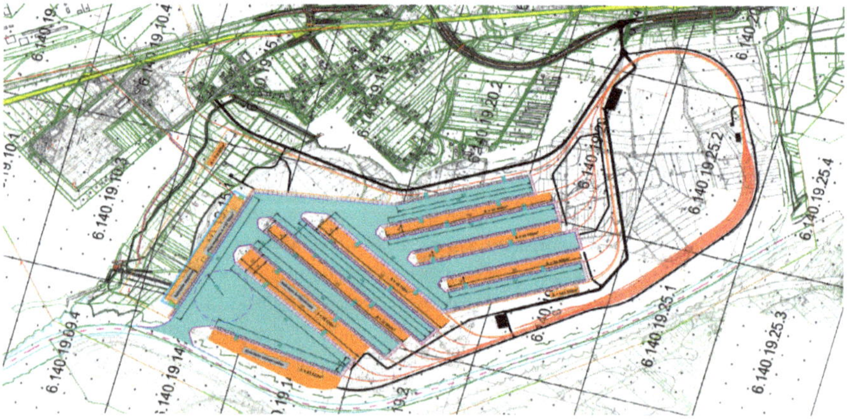

Fig. 1. (*continued*)

The most advanced solution is Conceptual Model C, which assumes the construction of 16 quays, including coal, universal, and container terminals. By extending the entrance channel to the port, it will be possible to connect an additional five port basins. Such a highly developed port, designed to perform numerous complex operations influenced by diverse and strongly interrelated external factors, should comprise multiple independent zones that operate in an organized, failure-free manner while complementing one another. Internal port solutions within individual zones are strictly tied to the spatial arrangement of the key structural and functional components that comprise the internal port infrastructure. The layout of individual elements must be preceded by highly detailed analyses, as incorrect planning and positioning would result in the inability to ensure efficient port operations and could expose the entire operational system to multiple conflicts and disruptions within functional subsystems. The primary determinants for estimating the final parameters of internal port facilities (port waters, port land area, and hydrotechnical structures) include: navigational conditions of the waterway, operational and maneuvering parameters of the serviced inland vessels, the type of required cargo-handling equipment, the planned handling capacity and throughput of the port, and the types of cargo-handling quays intended for implementation. The technical and technological equipment, in turn, depends on the adopted solutions within the planned functional zones. Defining the implementation scope is a fundamental condition for a properly designed logistics center, ensuring optimal utilization of location advantages, infrastructure, and the transport network's capacity.

3 Research Methodology

The research methodology integrates literature, legal, technological, and logistical analyses within a systemic and modular framework, enabling the development of a comprehensive and universal conceptual model of an inland river port as a foundation for creating modal logistics centers. The analysis of international scholarly articles published in Journal of Transport Geography, Transport Reviews, and Research in Transportation Business & Management concerning port operations, modal integration, and sustainable development [11, 14], together with findings from other source materials, provides the basis for identifying the key structural, technological, and organizational components of the port. The analysis of legal acts and guidelines determines the design and implementation process of the port, including administrative procedures and environmental decisions. Consideration of legal and environmental requirements enables the development of a model compliant with applicable standards and regulations [20, 22, 23]. The identification of external factors (geotechnical, hydrological, and environmental conditions) and internal factors (functional requirements, throughput capacity, and operational automation) enables the definition of port modules and their interrelations.

The recognition of legal and environmental requirements constitutes an essential complement to the analysis of an intermodal port. The integration of these determinants with structural, technological, and logistical aspects makes it possible to: ensure investment compliance with applicable regulations, optimize location and infrastructure layout, implement solutions minimizing environmental impact, and prepare the port as a modern, safe, and sustainable logistics hub within the region. In the analysis of logistical and organizational processes, particular importance is attached to identifying the port's organizational structures, the division of roles (investor, designer, contractor, supervisor), scheduling of works, and risk management. The findings allow for the development of a coherent scheme of both the construction and operational processes of the port. The integration of results from literature, legal, structural-technological, and logistical analyses into a modular and universal conceptual model constitutes the study's core outcome. The model is scalable and adaptable to various inland port locations and functional profiles. Block diagrams and process flow charts are applied to visualize relationships between investment stages and port modules. Model validation and verification involve comparison with existing inland ports in Europe and Poland (e.g., Duisburg, Rotterdam, Wrocław, Szczecin). The verification process includes assessment of compliance with functional, legal, and environmental requirements. The model enables the simulation of operational scenarios, allowing their optimization and adaptation to real investment conditions. Figure 2 presents the graphical model of the research methodology.

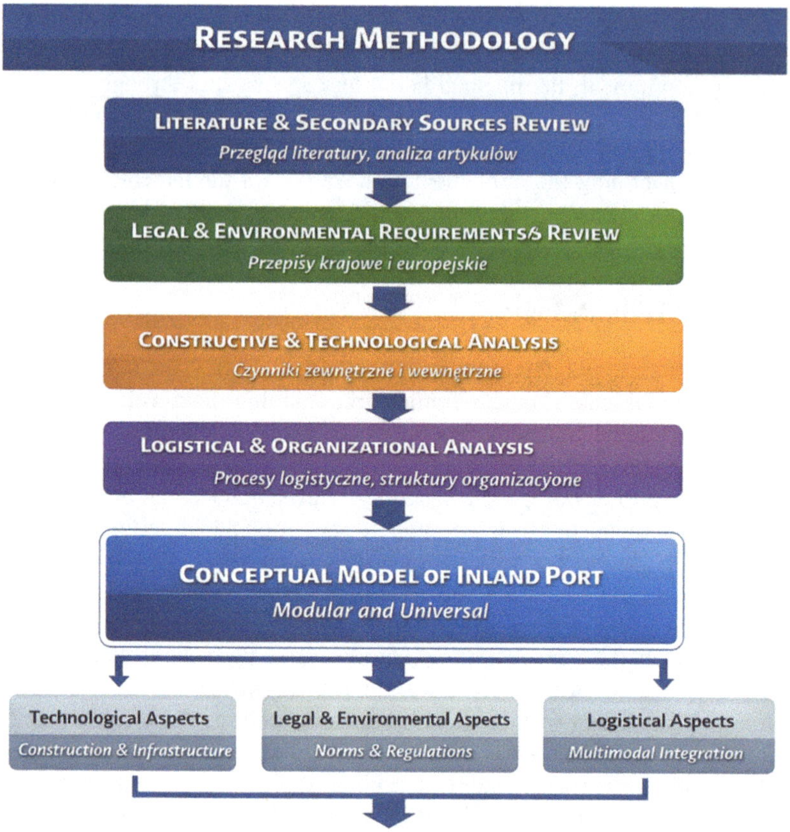

Fig. 2. Research methodology for the construction of an inland port [author's work].

4 Conceptual Model for the Construction of an Inland River Port

The proposed conceptual model for the preparation and organization of the construction process for an inland river port is a universal, modular framework that can be adapted to various locations and port typologies while maintaining full functionality as a logistics hub. The model integrates three principal domains: organizational and logistical aspects, structural and technological components, as well as legal and environmental requirements:

a) **Preliminary and contextual analysis.** The model (see Fig. 3) begins with a comprehensive analysis of the port location and its context, including assessments of hydrological and geotechnical parameters, river accessibility, and potential environmental risks such as flooding, nature conservation constraints, and noise emissions [11, 14]. Simultaneously, investor requirements are identified, including the intended port capacity, the types of cargo to be handled, and the potential for integration with other modes of transport [1].

b) **Structure of the construction process.** The model defines the principal stages of the construction process: planning and development of the port concept, technical and construction design, acquisition of administrative and environmental permits, execution of construction works, and commissioning of the port, including multimodal integration. Each stage incorporates schedule management, quality control, safety management, and investment risk assessment [9, 15].
c) **Structural and technological aspects.** The model identifies external factors that influence the selection of structural solutions, such as soil conditions, water flow dynamics, environmental impacts, and climatic requirements. At the same time, internal factors are considered, including port functionality, terminal throughput, cargo handling systems, and operational automation [2, 3]. Technical and technological components include quays, storage yards, warehouses, transport systems, and information systems supporting port operations, thereby ensuring the reliability and efficiency of the logistics center [12, 14, 24].
d) **Multimodal Integration.** The model enables planning of port connections for road, rail, and inland waterway transport, the design of cargo-handling and warehousing systems, and the analysis of goods flows to optimize logistics processes [1, 2]. The modular approach allows infrastructure to scale based on operational requirements and local conditions.
e) **Legal and environmental requirements.** An integral component of the model is the analysis of legal and environmental frameworks, including construction law, environmental protection regulations, water management legislation, and sustainable development standards applicable in Poland and the European Union [17, 20–23]. These requirements determine the planning, design, and implementation of the investment, including administrative procedures, environmental decisions, and environmental impact assessments [13, 16].
f) **Development of the Conceptual Model.** By integrating all the aforementioned domains, the model enables comprehensive management of the construction process—from location analysis, feasibility studies, and design, to construction execution and port operation (see Fig. 3). The proposed approach supports the development of efficient, reliable, and sustainable logistics centers while ensuring compliance with the highest structural, technological, and operational standards [9, 15].

The model may be further developed into a phased (stage-based) structure by assigning specific activities, stage deliverables, and control points (milestones) (see Fig. 4).

Strategic conclusions based on the extended model (see Fig. 4) are following:

a) **Necessity of phased investment implementation.** The port should be developed in a modular manner (e.g., initially as a general cargo terminal, followed by a container terminal), thereby reducing financial risk.
b) **Priority of multimodal integration.** The economic efficiency of the port depends on effective rail and road connections. Hydrotechnical investment without land-side integration will not ensure profitability.
c) **Early initiation of environmental procedures.** The greatest schedule risk is associated with obtaining environmental decisions and water-law permits; therefore, these procedures should be initiated in parallel with the conceptual phase.

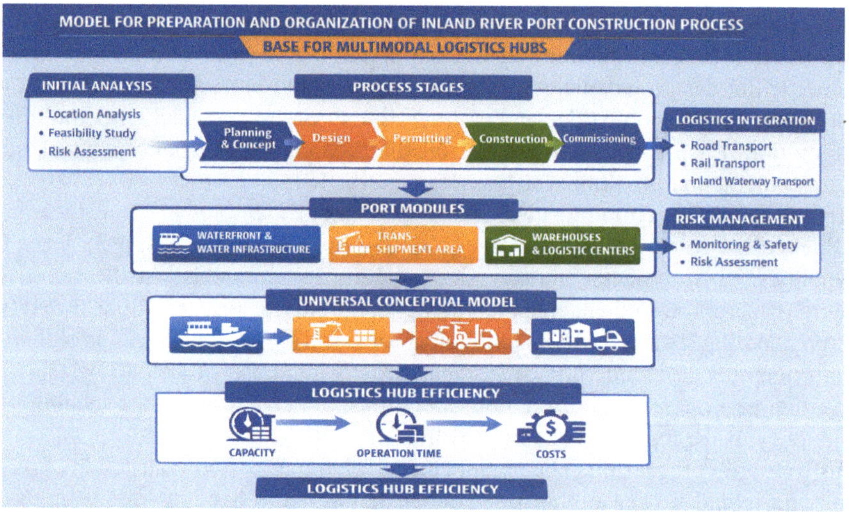

Fig. 3. Model for the preparation and organization of the inland river port construction process [author's work].

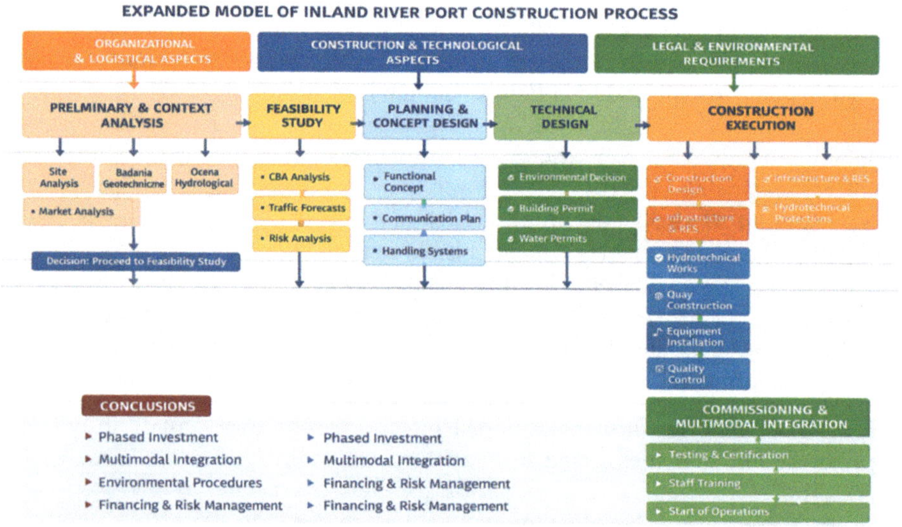

Fig. 4. Extended model of the inland river port construction process [author's work].

d) **Securing financing prior to the execution phase.** The feasibility study should clearly define the financial model and the level of EU co-financing.
e) **Implementation of digital tools (BIM + Digital Twin).** This enables the reduction of design errors, cost optimization, and facilitation of subsequent operational management.

f) **Climate risk management.** Hydrological changes (droughts, floods) should be incorporated into the design of technical depth parameters and quay structures.

5 Summary

The proposed conceptual model for the preparation and organization of the construction process of an inland river port incorporates both structural-engineering and technical-technological aspects that are crucial for the functioning of the port as a logistics center. The model distinguishes between external factors influencing the selection and application of structural solutions (e.g., geotechnical and hydrological conditions, river accessibility, environmental impacts, climatic and transport conditions) and internal factors related to the functional requirements of the port that determine the operational reliability of the logistics hub (e.g., terminal capacity, vessel berthing time, storage and handling systems, automation of operations). The model also includes technical and technological solutions such as quay and riverbank reinforcement structures, storage and transshipment systems, gantry cranes, internal transport systems, IT systems for traffic and port operations management, and technologies enabling efficient integration with the multimodal transport network. Owing to its modular approach, the model enables the correlation of design and technological decisions with the functional requirements of the port and future logistics centers. An important complement to the model is the analysis of legal and environmental conditions at both national and European levels, including regulations concerning construction, environmental protection, water management, and port operational safety. These conditions constitute additional constraints and guidelines in planning and implementing the construction process, requiring consideration of administrative procedures, environmental decisions, environmental impact assessments, and sustainable development standards. The planning and implementation of the intermodal port in Opole city require consideration of a number of significant legal and environmental requirements that form key determinants of the construction process. These include national regulations, such as the Construction Law and the Water Law, as well as ordinances governing technical conditions for port structures and environmental protection standards. The identification of environmental requirements includes, i.a., the protection of surface waters, the reduction of dust emissions during the handling of coal and other bulk cargo, the protection of natural habitats, the monitoring of noise and vibration, and the mitigation of flood risk. In parallel, relevant European Union directives must be taken into account, including the EIA (Environmental Impact Assessment), the Water Framework Directive, and the Habitats and Birds Directives [17, 18]. The integration of these requirements with structural, technological, and logistical aspects enables optimization of the location of quays, storage yards, warehouses, railway tracks, and internal roads, ensuring a safe, functional, and sustainable port. Consideration of legal and environmental factors also enables the development of a port that is consistent with transport policy objectives and sustainable development standards at both the national and European levels. By integrating all these elements, the model enables comprehensive planning and organization of the construction process, ensuring compliance with the highest structural, technological, safety, and operational efficiency standards, which is essential for the development of modern and reliable multimodal logistics centers.

From both scientific and practical perspectives, it is recommended to:

- develop a decision tree scheme for investors,
- prepare a risk matrix model,
- develop an indicator-based model for assessing inland port efficiency,
- conduct logistics flow simulations (e.g., scenario-based modeling).

After defining the management methodology for the analyzed infrastructure project—covering the organizational structure of the investment process as well as the identification of the design and construction stages of the intermodal port—the next stage of the research involves conducting a variant analysis using multi-criteria methods. This approach results from the multi-aspect nature of port investments, which require the simultaneous consideration of technical, logistical, economic, and environmental factors. In the scientific literature, the dominant approaches in location and design analyses are MCDM (Multi-Criteria Decision Making) methods, such as AHP, TOPSIS, VIKOR, and hybrid approaches that combine expert evaluation with synthetic modeling techniques [2, 9, 27–30]. The multi-criteria analysis will serve as a decision-support tool by enabling the systematic and quantitative comparison of several conceptual variants. The application of MCDM methods will enable a synthetic assessment of variants using defined criteria and weights that reflect their significance to the port's operational efficiency. Conducting the multi-criteria analysis will constitute the next stage, integrating the previously performed location, technical, and organizational analyses to identify the optimal variant from the perspectives of both investment implementation and the facility's future operation. This will be the subject of subsequent research and analyses.

References

1. Bedoya-Maya F et al (2025) Cargo consolidation in port-hinterland container transport: a spatial economic assessment for inland waterways. Res Transp Bus Manag 59:101254. https://doi.org/10.1016/j.rtbm.2024.101254
2. Caris A, Limbourg S, Macharis C, van Lier T, Cools M (2014) Integration of inland waterway transport in the intermodal supply chain: a taxonomy of research challenges. J Transp Geogr 41:126–136. https://doi.org/10.1016/j.jtrangeo.2014.08.022
3. Chang Z, Weng J, Qi Z, Yang D (2019) Assess economic and environmental trade-off for inland port location. Int J Shipp Transp Logist 11(2–3):234–261. https://doi.org/10.1504/IJSTL.2019.099277
4. Czubak A, Rak A (2024) Planning and organization of the construction process on the example of an inland intermodal river port. Zeszyty Naukowe Akademii Górnośląskiej 14(2):16–25. https://doi.org/10.53259/2024.02.02
5. Czubak A, Rak A (2025) Construction planning model and transport function of an intermodal river port as an important element of 21st century transport. In: Lewczuk K, Rosiński A, Siergiejczyk M (eds) Monograph: transport systems in the era of automation and digitalization. Publishing House of the Warsaw University of Technology, pp 21–40
6. Jarzębińska T (2007) The role of inland ports in integration of polish waterways with the European network. Pol Marit Res 14(2007/S1):30–33
7. Kotowska I, Mańkowska M, Pluciński M (2019) Planning the development of inland shipping in the seaport hinterland transport: a case study of the Oder River in Poland. Sci J Marit Univ Szczecin 58(130):84–92

8. Polska M (2019) Feasibility study of alternative options for the construction of an Inland intermodal port on the Oder river with coal and other cargo handling capability [Report prepared for PGE GiEK S.A., Elektrownia Opole Branch]. Multiconsult Polska Sp. z o.o., ul. Bonifraterska 17, 00–203 Warszawa
9. Notteboom T (2016) The adaptive capacity of container ports in an era of mega vessels: the case of upstream seaports Antwerp and Hamburg. J Transp Geogr 54:295–309. https://doi.org/10.1016/j.jtrangeo.2016.06.002
10. Notteboom T, Yang D, Xu H (2020) Container barge network development in inland rivers: a comparison between the Yangtze River and the Rhine River. Transp Res A Policy Pract 132:587–605. https://doi.org/10.1016/j.tra.2019.10.014
11. Rodrigue JP, Debrie J, Fremont A, Gouvernal E (2010) Functions and actors of inland ports: European and north American dynamics. J Transp Geogr 18(4):519–529. https://doi.org/10.1016/j.jtrangeo.2010.03.008
12. Sdoukopoulos E, Boile M (2020) Port-hinterland concept evolution: a critical review. J Transp Geogr 86:102775. https://doi.org/10.1016/j.jtrangeo.2020.102775
13. Shayan NF, Mohabbati-Kalejahi N (2026) Inland port sustainability: systematic review uncovering terminologies, topics, methodologies, and geographical scopes. Res Transp Bus Manag 64:101512. https://doi.org/10.1016/j.rtbm.2025.101512
14. Witte P, Wiegmans B, Roso V, Hall PV (2020) Moving beyond land and water: understanding the development and spatial organization of inland ports. J Transp Geogr 84:102676. https://doi.org/10.1016/j.jtrangeo.2020.102676
15. Witte P, Wiegmans B, Ng AKY (2019) A critical review on the evolution and development of inland port research. J Transp Geogr 74:53–61. https://doi.org/10.1016/j.jtrangeo.2018.11.001
16. Zhang L, Liu Q, Mathisen TA, Ge YE (2026) Navigating environmental sustainability in inland waterway transport: emission challenges and solutions. Mart Policy Manag 53(4):625–649. https://doi.org/10.1080/03088839.2025.2523589
17. European Committee for Standardization (CEN/TC 15, 147, 320) – European standards concerning port and cargo-handling infrastructure
18. European Commission (2021) EU inland waterways transport strategy
19. Minister of Infrastructure / Water Management (2018–2022) Sectoral reports and analyses on the modernization of the Oder Inland Waterway
20. Polish Standards (PN-EN) concerning quay structures, storage yards, and operational loads
21. Regulation of the Minister of Infrastructure of 12 March 2020 on the technical requirements to be met by port structures and hydraulic engineering facilities
22. Act of 7 July 1994 – Building Law (Journal of Laws 2023, item 1145, as amended)
23. Act of 20 July 2017 – Water Law (Journal of Laws 2023, item 205)
24. Kulczyk J, Winter J (2003) Śródlądowy transport wodny. Wrocław University of Science and Technology Publishing House
25. Belton V, Stewart TJ (2002) Multiple criteria decision analysis: an integrated approach. Springer-Science+Business Media, B.V. (originally published by Kluwer Academic Publishers). https://doi.org/10.1007/978-1-4615-1495-4
26. Hwang CL, Yoon K (1981) Multiple Attribute Decision Making: Methods and Applications. A State-of-the-Art Survey. Springer. https://doi.org/10.1007/978-3-642-48318-9
27. Triantaphyllou E (2000) Multi-criteria decision making methods: a comparative study. Springer-Science+Business Media, B.V. (originally published by Kluwer Academic Publishers). https://doi.org/10.1007/978-1-4757-3157-6
28. Notteboom TE, Rodrigue JP (2005) Port regionalization: towards a new phase in port development. Marit Policy Manag 32(3):297–313. https://doi.org/10.1080/03088830500139885

29. Behzadian M et al (2012) A state-of the-art survey of TOPSIS applications. Expert Syst Appl 39(17):13051–13069. https://doi.org/10.1016/j.eswa.2012.05.056
30. Tighnavard Balasbaneh A, Aldrovandi S, Sher W (2025) A systematic review of implementing multi-criteria decision-making (MCDM) approaches for the circular economy and cost assessment. Sustainability 17(11):5007. https://doi.org/10.3390/su17115007

Author Index

Z. Perkowski et al. (Eds.): ECCE 2026, LNCE 929, p. 277, 2027.
https://doi.org/10.1007/978-3-032-39515-3